Contaminated Sediments in Freshwater Systems

Contaminated Sediments in Freshwater Systems

Frank R. Spellman

CRC Press
Taylor & Francis Group
Boca Raton London New York

CRC Press is an imprint of the
Taylor & Francis Group, an **informa** business

CRC Press
Taylor & Francis Group
6000 Broken Sound Parkway NW, Suite 300
Boca Raton, FL 33487-2742

Printed on acid-free paper
Version Date: 20160414

International Standard Book Number-13: 978-1-4987-7517-5 (Hardback)

Library of Congress Cataloging-in-Publication Data

Names: Spellman, Frank R.
Title: Contaminated sediments in freshwater systems / Frank R. Spellman.
Description: Boca Raton : Taylor & Francis, 2016. | "A CRC title." | Includes bibliographical references and index.
Identifiers: LCCN 2016016588 | ISBN 9781498775175 (alk. paper)
Subjects: LCSH: Contaminated sediments. | River sediments. | Stream chemistry. | Water--Pollution. | Water quality.
Classification: LCC TD427.S33 S64 2016 | DDC 628.1/68--dc23
LC record available at https://lccn.loc.gov/2016016588

Visit the Taylor & Francis Web site at
http://www.taylorandfrancis.com

and the CRC Press Web site at
http://www.crcpress.com

Printed and bound in the United States of America by
Edwards Brothers Malloy on sustainably sourced paper

Contents

SECTION II Sediment Damage

SECTION III Sediment Sampling

Preface

This book is dedicated to environmental professionals, like me, or students who must wade into the frigid waters of the Yellowstone, Shoshone, Columbia–Snake, or Flathead rivers; Bitch Creek in Idaho; or the somewhat warmer waters of the Virgin River in Utah to sample, assess, and evaluate the condition of the water body using biological surveys and other direct measurements, as well as the collection of resident biota, sediments, and interstitial water.

"Wait a minute—freshwater biota sampling?"

The title of this book, *Contaminated Sediments in Freshwater Systems*, suggests that its intended purpose is to help engineers, scientists, environmental practitioners, administrative managers, students, and general readers better understand contaminated sediments. And that is, indeed, what the book, *Contaminated Sediments in Freshwater Systems*, is all about.

"Wait a minute—contaminated sediments? Why should we be concerned about contaminated sediments? Are you kidding me?"

Yes, contaminated sediments, with no kidding involved. Protecting sediment quality is an important part of restoring and maintaining the biological integrity of our nation's waters as well as protecting aquatic life, wildlife, and human health. Sediment is an integral component of freshwater surface water ecosystems that provides habitat, feeding, spawning, and rearing areas for many aquatic organisms.

Assessment of freshwater sediments is important and is intended to determine whether chemical and biological concentrations in the sediments are sufficient to cause adverse effects on either aquatic organisms or organisms higher in the food chain, including humans. Sediment serves as a reservoir for contaminants and therefore a potential source of pollutants to the water column, organisms, and ultimately human consumers of those organisms. These pollutants can arise from a number of sources, including municipal and industrial discharges, urban and agricultural runoff, and atmospheric deposition.

One of the main objectives of this book is to describe proven, scientifically sound methods for assessing sediment quality. The description of sediment assessment methods in this book integrates physical, chemical, and biological information. This kind of background information helps responsible parties or concerned environmental professionals make decisions regarding the possible need for contaminated sediment remediation based on a preponderance of evidence. Doing so involves a particular thought process and methodology.

This text is not limited in its potential for other users. For example, this work can be utilized by water practitioners to gain valuable insight into the substance they work so hard to collect, treat, supply, reuse, or discharge for its intended purpose, but it can just as easily provide important information for policymakers who may be tasked with making decisions concerning water quality and resource utilization. Consequently, this book will serve a varied audience that includes students, lay personnel, regulators, technical experts, attorneys, business leaders, and concerned citizens.

To maximize the usefulness of this text, the material is presented in plain English and in as simplified and concise a format as possible. Moreover, this text is accessible to those who have no experience with water science and sedimentology. Readers who work through the text systematically will develop an understanding of the importance of contaminated sediments and the significance of their impact on our environment, thus adding a critical component to the reader's professional knowledge.

A final word. This book is not your grandfather or father's typical technical book. The reader is advised that this book is presented in the author's typical conversational style, format, and tone. The text follows a pattern that is nontraditional; that is, the approach used here is based on real-world experience and proven parameters, not on theoretical gobbledygook. It is hoped that this book can help describe, define, and identify sediment contamination problems. Understanding the problem is always the first step toward achieving prevention and, if necessary, appropriate remediation procedures. The text is clearly written and user friendly and is presented in a conversational style. As with all of my other technical works to date, I simplify content because my intent is to ensure that there is no failure to communicate with the reader. Failure to communicate with anyone is never an option with me.

Prologue

In recent years, concern about the health and vitality of aquatic ecosystems has grown in the United States. One of the principal reasons for this is the realization that many toxic and bioaccumulative chemicals that are found in only trace amounts in water can accumulate to elevated levels in sediments. The net accumulation of a substance by an organism can occur as a result of uptake from all environmental sources. Substances of concern include metals, polycyclic aromatic hydrocarbons (PAHs), polychlorinated biphenyls (PCBs), chlorophenols, organochlorine (OC) pesticides, and polybrominated diphenyl ethers (PBDEs). Some of these pollutants, such as OC pesticides and PCBs, were released into the environment long ago. The use of many of these substances has been banned in the United States for more than 30 years; nevertheless, these chemicals continue to persist in the environment. Chapter 8, for example, describes the 2014 Gold King Mine spill, which affected surface waters in Colorado, New Mexico, and Utah and produced the effect known as yellow boy.

Contaminants enter our waters every day in the form of soil particles dislodged by raindrops and traveling as passengers of runoff into streams, rivers, lakes, or oceans; these contaminants are deposited in these bodies of water as sediments. Although rivers and streams have always transported enormous quantities of sediment to the sea, their sediment loads today are greater than ever; in fact, by weight, sediments are the most abundant water pollutant. Soils stripped of vegetation by crop cultivation, timber cutting, stripmining, overgrazing, road building, and other construction activities are subject to high rates of erosion. When eroded, sediments by the millions of tons are deposited into aquatic systems, muddying streams and rivers.

The obvious result of soil erosion is the loss of valuable agricultural soils, but other problems are associated with the wearing down of soil, as well. Eroded soil particles eventually fill lakes, ponds, reservoirs, harbors, navigation channels, and river channels. As a result, the accumulation of sediments greatly reduces the attractiveness of lakes and reservoirs, which causes them to lose recreational value. Sedimentation also impedes navigation, covers bottom-dwelling organisms, eliminates valuable fish-spawning areas, and reduces the light penetration necessary for photosynthesis. Another problem with erosion is that soils eroded from farmlands sweep nutrients in the form of nitrogen and phosphorus into surface waters. In small quantities, these nutrients are not a problem; however, a dramatic increase in the sediment load can cause significant problems—ecological changes, in fact.

Although the Chesapeake Bay is not a freshwater system, major freshwater rivers in the region help to form the Bay. For this reason and for illustrative purposes, I have included here a discussion about the ecological changes that have occurred in the Chesapeake Bay that have been partly caused by the influx of nutrients, sediments, and contaminated sediments.

WHAT'S WRONG WITH THE CHESAPEAKE BAY?

Not all that long ago, maybe less than 50 years ago, inhabitants of towns on the Eastern Shore of the Chesapeake Bay often went wading in clear, knee-deep waters to catch crabs. The crabbers made their way through the lush, waving grasses on the Bay's bottom, carrying crab pot nets attached to long poles and dragging along containers tied by a string to their waists. The crabbers waited until the crabs scampered out of the grasses, then netted them and flipped them into their containers, never breaking stride as they continued to scoop up scampering crabs. The water was so clear that the crabbers could see their own feet. Today, the clear water of the past has been replaced with brown and turbid water. The crabs have moved on to greener pastures and cleaner waters. The lush, thick grasses that tickled the crabber's feet are gone—and so are the crabs. In less than 50 years, submerged grasses have vanished from many parts of the Upper and Middle Bay. Scientists and environmentalists believe that they are beginning to understand why. The answers lie in assaults on the Chesapeake, many of them caused by the heavy hand of humans. The ecology of the Bay has changed. Some scientists, ecologists, and other environmental specialists suspect the Bay is dying.

What is going on with the Chesapeake Bay?

The answer is complex. Actually, over the last 50 years, many different groups and individuals have proposed many answers, most of which involve much speculation as to what is causing the problem. Here is an example of one of these "answers"—one that isn't really the answer to the problem at all.

Environmental policymakers in the Commonwealth of Virginia came up with what is called the Lower James River Tributary Strategy, which suggested that nitrogen from the Lower James River and other tributaries was the possible culprit responsible for contaminating the Lower Chesapeake Bay region. Nitrogen is normally a nutrient. In excess, however, nitrogen is a pollutant. Some theorists jumped on nitrogen as being the cause of a decrease in the oyster and other aquatic organism populations in the Lower Chesapeake Bay region. Oysters, like crabs, are important to the lower region, for both economical and environmental reasons. From an environmental point of view, oysters are important to the Lower Chesapeake Bay region because, in the past, oysters' life-cycle habits worked to maintain relatively clean Bay water. Oysters are filter feeders. They suck in water as well as its accompanying nutrients and other substances. The oyster sorts out the ingredients in the water and utilizes the nutrients it needs to sustain life. Impurities, including pollutants, are aggregated and excreted by oysters back into the James River.

Years ago, when oysters thrived in the Lower Chesapeake Bay they were able to take in turbid Bay water and turn it almost clear in a matter of about three days. Of course, this is not the case today because the oysters are almost all gone. Where did they go? Who knows? One thing we know for certain is that oysters are no longer thriving, no longer colonizing the Lower Chesapeake Bay region in the numbers they did in the past. They are no longer providing economic stability for watermen; they are no longer cleaning the Bay.

Some "experts" were sure that the problem had to be nutrient contamination, and nitrogen was the culprit. Wrong! Most shots taken from the hip miss the target. The proponents of nutrient contamination missed their shot.

A regional sanitation authority and a local university in the Lower Chesapeake Bay region formed a study group to formally, professionally, and scientifically study the problem. Over a 5-year period, using biological nutrient removal (BNR) techniques at a local wastewater treatment facility, they determined that the effluent leaving the treatment plant and entering the Lower James River consistently contained below 8 mg/L of nitrogen for 5 consecutive years.

So, has the water in the Chesapeake Bay become cleaner, clearer? Have the oysters returned? The answer to both of these questions is no, not really. But didn't some environmentalists, regulators, and other well-meaning "experts" state that the problem was nitrogen? If nitrogen levels have been reduced in the Lower James River, then shouldn't the oysters start thriving, colonizing, and cleaning the Lower Chesapeake Bay again? You might think so, but they are not, even though the nitrogen level in the wastewater effluent has been significantly lowered through treatment and a major point source contributor of nitrogen was reduced in the Lower Chesapeake Bay.

If the nitrogen level has been reduced, then where are the oysters? A more important question, though, is what is the real problem? The truth is that no one at this point can provide a definitive answer to this question, but a number of questions need to be answered before another theory on how to clean up the Lower Chesapeake Bay is proposed: (1) Is nitrogen from the Lower James River and other tributaries that feed the Chesapeake Bay having an impact on the Bay (and the oysters—and maybe the crabs)? (2) Is there evidence of low dissolved oxygen in the Lower James River? (3) Although concentrations of nitrogen in the Lower James River exist, are there corresponding high levels of plankton (chlorophyll *a*)? (4) Wouldn't removing nitrogen simply for the sake of removing nitrogen produce no environmental benefits, be very expensive, and divert valuable resources from other significant environmental issues?

Back to the problem of the decreased oyster (and crab) population in the Lower James River/Chesapeake Bay region. Why have the oyster and crab populations decreased? One theory states that because the tributaries feeding the Lower Chesapeake Bay (including the James River) carry megatons of sediments into the Bay, they add to the Bay's turbidity problem. When waters are highly turbid, oysters do the best they can to filter out the sediments but eventually their numbers decrease and they fade into the abyss. A similar fate awaits the crab. Highly turbid waters do not allow sunlight to penetrate the murky water. Without sunlight, the seagrasses will not flourish. Without the seagrasses, the crab population declines.

Is this the answer? Is the problem with the Lower Chesapeake Bay and its oyster population and with the Eastern Shore and its crab population related to turbidity? Only solid, legitimate, careful scientific analysis may provide the answer. One thing is certain, though. Before we leap into ill-advised decisions that are not necessarily based on sound science but feel good, we need to step back and size up the situation. Such a sizing-up procedure can be correctly accomplished only through the use of scientific methods.

Obviously, that we need to stop the pollution of our surface water bodies is not in question. However, shouldn't we replace the timeworn and frustrating position that "we must start somewhere" with good common sense and legitimate science?

The bottom line is we shouldn't do anything to our environment until science supports the tampering—and the investment. It is important to point out that from 1984 through 2016, because of reductions made in nutrient and sediment deposition in the Lower Chesapeake Bay Regions, the Bay grasses have rebounded. Will the oyster and crab populations also rebound? Only time will tell.

You could not step twice into the same rivers, for other waters are ever flowing on to you.

—Heraclitus of Ephesus

Author

Frank R. Spellman, PhD, is a retired adjunct assistant professor of environmental health at Old Dominion University, Norfolk, Virginia, and the author of more than 100 books covering topics ranging from concentrated animal feeding operations (CAFOs) to all areas of environmental science and occupational health. Many of his texts are readily available online, and several have been adopted for classroom use at major universities throughout the United States, Canada, Europe, and Russia; two have been translated into Spanish for South American markets. Dr. Spellman has been cited in more than 850 publications. He serves as a professional expert witness for three law groups and as an incident/accident investigator for the U.S. Department of Justice and a northern Virginia law firm. In addition, he consults on homeland security vulnerability assessments for critical infrastructures, including water/wastewater facilities, and conducts pre-Occupational Safety and Health Administration and Environmental Protection Agency audits throughout the country. Dr. Spellman receives frequent requests to co-author with well-recognized experts in several scientific fields; for example, he is a contributing author to the prestigious text *The Engineering Handbook*, 2nd ed. Dr. Spellman lectures on wastewater treatment, water treatment, and homeland security, as well as on safety topics, throughout the country and teaches water/wastewater operator short courses at Virginia Tech in Blacksburg. In 2011, he traced and documented the ancient water distribution system at Machu Picchu, Peru, and surveyed several drinking water resources in Amazonia, Ecuador. He has also studied and surveyed two separate potable water supplies in the Galapagos Islands, in addition to studying Darwin's finches while there. Dr. Spellman earned a BA in public administration, a BS in business management, an MBA, and both an MS and a PhD in environmental engineering.

Acronyms, Abbreviations, and Conversion Factors

ACRONYMS AND ABBREVIATIONS

°C	Degrees centigrade or Celsius
°F	Degrees Fahrenheit
μ	Micron
μg	Microgram
μm	Micrometer
A/O	Anoxic/oxic
A2/O	Anaerobic/anaerobic/oxic
AC	Alternating current
ACEE	American Council for an Energy Efficient Economy
Al_3	Aluminum sulfate (alum)
Amp	Amperes
ANAMMOX	Anaerobic ammonia oxidation
APPA	American Public Power Association
AS	Activated sludge
ASCE	American Society of Civil Engineers
ASE	Alliance to Save Energy
AT 3	Aeration tank 3 process
Atm	Atmosphere
AVS	Acid volatile sulfide
AWWA	American Water Works Association
BABE	Bio-augmentation batch enhanced
BAF	Biological aerated filter
BAR	Bioaugmentation reaeration
BASIN	Biofilm activated sludge innovative nitrification
BEP	Best efficiency point
bhp	Brake horsepower
BNR	Biological nutrient removal
BOD	Biochemical oxygen demand
BOD/TKN	Ratio of biochemical oxygen demand to total Kjeldahl nitrogen
BOD/TP	Ratio of biochemical oxygen demand to total phosphorus
BPR	Biological phosphorus removal
CANON	Completely autotrophic nitrogen removal over nitrate
CAS	Cyclic activated sludge
CBOD	Carbonaceous biochemical oxygen demand
CCCSD	Central Contra Costa Sanitary District
CEC	California Energy Commission

CEE	Consortium for Energy Efficiency
cfm	Cubic feet per minute
CFO	Cost flow opportunity
cfs	Cubic feet per second
CHP	Combined heat and power
Ci	Curie
COD	Chemical oxygen demand
CP	Central plant
CV	Coefficient of variation
CWSRF	Clean Water State Revolving Fund
DAF	Dissolved-air flotation unit
DCS	Distributed control system
DO	Dissolved oxygen
DOE	Department of Energy
DON	Dissolved organic nitrogen
DSIRE	Database of State Incentives for Renewables and Efficiency
EBPR	Enhanced biological phosphorus removal
ECM	Energy conservation measure
EPA	Environmental Protection Agency
EPACT	Energy Policy Act
EPC	Energy performance contracting
EPRI	Electric Power Research Institute
ESCO	Energy services company
$FeCl_3$	Ferric chloride
FFS	Fixed-film system
GAO	Glycogen accumulating organism
GBMSD	Green Bay (Wisconsin) Metropolitan Sewerage District
gpd	Gallons per day
gpm	Gallons per minute
H_2CO_3	Carbonic acid
HCO_3^-	Bicarbonate
HDWK	Headworks
hp	Horsepower
HRT	Hydraulic retention time
Hz	Hertz
I&C	Instrumentation and control
I&I	Inflow and infiltration
IFAS	Integrated fixed-film activated sludge
IOA	International Ozone Association
IUVA	International Ultraviolet Association
kWh	Kilowatt hour
kWh/year	Kilowatt hours per year
LPHO	Low pressure high output
M	Mega
M	Million
MBR	Membrane bioreactor

MG	Million gallons
mg/L	Milligrams per liter (equivalent to parts per million)
MGD	Million gallons per day
MLE	Modified Ludzack–Ettinger process
MLSS	Mixed liquor suspended solids
MPN	Most probable number
MW	Molecular weight
N	Nitrogen
NAESCO	National Association of Energy Service Companies
NEMA	National Electrical Manufacturers Association
NH_4	Ammonium
NH_4-N	Ammonia nitrogen
NL	No limit
NPDES	National Pollutant Discharge Elimination System
NYSERDA	New York State Energy Research and Development Authority
O&M	Operations and maintenance
ORP	Oxidation–reduction potential
Pa	Pascal
PAO	Phosphate accumulating organism
PG&E	Pacific Gas and Electric
PID	Phased isolation ditch
PLC	Programmable logic controller
PO_4^{3-}	Phosphate
POTW	Publicly owned treatment works
PSAT	Pump system assessment tool
psi	Pounds per square inch
psig	Pounds per square inch gauge
RAS	Return activated sludge
rpm	Revolutions per minute
SBR	Sequencing batch reactor
SCADA	Supervisory Control and Data Acquisition
SCFM	Standard cubic feet per minute
SRT	Solids retention time
TDH	Total dynamic head
TKL	Total Kjeldahl nitrogen
TMDL	Total maximum daily load
TN	Total nitrogen
TP	Total phosphorus
TSS	Total suspended solids
TVA	Tennessee Valley Authority
UV	Ultraviolet light
UVT	Ultraviolet light transmittance
VFD	Variable frequency drive
VSS	Volatile suspended solids
W	Watt
WAS	Waste activated sludge

WEF	Water Environment Federation
WERF	Water Environment Research Foundation
WMARSS	Waco Metropolitan Area Regional Sewer System
WPCP	Water pollution control plant
WRF	Water Research Foundation
WSU	Washington State University
WWTP	Wastewater treatment plant

CONVERSION FACTORS

Inch/Pound to SI

Multiply	by	To Obtain
Cubic feet per second (ft^3/s)	0.02832	Cubic meters per second
Cubic yards (yd^3)	0.7646	Cubic meters
Feet (ft)	0.3048	Meters
Feet per foot (ft/ft)	1.0	Meters per meter
Feet per second (ft/s)	0.3048	Meters per second
Feet per squared second (ft/s^2)	0.305	Meters per squared second
Inches (in.)	2.54	Centimeters
Inches (in.)	25.4	Millimeters
Miles (mi)	1.609	Kilometers
Pascals (Pa)	1.0	Newtons per square meter
Pounds per cubic foot (lb/ft^3)	0.00000624	Milligrams per liter
Pounds per foot-squared second ($lb/ft{\cdot}s^2$)	1.488	Newtons per square meter
Square miles (mi^2)	2.590	Square kilometers
Tons	907.185	Kilograms
Tons	0.907	Metric tons
Tons per day (ton/d)	0.01050	Kilograms per second

SI to Inch/Pound

Multiply	by	To Obtain
Grams (g)	0.03527	Ounces, avoirdupois
Kilograms per cubic meter (kg/m^3)	0.063	Pounds per cubic feet
Kilograms per meter per squared second ($kg/m{\cdot}s^2$)	1.0	Newtons per square meter
Meters (m)	3.281	Feet
Meters per second (m/s)	3.281	Feet per second
Meters per second squared (m/s^2)	3.281	Feet per second squared
Millimeters (mm)	0.039	Inches
Newtons per square meter (N/m^2)	0.6719	Pounds per foot squared per second
Newtons per square meter (N/m^2)	1.0	Pascal (Pa)
Square meters per second (m^2/s)	1.076	Square feet per second

Section I

Sediment Fundamentals

1 Transformation from Mountain to Sediment

HOW MOUNTAINS BECOME SEDIMENT

Early in the spring, on a snow- and ice-covered high alpine meadow, the water cycle can be readily observed. The main component of the cycle—water—has been held in reserve, literally frozen during the long, dark winter months. But, when the longer, warmer spring days arrive, the sun is higher, more direct, and of longer duration, and the frozen masses of water respond to the increased warmth. The melt begins with a single drop, then two, then three, then more and more until the beginning of a high-gradient rill is formed. As the snow and ice melt, the drops join a vigorous chorus that seems to have no end; they fall from their ice-bound lip to the bare rock and soil terrain below. A chunk of Laramide orogeny period rock, a 4113-mm boulder that formed 80 to 55 million years ago (Ma) during the late Cretaceous to Paleocene, finally succumbs to gravity and falls with the cascading drops of melt water. The snowmelt and the boulder meet on terrain that is not like glacial till, the unconsolidated, heterogeneous mixture of clay, sand, gravel, and boulders that have been dug out, ground out, and exposed by the force of a huge, slow, and inexorably moving glacier. Instead, this rocky terrain is exposed to the falling drops of snowmelt because of a combination of wind and the tiny but powerful force of the drops of water, season after season. The thunderous and earth-shaking encounter of the falling boulder with the surface below removes any lingering groundcover not yet removed by the snowmelt drops. The boulder breaks and disintegrates into a couple of smaller 512- to 256-mm-sized boulders (a process that will continue as headward erosion and bifurcation work in harmony to reduce a giant down to sediment). The combination of falling snowmelt drops and boulders removes any possible semblance of soil or earthy flesh and exposes the underlying, intimate bones of the globe.

Gradually, the single drops increase to a small rush. They join to form a splashing, rebounding, helter-skelter cascade, many separate rivulets that trickle and sparkle in the bright sunlight and then run their way down the face of the granite mountain. A few of the smaller boulders continue their downward journey and also fall and run their way down the face of the mountain. At an indented ledge halfway down the mountain slope, a pool forms whose beauty, clarity, phantom blue eye, and sweet iciness provide the visitor with an incomprehensible, incomparable gift—a blessing from Earth.

The mountain pool fills slowly, tranquil under the sunlit blue sky, reflecting the pines, snow, and sky that surround it, offering an open invitation to lie down and drink and to peer into the deep, glass-clear waters, so clear that it seems possible to reach down over 50 feet and touch the very bowels of the mountain. The pool has no transition from shallow margin to depth; it is simply deep and pure. But the pool's depth is not permanent. Like all things in Nature, the pool changes. The pool changes in size and depth not because of the onslaught of melt water but because of the boulders that now and then impact it with a gigantic splash. Many of the boulders sink to the bottom, raising the height of the floor level of the pool and reducing the depth of the pool, causing more snowmelt water to pour over the edge.

As the pool fills with more melt water, the boulders settle in its depths, and the pool's surface is glasslike again, we wish to freeze time, to hold this place and this pool in its perfect state forever; it is such a rarity to us in our modern world. But, Mother Nature calls, ever prodding, urging, pushing. For a brief instant, the water laps in the breeze against the outermost edge of the ridge, then a trickle flows over the rim, and then a smaller cobble (122 mm), teeter-totters on the edge. Eventually, the giant hand of gravity reaches out and tips the overflowing melt and swaying cobble onward, and they continue their downward journey, following the path of least resistance to their next destination, several thousand feet below.

When the overflow hits the angled, broken rocks below, it bounces, bursts, and mists its way against steep, V-shaped walls that form a small, deep valley, carved out over time by water and the forces of Earth, still high in altitude but with a rock-strewn bed bent downward, toward the sea. The smaller cobble, unlike the growing stream of melt water, is decreasing in size due to the several violent collisions it has endured and is now coarse gravel (25 mm) as it accompanies the melt water downward.

Within the valley confines, the melt water has grown from drops to rivulets to a small formation of flowing water. The former boulder is still coarse gravel size. And flow and tumble along they do, through what is at first a narrow, V-shaped valley that widens to form more of a U-shape. The journey continues, as the water and cobble pick up speed and spill and tumble over massive stream-bound boulders, but then the flow slows.

At a larger, but shallower pool, waters from higher elevations have joined the main body, and our coarse gravel is now fine gravel (5 mm) that has come from the hillsides, from crevices, from springs, from rills, from mountain creeks. At the influent poolsides, all appears peaceful, quiet, and restful, but not far away, at the effluent end of the pool, gravity takes control again. The overflow of melt water and fine gravel pour over the jagged lip and cascade downward several hundred feet, with the waterfall carrying its load toward a violent, crushing, mist-filled landing.

The water separates and joins again and again, forming a deep, furious, wild stream that calms gradually as it continues to flow over lands that are less steep. The gravel piggybacks the spinning water flow, which eventually widens into pools overhung by vegetation and surrounded by tall trees. The pure, crystalline waters have become progressively more discolored on their downward journey; they are now stained brown with humic acid and tannins and are literally filled with suspended noncohesive sediments. The once pure stream is now muddy. The fine gravel has become very fine (2 mm) and noncohesive, and it continues to roll, skid, and skip along the bedlayer.

The mass divides and flows in different directions, over different landscapes. Small streams divert and flow off into open country. Different soils work to retain or speed up the waters, and in some places the waters spread out into shallow swamps, bogs, marshes, fens, or mires. Other streams pause long enough to fill deep depressions in the land and form lakes. For a time, the water pauses and slows in its journey to the sea, but only for the short term, because lakes are only a temporary resting place in the water cycle. The water will eventually move on due to evaporation or seepage into groundwater or by forming another stream. The splintered, broken, crushed, and somewhat smoother and rounded sediments continue to be carried by the flow of water. Many of the larger pieces settle at the openings to lakes and ponds, while smaller pebbles spread out to deeper reaches, and sand and silt deposit in deeper areas closer to the stream outlet. A smaller amount of the rocky load continues to be carried along with the main flow as it reforms the river once again. Leaving some of itself behind, the water continues to flow but changes and forms another river, which braids its way through the landscape and heads for the sea. As it continues its downward journey toward sea level, the flow of water slows down; the river bottom changes from rock and stone to silt and clay. Plants begin to grow, stems thicken, and leaves broaden. The river is now full of life and the nutrients needed to sustain life, but still it courses onward to its destination, finally spilling into the sea.

What happened to the very fine gravel? Did it disappear? Did it spill into the sea?

Well, yes and no. Some reached the sea as sand (1 mm).

The rest of the sediment?

The other sediment settled along, on, and in the river beds during the course of the journey.

Did it settle permanently, forever?

No. The fine gravel and sand, remnants of the mighty boulder that fell from the mountain, are now settled but not permanently. Remember, nothing in nature is permanent. We are all guests, short-term guests. And the residence of the settled sand (0.125 mm) is temporary. The elements of shear forces, measureable amounts of motion caused by critical shear stress, hyporheic exchange, bioturbation, groundwater upwelling, sediment resuspension, and benthic infaunal activity all work alone or in harmony to lift and move the sediment to different locations. The inexorable influence of erosion will convert some sediments to a cohesive form while other sediments will remain noncohesive, at least for the time being. In their new locations these sediments will rest or move on, perhaps becoming contaminated. With the passage of time and flow, these contaminated sediments will be biotransformed. At least that is our expectation … our hope.

This is how it works. This is how mountains become sediments. This is how Nature works.

ANOTHER PERSPECTIVE

Let's approach the formation of sediment from a different perspective. The end product produced in the following process is much like the sediment produced by the fluvial processes described above, but water is not the main factor in this particular account. If modern humans could be transported back in time, we would instantly recognize a massive structure rising before us, even though we might be taken aback at what we are seeing: a youthful mountain range with considerable mass, steep sides, and a height that could certainly reach beyond any cloud. One particular peak—the tallest, most massive one—would grab our attention, with its polyhedron-shape, polygonal base, and triangular faces culminating in a single sharp-tipped apex. We would think it was comparable in shape, though larger in size, to the largest of the great Egyptian pyramids, even though the pyramids were originally covered in a sheet of limestone, not the thick, perpetual sheet of solid ice and snow covering this mountain peak.

Returning to the present day, if we were to walk this same site the changes would be obvious and startling—and entirely relative to time. What stood as an incomparable mountain peak eons ago has today lost its ancient majesty. In fact, we wouldn't give it a second thought as we walked across its remnants and through the vegetation growing from its pulverized and amended remains.

Over 300 million years ago, the pyramid-shaped mountain peak stood in full, unchallenged splendor above the clouds, wrapped in a cloak of ice, a mighty fortress of stone seemingly vulnerable to nothing, standing higher than any mountain ever stood—or ever would stand—on Earth. And so it stood, for millions upon millions of passings of the Earth around the sun. Born when Mother Earth took a deep breath, the pyramid-shaped peak stood tall and undisturbed until millions of years later when Mother Earth stretched. Today we would call this stretch a massive earthquake—humans have never witnessed one of such magnitude. Rather than registering on the Richter scale, it would have destroyed it. But when this massive earthquake shattered the Earth's surface, nothing we would call intelligent life lived on Earth—and it's a good thing.

During this massive upheaval, the peak shook to its very foundations, and the initial shockwave and the hundred-plus aftershocks fractured its solid granite structure. This fracture was so massive that each aftershock widened it and loosened the base foundation of the peak itself. Only 10,000 years later (a few seconds relative to geologic time), the fracture's effects totally altered the shape of the peak forever. During a horrendous windstorm, one of an intensity known only in Earth's earliest days, a sharp tremor (emanating from deep within the Earth and shooting up the spine of the mountain itself, up to the very peak) widened the gaping wound still more.

Decades of continued tremors and terrible windstorms passed (no present-day structure could withstand a blasting from such a wind), and finally the highest peak of the time, of all time, fell. It broke off completely at its base, and, following the laws of gravity (as effective and powerful a force then as today, of course), tumbled and fell more than 20,000 feet, straight down. It collided with the expanding base of the mountain range, the Earth-shattering impact destroying several thousand acres. It finally came to rest (what remained intact) on a precipitous ledge, at 15,000 feet in elevation. The pyramid-shaped peak, much smaller now, sat precariously perched on the precipitous ledge for about 5 million years.

Nothing, absolutely nothing, is safe from time. The most inescapable natural law is that of entropy. Time and entropy mean change and decay—harsh, sometimes brutal, but always inevitable. The bruised, scarred, truncated, but still massive rock form, once a majestic peak, was now a victim of Nature, which, with its chief ally, time, at its side, works to degrade anything and everything that has substance and form. For better or for worse, in doing so, Nature is ruthless, sometimes brutal, and always inevitable—but never without purpose.

While resting on the ledge, the giant rock, over the course of that 5 million years, was exposed to constantly changing conditions. For several thousand years, Earth's climate was unusually warm—almost tropical—everywhere. Throughout this relatively warm era, the rock was not covered with ice and snow, but instead baked in intense heat, steamed in hot rain, and seared in the gritty, heavy windstorms that arose and released their abrasive fury, sculpting the rock's surface each day for more than 10,000 years.

Then came a pause in the endless windstorms and upheavals of the young planet, a span of time when the weather was not furnace hot or arctic cold, but moderate. The rock was still exposed to sunlight but at lower temperatures, to rainfall at increased levels, and to fewer windstorms of increased fury. The climate remained so for some years until the cycle repeated itself over and over—arctic cold, moderately warm, furnace hot.

During the last of these cycles, the rock, considerably affected by physical and chemical exposure, was reduced in size and shape even more. Considerably smaller now than when it landed on the ledge, a mere pebble compared to its former size, it fell again, 8000 feet to the base of the mountain range, where it came to rest on a bed of talus. Reduced in size still more, it remained on its sloping talus bed for many more thousands of years.

Somewhere around 15,000 BC, the rock form, continuously exposed to chemical and mechanical weathering, its physical structure weakened by its long-ago falls, fractured and broke into rocks of ever-decreasing size until the largest intact fragment left from the original rock was no bigger than a four-bedroom house. As time rolled on, until about the time the Egyptians were building their pyramids, this rock continued to be reduced in size, to roughly ten feet square. Over the next thousand years, it continued to shrink to the size of a beach ball due to wearing, crumbling, and flaking away; it was surrounded by fragments of its former self. The rock became covered with moss and lichen, with a web of fissures, tiny crevices, and fractures woven throughout.

Over the next thousand or so years, via *bare rock succession*, what had once been the mother of all mountain peaks, the highest point on Earth, had been reduced to nothing more than a handful of soil. How did this happen? What is bare rock succession? If a layer of soil is completely stripped off land by natural means (e.g., water, wind), by anthropogenic means (tillage plus erosion), or by cataclysmic occurrence (a massive landslide or earthquake), only after many years can a soil-denuded area return to something approaching its original state or can a bare rock be converted to soil. But, given enough time—perhaps a millennium—the scars heal over, and a new, virgin layer of soil forms where only bare rock once existed. The series of events that take place in this restoration process are known as bare rock succession, which is indeed a true succession with identifiable stages succeeding the stages that existed earlier.

Bare rock, no matter how it is laid open to view, is exposed to the atmosphere. The geologic processes that cause weathering break down the surface into smaller and smaller fragments. Many forms of weathering exist, and all effectively reduce bare rock surfaces to smaller particles or chemicals in solution.

FIGURE 1.1 The first stage of bare rock succession: lichens growing on bare rock. (Photograph by author.)

Lichens appear to cover the bare rock first. These hardy plants grow on the rock itself (see Figure 1.1). They produce weak acids that assist in the slow weathering of the rock surface. The lichens also trap wind-carried soil particles, which eventually produce a very thin soil layer—a change in environmental conditions that gives rise to the next stage in bare rock succession. Mosses replace lichens, thriving in the meager soil the lichens and weathering provide. They produce a larger growing area and trap even more soil particles, providing a more moist bare rock surface. The combination of more soil and moisture establishes abiotic conditions that favor the next succession stage.

Now the seeds of herbaceous plants invade what was once bare rock. Grasses and other flowering plants take hold. Organic matter provided by the dead plant tissue is added to the thin soil, while the rock still weathers from below. More and more organisms join the community as it becomes larger and more complex. By this time, the plant and animal community is fairly complicated. The next major invasion is by weedy shrubs that can survive in the amount of soil and moisture present. As time passes, the process of building soil speeds up as more and more plants and animals invade the area. Soon trees take root and forest succession is evident. Many years are required, of course, before a climax forest will grow here, but the scene is set for that to occur.

Today, only the remnants of the former, incomparable, pyramid-shaped peak are left, in the form of sediment—sediment packed full of organic humus, sediment that looks like rocky road chocolate (pebbled mud) when wet. When the sediment is dry, though, most people would think of it simply as a handful of pebbles, just plain old dirt. And that is what it is.

2 Setting the Stage

Probably the best way to understand where this book is headed, to get a really good feel for it, to get to the heart of what the message is all about is to read the following by Rachel Carson (1962):

> We poison the caddis flies in a stream and the salmon runs dwindle and die. We poison the gnats in a lake and the poison travels from link to link of the food chain and soon the birds of the lake margins become victims. We spray our elms and the following springs are silent of robin song, not because we sprayed the robins directly but because the poison traveled, step by step, through the now familiar elm leaf–earthworm–robin cycle. These are matters of record, observable, part of the visible world around us. They reflect the web of life—or death—that scientists know as ecology.

As Carson pointed out, what we do to any part of our environment has an impact on other parts. In other words, there is an interrelationship between the parts that make up our environment. Probably the best way to state this interrelationship is to point out that on Earth it is all about specific interactions that exist between organisms and their living and nonliving environment.

BENTHIC HABITAT

Before setting the stage for the discussion that follows, a brief look at the benthic habitat is called for. The benthic habitat is found in a streambed, or benthos. A streambed is comprised of various physical and organic materials where erosion and deposition are continuous characteristics. Erosion and deposition may occur simultaneously and alternately at different locations in the same streambed. Where channels are exceptionally deep and taper slowly to meet the relatively flattened streambed, habitats may form on the slopes of the channel. These habitats are referred to as *littoral habitats*. Shallow channels may dry up periodically in accordance with weather changes. The streambed is then exposed to open air and may take on the characteristics of a wetland.

Silt and organic materials settle and accumulate in the streambed of slowly flowing streams. These materials decay and become the primary food resource for the invertebrates inhabiting the streambed. Productivity in this habitat depends on the breakdown of these organic materials by herbivores. Bottom-dwelling organisms do not use all of the organic materials; a substantial amount becomes part of the streambed in the form of peat.

In faster moving streams, organic materials do not accumulate so easily. Primary production occurs in a different type of habitat found in the riffle regions with shoals and rocky regions for organisms to adhere to. Plants that can root themselves into the streambed dominate these regions. By plants, we are referring mostly to forms of algae, often microscopic and filamentous, that can cover rocks and debris that have settled into the streambed during summer months.

Note: If you have ever stepped into a stream, the green, slippery slime on the rocks you encountered in the streambed is representative of this type of algae.

Although the filamentous algae seem well anchored, strong currents can easily lift the algae from the streambed and carry it downstream, where it becomes a food resource for low-level consumers. One factor that greatly influences the productivity of a stream is the width of the channel; a direct relationship exists between stream width and richness of bottom organisms. Bottom-dwelling organisms are very important to the ecosystem, as they provide food for other, larger benthic organisms through consuming detritus.

BENTHIC PLANTS AND ANIMALS

Vegetation is not common in the streambed of slow-moving streams; however, vegetation may anchor along the banks. Algae (mainly green and blue–green) as well as common types of water moss attach themselves to rocks in fast-moving streams. Mosses and liverworts often climb up the sides of the channel onto the banks, as well. Some plants similar to the reeds of wetlands with long stems and narrow leaves are able to maintain roots and withstand the current. Aquatic insects and invertebrates dominate slow-moving streams. Most aquatic insects are in their larval and nymph forms such as the blackfly, caddisfly, and stonefly. Adult water beetles and waterbugs are also abundant. Insect larvae and nymphs provide the primary food source for many fish species, including American eel and brown bullhead catfish. Representatives of crustaceans, rotifers, and nematodes (flat worms) are sometimes present. The abundance of leeches, worms, and mollusks (especially freshwater mussels) varies with stream conditions but generally favors low-phosphate conditions. Larger animals found in slow-moving streams and rivers include newts, tadpoles, and frogs. The important characteristic of all life in streams is adaptability to withstand currents.

BENTHIC MACROINVERTEBRATES

The emphasis on aquatic insect studies, which have expanded exponentially in the last several decades, has been largely ecological. Freshwater macroinvertebrates are ubiquitous; even polluted waters contain some representative of this diverse and ecologically important group of organisms. Benthic macroinvertebrates are aquatic organisms without backbones that spend at least a part of their life cycle on the stream bottom. Examples include aquatic insects, such as stoneflies, mayflies, caddisflies, midges, and beetles, as well as crayfish, worms, clams, and snails. Most hatch from eggs and mature from larvae to adults. The majority of the insects spend their larval phase on the river bottom and, after a few weeks to several years, emerge as winged adults. The aquatic beetles, true bugs, and other groups remain in the water as adults. Macroinvertebrates typically collected from the stream substrate are either aquatic larvae or adults. In practice, stream ecologists observe indicator organisms and their responses to determine the quality of the stream environment. A number of methods can be used for determining water quality based on biologic characteristics. A wide variety of indicator organisms (biotic groups) are used for biomonitoring. The most often used include algae, bacteria, fish, and macroinvertebrates.

In stream ecology studies, benthic macroinvertebrates are studied for a number of reasons. Simply, they offer a number of advantages:

1. They are ubiquitous, so they are affected by perturbations in many different habitats.
2. They are species rich, so the large number of species produces a range of responses.
3. They are sedentary, so they stay put, which allows determination of the spatial extent of a perturbation.
4. They are long-lived, which allows temporal changes in abundance and age structure to be followed.
5. They integrate conditions temporally, so like any biotic group they provide evidence of conditions over long periods.

In addition, benthic macroinvertebrates are preferred as bioindicators because they are easily collected and handled by samplers; they require no special culture protocols. They are visible to the naked eye and samplers easily distinguish their characteristics. They have a variety of fascinating adaptations to stream life. Certain benthic macroinvertebrates have very special tolerances and thus are excellent specific indicators of water quality. Useful benthic macroinvertebrate data are easy to collect without expensive equipment. The data obtained by macroinvertebrate sampling can serve to indicate the need for additional data collection, possibly including water analysis and fish sampling.

WATER QUALITY AND SEDIMENTS

Traditionally, concerns relative to the management of aquatic resources in freshwater systems have focused primarily on water quality. The focus has been on questions or concerns related to, for example, what was in the water that you just drank? Was it only hydrogen and oxygen atoms? More importantly, was it safe for drinking? All water is of a certain quality, which we cannot determine solely by looking at it. The real question is what does "water quality" really mean? Water loaded with sediments such as dirt and grime might work fine for a lettuce plant but would you want to drink it? Some would think this question relative—relative to whether or not you are dying of thirst. Anyway, water quality can be thought of as a measure of the suitability of water for a particular use based on selected physical, chemical, and biological characteristics. The bottom line on water quality—that is, on the quality of water you are about to drink or use—is a matter of judgment. Like pollution, water quality, in many instances, can be a judgment call.

Early aquatic resource management efforts were often directed toward ensuring the potability of surface water sources. Subsequently, the scope of these management initiatives expanded to include protection of instream (i.e., fish and aquatic life), agricultural, industrial, and recreational water uses. Although initiatives undertaken in the past several decades have unquestionably improved water quality conditions, a growing body of evidence indicates that management efforts directed solely at the attainment of surface water quality criteria may not provide an adequate basis for protecting the designated uses of aquatic ecosystems (USEPA, 2002).

In recent years, concerns relative to the health and vitality of aquatic ecosystems have begun to surface in North America. One of the principal reasons why is that many toxic and bioaccumulative chemicals can accumulate to elevated levels in sediments; these chemicals include metals, polycyclic aromatic hydrocarbons (PAHs), polychlorinated biphenyls (PCBs), chlorophenols, organochlorine (OC) pesticides, and polybrominated diphenyl ethers (PBDEs), which are found in only trace amounts in water. Some of these pollutants, such as OC pesticides and PCBs, were released into the environment long ago. The use of many of these substances has been banned in North America for more than 40 years; nevertheless, these chemicals continue to persist in the environment. Other contaminants enter our waters every day from industrial and municipal discharges, urban and agricultural runoff, and atmospheric deposition from remote sources. Due to their physical and chemical properties, many of these substances tend to accumulate in sediments. In addition to providing sinks for many chemicals, sediments can also serve as potential sources of pollutants to the water column when conditions change in the receiving water system (e.g., during periods of anoxia, after severe storms).

Information from a variety of sources indicates that sediments in aquatic ecosystems throughout North America are contaminated by a wide range of toxic and bioaccumulative substances, including metals, PAHs, PCBs, OC pesticides, a variety of semivolatile organic chemicals (SVOCs), and polychlorinated dibenzo-*p*-dioxins and furans (PCDDs and PCDFs). Contaminated sediments pose a major risk, for example, to the beneficial uses of aquatic ecosystems throughout the Great Lakes. The imposition of fish consumption advisories has adversely affected commercial, sport, and food fisheries in many areas. In addition, degradation of the benthic community and other factors have adversely affected fish and wildlife populations. Furthermore, fish in many of these areas often have higher levels of tumors and other abnormalities than fish from reference areas. Contaminated sediments have also threatened the viability of many commercial ports because of restrictions placed on dredging navigational channels and the disposal of dredged materials. Overall, contaminated sediments have been linked to beneficial use impairments in Canada and the United States (USEPA, 2002).

KEY TERMS AND DEFINITIONS

Although a glossary of terms and definitions is included at the end of the book, experience has shown that defining basic, fundamental, and important terms at the beginning of a technical presentation is paramount to understanding the technical information that follows. Many of the terms used in the text that follows are presented here first to provide a map to understanding:

Activity—A dimensionless quantity expressing the escaping tendency of a component in a system relative to that in a pure state of the same component at the same temperature and total pressure. It is the ratio of the partial pressures (or, more precisely, the fugacities) of the dissolved component in the system to that of the component at standard state.

Activity coefficient—A dimensionless quantity that corrects for the deviation of the partial pressure (or fugacity) of a component in solution from that value defined by Raoult's law. If the observed partial pressure is greater than that predicted by Raoult's law, the activity coefficient is less than one. Most organic solutes in water show positive deviation from ideality; that is, the activity coefficient is greater than one. This term provides an approximation of how much interaction exists between molecules at higher concentrations.

Adsorption—The interphase accumulation or concentration of substances at a surface or interface. The process can occur at an interface between any two phases, such as liquid–liquid, gas–liquid, gas–solid, or liquid–solid interfaces. The material being concentrated or adsorbed is the adsorbate, and the adsorbing phase is termed the adsorbent.

Advection—The transport of particles due to the motion or velocity of the fluid.

Anthropogenic compounds—Compounds that are produced as a result of the activities of humans as contrasted with compounds formed by the actions of natural forces and events.

Bedload—Sediment that is transported in a stream by rolling, sliding, or skipping along the bed and very close to it (i.e., within the bedlayer).

Bedload layer—A thin layer through which the bedload discharges, commonly assumed to be only a few grain diameters thick.

Benthic—Of the seafloor, or pertaining to organisms living on or in the seafloor.

Bioturbation—Mixing processes in sediment layer caused by the activity of biological organisms.

Boundary layer—The thin layer of fluid next to a solid boundary (e.g., bottom of an estuary) where friction is very important.

Bulk density—The total mass density of sediment and water in a given volume of sediment bed material.

Coagulation—A process where charge neutralization of discrete colloids occurs through interactions with available counterions (i.e., the mobile ions in ion exchange).

Cohesive—Description of sediments, generally less than 200 μm in diameter, that tend to stick together and resist separation.

Colloids—Operationally defined as discrete particles with at least one characteristic dimension in the micrometer to nanometer range.

Critical shear stress—The shear stress at which sediments begin to exhibit a measurable amount of motion.

Diagenesis—The gradual and successive chemical and physical changes that take place in sediment previous to or during its consolidation.

Dissolved—Operationally defined as discrete particles less than 0.45 μm in their maximum dimension.

Epibenthic—Pertaining to organisms living near the seafloor.

Fetch—Distance of water over which the wind blows in essentially a constant direction.

Flocculation—Aggregation of natural discrete colloids into larger masses by the mixing action of water.

Fluvial—Pertaining to rivers or streams.

Flux—The rate of flow of a physical substance (e.g., water or sediments) through a given area.

Fulvic acids—The fraction of organic matter that remains dissolved in solution after sequential extraction of the sample with alkali and then acid.

Humic acids—The fraction of organic matter that remains dissolved in solution after extraction with alkali but that precipitates from solution upon further extraction with acid.

Humins—The fraction of organic matter that is not soluble in either alkali or acid.

Interstitial water—Water occurring in the small openings, spaces, and voids between particles of unconsolidated materials in that portion of the vadose water zone between the root zone and the water table. The water is held in place by entrapment, ionic attraction, and capillary or adhesive forces, rather than from upward pressure components of saturation.

Lipophilic—The characteristic of solutes or solvents that are readily miscible in other organic solvents such as lipids. Lipids as a heterogeneous group of substances include fatty acids, neutral fats, phosphatides, glycolipids, aliphatic alcohols and waxes, terpenes, and steroids. Lipids are categorized by their extractability in nonpolar organic solvents such as chloroform carbon tetrachloride, benzene, ether, carbon disulfide, and petroleum ether.

Miscible—The ability of two or more substances to mix and to form a single, homogeneous phase.

Noncohesive—Description of sediments, generally more than 200 μm in diameter, that exhibit no tendency toward resisting separation.

Particulate—Operationally defined as discrete particles greater than 0.45 μm in their maximum dimension.

Partition—The distribution of a compound between two different bulk phases, usually by a solubilization process.

Polarity—Local charge groupings occurring as a result of geometric asymmetries between atoms of a given molecule. This chemical characteristic can lead to incompatibilities between liquids of different polarities. The relative polarity of large macromolecules, such as the humic and fulvic acids, is measured by comparing the oxygen-to-carbon ratios of the molecules. The larger the magnitude of the ratio, the greater the relative polarity of the compound.

Porewater (or interstitial water)—Subsurface water in an interstice, or pore (Bates and Jackson, 1984). Sediment interstitial water, or porewater, is defined as the water occupying the spaces between sediment particles. Contaminants in the interstitial water and in the solid phase are expected to be at thermodynamic equilibrium. This makes interstitial waters useful for assessing contaminant levels and associated toxicity (USEPA, 2001).

Porosity—The ratio of openings (voids or pores) to the total volume of a soil or rock. Porosity is expressed either as a decimal fraction or as a percentage (Heath, 1983). The porosity of a rock is its property of containing interstices (Meinzer, 1923).

Sediment—Fragmented material that originates from the disintegration of rocks and is transported by, suspended in, or deposited by water or air or is accumulated in beds by other natural processes; it is detached fragmental matter that originates from either chemical or physical weathering of rocks and minerals and is transported by, suspended in, or deposited by water or air or is accumulated in beds by other natural agencies (Osterkamp, 2008).

Shear stress—The force due to friction exerted on a unit area of the sediment bed due to a moving water mass.

Solute—The compound dissolved in solution.

Sorption—General expression for a process in which a component moves from one phase to be accumulated in another. The material being sorbed is the sorbate, and the sorbing phase is termed the sorbent. Sorption is used when it is not certain if the accumulating mechanism is a partitioning, an adsorption, or an absorption process.

Surface microlayer—Upper, very thin layer of a surface water body characterized by high surface tension and some physical and chemical properties distinguishable from the bulk solution.

Surficial sediment layer—Upper sediment layer with a thickness that fluctuates depending on the disposition rates of new sediment and the degrees of resuspension generated by boundary layer turbulence. In this book, the surficial sediment layer is always considered to be submerged.

Suspended load—Sediment particles maintained in the water column by turbulence and carried with the flow of water.

Turbulent diffusion—The movement and dispersal of a mass in the water column due to random turbulent motions in the flow.

REFERENCES AND RECOMMENDED READING

AGI. (1980). *Glossary of Geology*. Falls Church, VA: American Geological Institute.

ASTM. (1980). Standard definitions of terms and symbols relating to soil and rock mechanics. In: *1980 Annual Book of ASTM Standards*, Part 19, pp. 1402–1419. Philadelphia, PA: American Society for Testing and Materials.

Bates, R.L. and Jackson, J.A. (1984). *Dictionary of Geological Terms*. New York: American Geological Institute.

Carson, R. (1962). *Silent Spring*. Boston: Houghton Mifflin.

Heath, R.C. (1983). *Basic Ground-Water Hydrology*, USGS Water Supply Paper 2220. Reston, VA: U.S. Geological Survey.

Meinzer, O.E (1923). *The Occurrence of Ground Water in the United States*, USGS Water Supply Paper 489. Reston, VA: U.S. Geological Survey.

NIH. (2015). *Dicofol*. Bethesda, MD: National Institutes of Health (https://pubchem.ncbi.nlm.nih.gov/compound/dicofol).

Osterkamp, W.R. (2008). *Annotated Definitions of Selected Geomorphic Terms and Related Terms of Hydrology, Sedimentology, Soil Science and Ecology.* USGS Open File Report 2008-1217. Reston, VA: U.S. Geological Survey.

Spellman, F.R. (2014). *The Science of Water*, 3rd ed. Boca Raton, FL: CRC Press.

USGS. (2009a). *Landslide Hazards Program Glossary.* Reston, VA: U.S. Geological Survey.

USGS. (2009b). *Glossary of Hydrologic Terms.* Portland, OR: U.S. Geological Survey, Oregon Water Science Center.

USEPA. (2001). Collection of interstitial water. In: *Methods of Collection, Storage and Manipulation of Sediments for Chemical and Toxicological Analyses—Technical Manual.* Washington, DC: U.S. Environmental Protection Agency, Office of Science and Technology.

USEPA. (2002). *A Guidance Manual to Support the Assessment of Contaminated Sediments in Freshwater Ecosystems*, Vols. I–III. Chicago, IL: Great Lakes National Program Office.

USEPA. (2006). Appendix A: codes used in water-quality processing system. In: *User's Manual for the National Water Information System of the U.S. Geological Survey*, USGS Open Report 2006-1145. Reston, VA: U.S. Geological Survey.

3 Surface Water Sediments

Science affects the way we think together.

—**Lewis Thomas, scientist**

INTRODUCTION

Before providing information on surface water sediments in general, it is important to define surface water itself, at least for the purposes of this book. Keep in mind that the focus herein is on freshwater surface systems. Saltwater systems and deposited sediments are important and involve discussions very similar to those presented here for freshwater surface water sediments, but our focus here is on surface water sediments.

SURFACE WATER

Approximately 40 million cubic miles of water cover or reside within the Earth. The oceans contain about 97% of all water on Earth. The other 3% is freshwater: (1) snow and ice on the surface of the Earth which contain about 2.25% of the water; (2) usable groundwater, which accounts for approximately 0.3%; and (3) surface freshwater, which is less than 0.5%. In the United States, for example, average rainfall is approximately 2.6 feet (a volume of 5900 km^3). Of this amount, approximately 71% evaporates (about 4200 cm^3), and 29% goes to stream flow (about 1700 cubic km^3).

Beneficial freshwater uses include manufacturing, food production, domestic and public needs, recreation, hydroelectric power production, and flood control. Total U.S. water withdrawals in 2005 were 410,000 million gallons per day; 85% of that was freshwater withdrawals, and surface water supplied 80% of all water withdrawals. Withdrawals for thermoelectric power generation, primarily coal, nuclear, and natural gas, were 201,000 million gallons per day. The remainder went toward domestic use, irrigation, livestock, aquaculture, mining, and various industrial uses (USGS, 2009). Historically, in the United States, water usage increased through 1980 but has declined somewhat. For example, in 1975, just under 350 billion gallons of freshwater were used per day. By 1980, about 375 billion gallons of freshwater were used each day. In 2010, it was estimated that just over 300 billion gallons of freshwater were used per day (Donnelly and Cooley, 2015).

The primary sources of freshwater include the following:

- Surface water from lakes, rivers, and streams
- Groundwater from springs, artesian wells, and drilled or dug wells
- Captured and stored rainfall in cisterns and water jars
- Desalinized seawater or brackish groundwater
- Reclaimed wastewater

Current federal drinking water regulations actually define three distinct and separate sources of freshwater: (1) surface water, (2) groundwater, and (3) groundwater under the direct influence of surface water (GUDISW). This last classification is the result of the Surface Water Treatment Rule (SWTR). The definition of what conditions constitute GUDISW, while specific, are not obvious.

As stated earlier, the focus in this book is on surface sources of freshwater, specifically because sediment deposition within surface waters, sediment transport within surface waters, and the contamination of sediments within surface waters are important issues for the overall environmental health of Earth. It is important to point out that surface waters are not uniformly distributed over the Earth's surface. In the United States, for example, only about 4% of the landmass is covered by rivers, lakes, and streams. The volumes of these freshwater sources depend on geographic, landscape, and temporal variations, as well as on the impact of human activities. Surface water is water that is open to the atmosphere and results from *overland flow* (i.e., *runoff* that has not yet reached a definite stream channel). Put a different way, surface water is the result of surface runoff. For the most part, however, *surface* (as used in the context of this text) refers to water flowing in streams and rivers, as well as water stored in natural or artificial lakes, manmade impoundments such as lakes made by damming a stream or river, springs that are affected by a change in level or quantity, shallow wells that are affected by precipitation, wells drilled next to or in a stream or river, rain catchments, and muskeg and tundra ponds.

Advantages and Disadvantage of Surface Water Supplies

The biggest advantage of using a surface water supply as a water source is that these sources are readily located; finding surface water sources does not demand sophisticated training or equipment. Many surface water sources have been used for decades and even centuries (in the United States, for example), and considerable data are available on the quantity and quality of the existing water supply. Surface water is also generally softer (i.e., not mineral laden), which makes its treatment much simpler.

The most significant disadvantage of using surface water as a water source is *pollution*. Surface waters are easily contaminated with microorganisms that cause waterborne diseases and chemicals that enter the river or stream from surface runoff and upstream discharges. Another problem with many surface water sources is *turbidity*, which fluctuates with the amount of precipitation. Increases in turbidity increase treatment costs and operator time. Surface water temperatures can be a problem because they fluctuate with ambient temperature, making consistent water quality production at a waterworks plant difficult. Drawing water from a surface water supply might also present problems; for example, intake structures may clog or become damaged from winter ice, or the source may be so shallow that it completely freezes in the winter. *Water rights* are another issue, in that removing surface water from a stream, lake, or spring requires a legal right. The lingering, seemingly unanswerable, question is who owns the water? Using surface water as a source means that the purveyor is obligated to meet the requirements of the Surface Water Treatment Rule and Interim Enhanced Surface Water Treatment Rule (IESWTR), which applies only to large public water systems (PWSs) serving more than 10,000 people. The IESWTR tightened controls on disinfection byproducts and turbidity and regulates *Cryptosporidium*.

Surface Water Hydrology

To properly manage and operate water systems, it is important to have a basic understanding of the movement of water and the factors that affect water quality and quantity—in other words, *hydrology*. A discipline of applied science, hydrology includes several components, such as the physical configuration of the watershed, the geology, soils, vegetation, nutrients, energy, wildlife, and the water itself. The area from which surface water flows is a *drainage basin* or catchment area. With a surface water source, this drainage basin is most often referred to, in nontechnical terms, as a *watershed* (when dealing with groundwater, we call this area a *recharge area*).

Note: The area that directly influences the quantity and quality of surface water is called the *drainage basin* or *watershed*.

When we trace on a map the course of a major river from its meager beginnings along its seaward path, it is readily apparent that its flow becomes larger and larger. Every tributary adds to its size, and between tributaries the river grows gradually

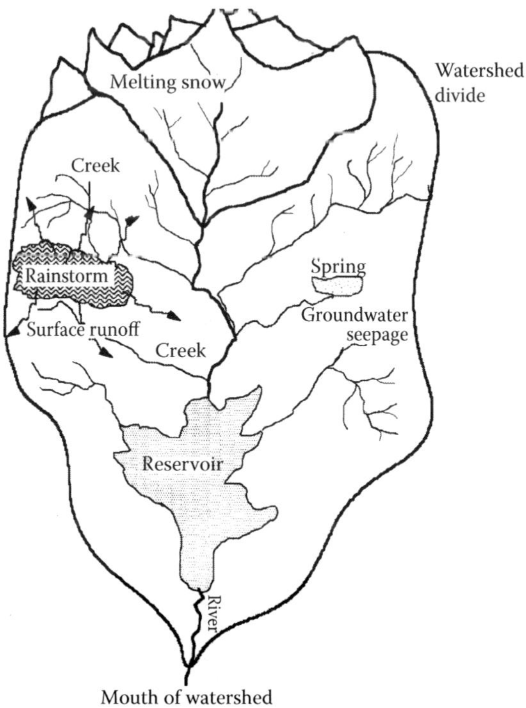

FIGURE 3.1 Typical watershed.

due to overland flow entering it directly (see Figure 3.1). Not only does the river grow, but its entire watershed or drainage basin, basically the land it drains into, also grows in the sense that it embraces an ever-larger area. The area of the watershed is commonly measured in square miles, sections, or acres. When taking water from a surface water source, knowing the size of the watershed is desirable.

Surface Water Quality

Surface waters should be of adequate quality to support aquatic life and be aesthetically pleasing, and waters used as sources of supply should be treatable by conventional processes to provide potable supplies that can meet the drinking water standards. Many lakes, reservoirs, and rivers are maintained at a quality suitable for swimming, water skiing, and boating as well as for drinking water. Whether the surface water supply is taken from a river, stream, lake, spring, impoundment, reservoir, or dam, surface water quality varies widely, especially in rivers, streams, and small lakes. These water bodies are not only susceptible to waste discharge contamination but also to "flash" contamination (can occur almost immediately and not necessarily over time). Lakes are subject to summer/winter stratification (turnover) and to algal blooms. Pollution sources range from runoff (agricultural, residential, and urban) to spills, municipal and industrial wastewater discharges, and recreational users, as well as from natural occurrences. Surface water supplies are difficult to protect from contamination and must always be treated.

SOIL VS. DIRT

> Soils are crucial to life on earth … soil quality determines the nature of plant ecosystems and the capacity of land to support animal life and society. As human societies become increasingly urbanized, fewer people have intimate contact with the soil, and individuals tend to lose sight of the many ways in which they depend upon soils for their prosperity and survival. The degree to which we are dependent on soils is likely to increase, not decrease, in the future. Of course, soils will continue to supply us with nearly all of our food and much of our fiber. On a hot day, would you rather wear a cotton shirt or one made of polyester? In addition, biomass grown on soils is likely to become an increasingly important source of energy and industrial feedstocks, as the world's finite supplies of petroleum are depleted over the coming century. The early signs of this trend can be seen in the soybean oil-based inks, the cornstarch plastics, and the wood alcohol fuels that are becoming increasingly important on the market.
>
> **Brady and Weil (1996, p. 2)**

In any discussion about sediments and their eventual contamination, we must initially describe, explain, and define exactly what soil is and why soil is as important to us as air and water. First, we must also clear up a major misconception about soil. People often confuse soil with dirt. Soil is not dirt. Dirt is misplaced soil—soil where we don't want it, contaminating our hands or clothes, tracked in on the floor.

Dirt we try to clean up and keep out of our environment. But *soil* is special—mysterious, critical to our survival, and, whether we realize it or not, essential to our existence. We have relegated soil to an ignoble position. We commonly degrade it—we consider only feces to be a worse substance, but soil deserves better.

Before we move on, let's take another look at that handful of "dirt". What do we really have in hand when we reach down and grab a handful of "dirt"? The point is that it isn't actually dirt but soil. What is soil, really? Perhaps no word causes more confusion in communications among various groups of laypersons and professionals—environmental scientists, environmental engineers, specialized groups of earth scientists, and engineers in general—than the word "soil." Why? From the professional's perspective, the problem lies in the reasons why different groups study soils.

Pedologists (soil scientists) are interested in soils as a medium for plant growth. Representing a corresponding branch of engineering soils specialists, *soil engineers* look at soil as a medium that can be excavated with tools. A *geologist*'s view of soil falls somewhere between that of pedologists and soil engineers—they are interested in soils and the weathering processes as past indicators of climatic conditions and in relation to the geologic formation of useful materials ranging from clay deposits to metallic ores.

To clear up this confusion, let's view that handful of soil from a different—but much more basic and revealing—perspective. Consider the following descriptions of soil to better understand what soil is and why it is critically important to us all:

1. A handful of soil is alive, a delicate living organism—as lively as an army of migrating caribou and as fascinating as a flock of egrets. Literally teeming with life of incomparable forms, soil deserves to be classified as an independent ecosystem or, more correctly stated, as many ecosystems.
2. When we reach down and pick up an handful of soil, exposing the stark bedrock surface, it should remind us, maybe startle some of us, that without its thin living soil layer Earth is a planet as lifeless as our own moon.

If you still prefer to call soil dirt, that's okay. Maybe you view dirt in the same way as E.L. Konigsburg's character Ethan does (Konigsburg, 1966, p. 64):

> The way I see it, the difference between farmers and suburbanites is the difference in the way we feel about dirt. To them, the earth is something to be respected and preserved, but dirt gets no respect. A farmer likes dirt. Suburbanites like to get rid of it. Dirt is the working layer of the earth, and dealing with dirt is as much a part of farm life as dealing with manure: neither is user-friendly, but both are necessary.

SOIL BASICS

Soil is the layer of bonded particles of sand, silt, and clay that covers the land surface of the Earth. Most soils develop multiple layers. The topmost layer (*topsoil*) is the layer in which plants grow. This topmost layer is actually an ecosystem composed of both biotic and abiotic components—inorganic chemicals, air, water, decaying organic material that provides vital nutrients for plant photosynthesis, and living organisms. Below the topmost layer (usually no more than a meter in thickness), is the *subsoil*, which is much less productive, partly because it contains much less organic matter. Below that is the *parent material*, the bedrock or other geologic material from which the soil is ultimately formed. The general rule of thumb is that it takes about 30 years to form one inch of topsoil from subsoil; it takes much longer than that for subsoil to be formed from parent material, the length of time depending on the nature of the underlying matter (Franck and Brownstone, 1992).

Soil Properties

From the environmental scientist's view (with regard to land conservation and remediation methodologies for contaminated soil remediation through reuse and recycling), four major properties of soil are of interest: soil texture, slope, structure, and organic matter. Soil texture (see Figure 3.2), or the relative proportions of the various soil separates in a soil, is a given and cannot be easily or practically changed significantly. It is determined by the size of the rock particles (sand, silt, and clay particles) or the soil separates within the soil. The largest soil particles are gravel, which consists of fragments larger than 2.0 mm in diameter. Particles between 0.05 and 2.0 mm are classified as sand. Silt particles range from 0.002 to 0.05 mm in diameter, and the smallest particles (clay particles) are less than 0.002 mm in diameter. Clays are composed of the smallest particles, but these particles have stronger bonds than silt or sand; once broken apart, though, they erode more readily. Particle size has a direct impact on erosion. Rarely does a soil consist of only one single size of particle; most are a mixture of various sizes. The *slope* (or steepness of the soil layer) is another given, important because the erosive power of runoff increases with the steepness of the slope. Slope also allows runoff to exert increased force on soil particles, which breaks them apart more readily and carries them farther away.

Soil structure (tilth) should not be confused with soil texture—they are different. In fact, in the field, the properties determined by soil texture may be considerably modified by soil structure. Soil structure refers to the combination or arrangement of primary soil particles into secondary particles (units or peds). Simply stated, soil structure refers to the way various soil particles clump together. Clusters of soil particles, called *aggregates*, can vary in size, shape, and arrangement; they combine naturally to form larger clumps called *peds*. Sand particles do not clump because sandy soils lack structure. Clay soils tend to stick together in large clumps. Good soil develops small friable (easily crumbled) clumps. Soil develops a unique, fairly stable structure in undisturbed landscapes, but agricultural practices break down the aggregates and peds, lessening erosion resistance.

The presence of decomposed or decomposing remains of plants and animals (*organic matter*) in soil improves not only fertility but also soil structure—especially the ability of soil to store water. Live organisms such as protozoa, nematodes, earthworms, insects, fungi, and bacteria are typical inhabitants of soil. These organisms work to either control the population of organisms in the soil or to aid in the recycling of dead organic matter. All soil organisms, in one way or another, work to release nutrients from the organic matter, changing complex organic materials into products that can be used by plants.

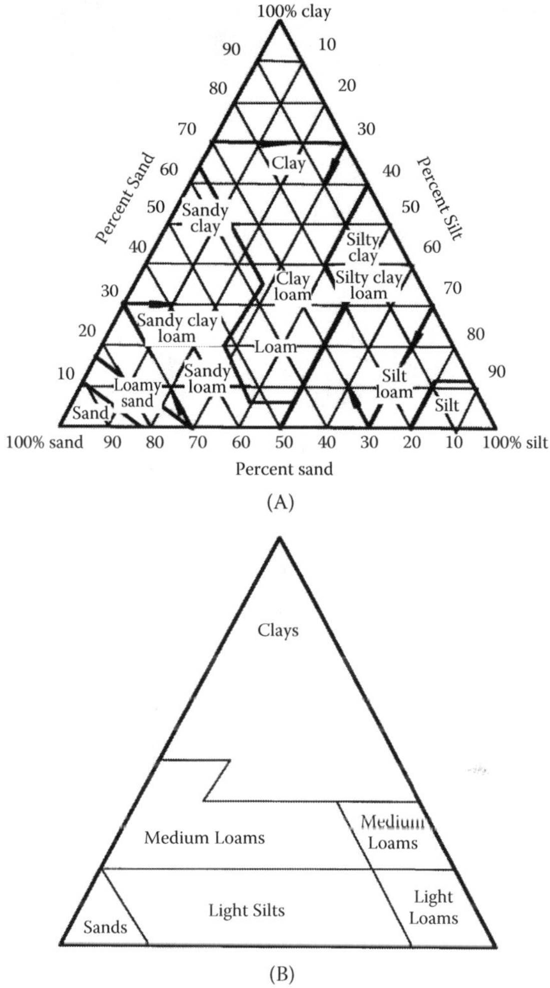

FIGURE 3.2 (A) Textural triangle similar to U.S. Department of Agriculture model; (B) broad groups of textural classes. (Adapted from Briggs, D. et al., *Fundamentals of the Physical Environment*, Routledge, London, 1997, p. 323.)

Soil Formation

Soil is formed as a result of physical, chemical, and biological interactions in specific locations. Just as vegetation varies among biomes, so do the soil types that support that vegetation. The vegetation of the tundra and that of the rain forest differ vastly from each other and from vegetation of the prairie and coniferous forest; soils differ in similar ways. In the soil-forming process, two related, but fundamentally different, processes are occurring simultaneously. The first is the formation of soil parent materials by weathering of rocks, rock fragments, and sediments. This set of processes is carried out in the zone of weathering. The end point is producing parent

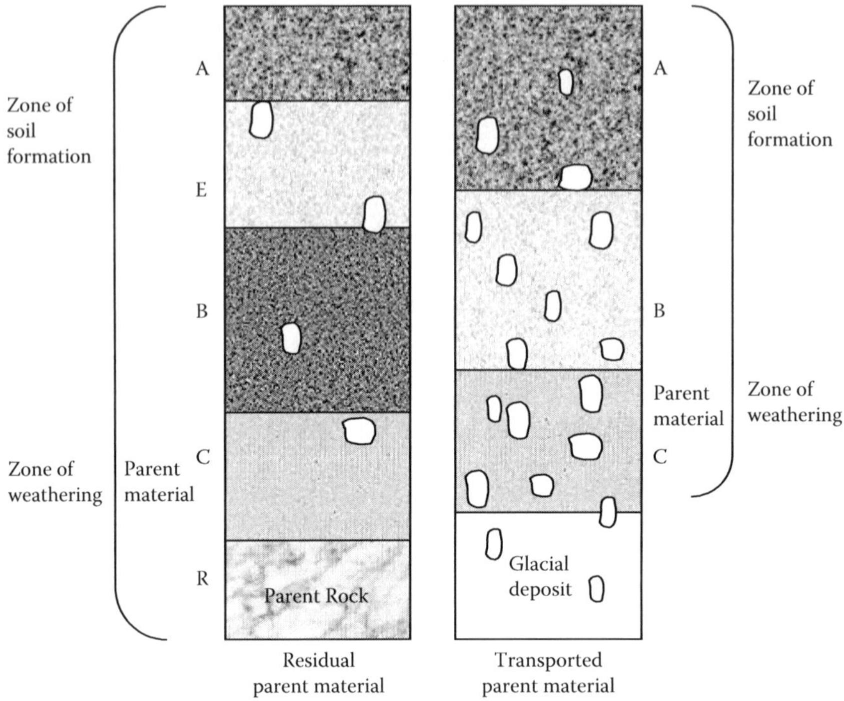

FIGURE 3.3 Soil profiles on residual and transported parent materials.

material for the soil to develop in and is referred to as C horizon material. It applies in the same way for glacial deposits as for rocks. The second set of processes is the formation of the soil profile by soil-forming processes, which gradually change the C horizon material into A, E, and B horizons. Figure 3.3 illustrates two soil profiles, one on hard granite and one on a glacial deposit.

Soil development takes time and is the result of two major processes: weathering and morphogenesis. *Weathering* (the breaking down of bedrock and other sediments that have been deposited on the bedrock by wind, water, volcanic eruptions, or melting glaciers) happens physically, chemically, or a combination of both. Physical weathering involves the breaking down of rock primarily by temperature changes and the physical action of water, ice, and wind. When a geographical location is characterized as having an arid desert biome, the repeated exposure to very high temperatures during the day followed by low temperatures at night causes rocks to expand and contract and eventually to crack and shatter. At the other extreme, in cold climates rock can crack and break as a result of repeated cycles of expansion of water in cracks and pores during freezing and contraction during thawing. Figure 3.4 shows another example of physical weathering in which a slot canyon is carved down to various-sized rocks that, with time, are reduced to soil particles in which various vegetation types spread their roots and grow; roots can exert enough pressure to enlarge cracks in solid rock, eventually splitting the rock. Plants such as mosses and lichens also penetrate rock and loosen particles.

FIGURE 3.4 Slot canyon carved by the Virgin River in Zion National Park, Utah. (Photograph by author.)

Bare rocks are also subjected to *chemical weathering*, which involves chemical attack and dissolution of rock. Accomplished primarily through oxidation via exposure to oxygen gas in the atmosphere, acidic precipitation (after having dissolved small amounts of carbon dioxide gas from the atmosphere), and acidic secretions of microorganisms (bacteria, fungi, and lichens), chemical weathering speeds up in warm climates and slows down in cold ones. Physical weathering and chemical weathering do not always (if ever) occur independently of each other; instead, they normally work in combination, and the results can be striking.

A classic example of the effect and power of their simultaneous actions can be seen in the ecological process known as *bare rock succession*, explained earlier in Chapter 1. An example of bare rock succession that can be seen today demonstrates just how effective and dramatic this weathering process can be. The Natural Bridge in Virginia (see Sidebar 3.1) illustrates the awesome power of physical and chemical processes, working in tandem, reshaping the Earth, and producing and transporting the particle material (eventually sediment particles) upon which and from which soil will eventually form.

SIDEBAR 3.1. NATURAL BRIDGE OF VIRGINIA

Thomas Jefferson stated that the Natural Bridge of Virginia (see Figure SB3.1.1) is "the most sublime of Nature's Works." The great stone causeway, situated a few miles west of the Blue Ridge Mountains in the heart of the great Appalachian Valley of western Virginia, has been proclaimed one of the natural wonders of the world. The proportions of the Natural Bridge are enormous. It is 90 feet long, and the width varies from 150 feet at one end to 50 feet at the other. The Natural Bridge is taller than Niagara Falls. The span contains approximately

FIGURE SB3.1.1 Natural Bridge in Virginia. (Photograph by author.)

450,00 cubic feet of rock. If we could weigh it, the mass would probably come in at about 36,000 tons (72,000,000 pounds). At its feet flows Cedar Creek, now only a small trickle in comparison to the massive, roaring flow of water it once was.

The usual question asked about the Natural Bridge is "How was it formed?" Many theories have been suggested as to the exact origin of the Natural Bridge. Thomas Jefferson held the theory that the Natural Bridge was formed by some sort of cataclysmic event (what he called "some great convulsion") and that its formation was relatively recent. (At the time, though, the Earth was believed to be only several thousands of years old. To Jefferson, that such an event could have occurred over the course of millions of years was inconceivable.)

Today we know that ideas about natural features such as the Natural Bridge change as we gain more knowledge. Spencer (1985) observed that, when talking about the exact age of the Natural Bridge, we must be careful to distinguish several important events; for example, the rocks that compose the bridge are early Ordovician (about 500 million years old). Toward the end of the Paleozoic Era (about 200+ million years ago), the internal forms of these rocks (the folds and breaks in the layers) were imposed during the Appalachian Mountain building process. Probably no more than a few million years ago, the formation of the stream drainage and the carving out of the bridge began.

With input from others, Jefferson later modified his "great convulsion" theory as the cause of the formation of the bridge. Jefferson, an astute student of science, was aware that other natural bridges on Earth had been formed by the work of water—the wearing action of water running through them—rather than by a convulsion of nature. He came to believe that these same wearing actions might have formed the Natural Bridge. One of Jefferson's friends, Francis W. Gilmer, put forward a detailed description of the origin of the Natural Bridge in 1816. Gilmer outlined his thinking on the subject in a paper he presented to the American Philosophical Society:

> ... instead of its being the effect of a sudden convulsion, or an extraordinary deviation from the ordinary laws of nature, it will be found to have been produced by the very slow operation of causes which have always, and must ever continue, to act in the same manner. ... the country above the bridge ... is calcareous. ... This rock is soluble in water to such a degree, as to be found in solution with all the waters of the country, and is so soft as to yield not only to its chemical agency, but also to its mechanical attrition. ... Here, as in calcareous countries generally, there are frequent and large fissures in the earth, which are sometimes conduits for subterraneous streams, called "sinking rivers."
>
> ... It is probable, then, that the water of Cedar Creek originally found a subterraneous passage beneath the arch of the present bridge ... The stream has gradually widened, and deepened this ravine to its present situation. Fragments of its sides also yielding to the expansion and contraction of heat and cold, tumbled down even above the height of the water. ... The stone and earth composing the arch of the bridge, remained there and nowhere else; because, the hill being of rock, the depth of rock was greatest above the surface of the water where the hill was highest, and this part being very thick, and the strata horizontal, the arch was strong enough to rest on such a base. ... Indeed, the very process by which the natural bridge was formed is still visibly going on; the water ... is excavating the rock, and widening the channel, which, after a long lapse of time, may become too wide to support the arch, and this wonder of our country will disappear.

Since the time of Gilmer's original theory about the origin of the Natural Bridge, all geologists who subsequently studied the Natural Bridge have agreed with his view that the bridge was formed by the action of running water diverted from the surface of the ground into a subterranean passage beneath the arch. They differ only in the details of this diversion. In 1893, for example, C.D. Walcott "suggested the bridge was once the site of a waterfall and at that time the valley floor of Cedar Creek was at a level close to the top of the canyon. Walcott postulated that the water somehow diverted just upstream from the waterfall into an underground passage that emptied out of the base of the falls. This would have left the span of dolomite between the diversion and edge of the waterfalls intact" (Spencer, 1985, p. 44).

Most of the others who have studied the origin of the Natural Bridge (e.g., Malott and Shrock, 1930; Woodward, 1936) favor ideas much closer to Gilmer's—that a surface stream was diverted into an opening in the Earth (a cave) from which water issued farther downstream. This underground flow formed a long, natural tunnel. Over time, the roof of this tunnel collapsed, leaving only the span of Natural Bridge (see Figure SB3.1.2).

Cedar Creek still flows beneath the Natural Bridge (see Figure SB3.1.3). It originates in the Allegheny Mountains and empties into the James River. The structure of the rock of which the Natural Bridge is made determined its location. The arch of the bridge, which is massive and compact, is formed from dolomite, and its structural integrity seems sound. However, in time, the bridge will fall into Cedar Creek; it will be gone, but just as Nature works to modify and eventually destroy the Natural Bridge, she also works to form other natural wonders of the world, as she must.

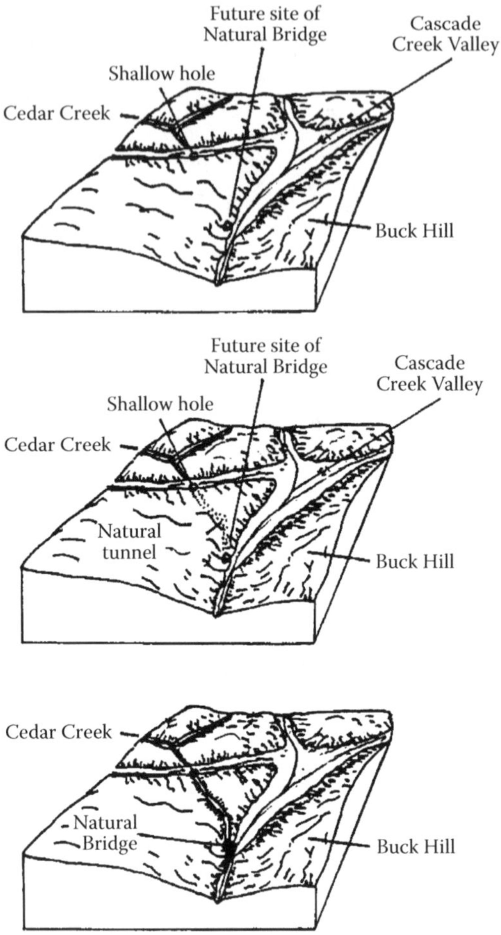

FIGURE SB3.1.2 How the Natural Bridge was formed over time: Beginning with a shallow hole, Cedar Creek formed a natural tunnel (as shown). Later, the tunnel roof caved in. Because the Natural Bridge was formed of hard dolomite, it remained in place.

FIGURE SB3.1.3 Cedar Creek as it appears today. Notice the inclined layers of limestone in the creek bed, rising up the slope.

The final stages of soil formation include the process of *morphogenesis*, or the production of a distinctive *soil profile* with its constituent layers, or *soil horizons* (see Figure 3.3). The soil profile (the vertical section of the soil from the surface through all its horizons, including C horizons) gives environmental scientists critical information. When properly interpreted, soil horizons can warn about potential problems with using the land and can tell much about the environment and history of a region. The soil profile allows us to describe, sample, and map soils.

Soil horizons are distinct layers, roughly parallel to the surface, which differ in color, texture, structure, and content of organic matter (see Figure 3.3). The clarity with which horizons can be recognized depends on the relative balance of the migration, stratification, aggregation, and mixing processes that take place in the soil during morphogenesis. In podzol-type soils, striking horizonation is quite apparent; in vertisol-type soils, the horizons are less distinct. When horizons have been evaluated, they are each assigned a letter symbol to reflect the genesis of the horizon (see Figure 3.3).

Certain processes work to create and destroy clear soil horizons. Various formations of soil horizons that tend to create clear horizons by vertical redistribution of soil materials include the leaching of ions in the soil solutions, movement of clay-sized particles, upward movement of water by capillary action, and surface deposition of dust and aerosols. Clear soil horizons are destroyed by mixing processes that occur because of organisms, cultivation practices, creep processes on slopes, frost heave, and swelling and shrinkage of clays—all part of the natural soil formation process.

The bottom line: For our purposes in this text, whether you call it dirt or soil, we call it sediment.

REFERENCES AND RECOMMENDED READING

Batie, S.S. (1983). *Soil Erosion: Crisis in America's Croplands?* Washington, DC: Conservation Foundation.

Birkeland, P.W. (1984). *Soils and Geomorphology*. New York: Oxford University Press.

Bohn, H.L., McNeal, B.L., and O'Connor, G.A. (1985). *Soil Chemistry*, 2nd ed. New York: John Wiley & Sons.

Brady, N.C. and Weil, R.R. (1996). *The Nature and Properties of Soils*, 11th ed. New York: Prentice Hall.

Briggs, D., Smithson, P., Addison, K., and Atkinson, K. (1997). *Fundamentals of the Physical Environment*, 2nd ed. New York: Routledge.

Donnelly, A. and Cooley, H. (2015). *Water Use Trends in the United States*. Oakland, CA: Pacific Institute (http://pacinst.org/publication/water-use-trends-in-the-united-states).

FitzPatrick, E.A. (1983). *Soils: Formation, Classification and Distribution*. London: Longman.

Franck, I. and Brownstone, D. (1992). *The Green Encyclopedia*. New York: Prentice Hall.

Konigsburg, E.L. (1996). *The View from Saturday*. New York: Scholastic Books.

Malott, C.A. and Shrock, R.R. (1930). Origin and development of Natural Bridge, Virginia. *American Journal of Science*, 19: 257–273.

Rowell, D.L. (2006). *Soil Science: Methods and Applications*. London: Longman.

Spencer, E.W. (1985). *Guidebook: Natural Bridge and Natural Bridge Caverns*. Lexington, VA: Poorhouse Mountain Studios.

Tomera, A.N. (1989). *Understanding Basic Ecological Concepts*. Portland, ME: J. Weston Walch, Publisher.

USGS. (2009). *Summary of Estimated Water Use in the United States in 2005*, Fact Sheet 2009–3098. Reston, VA: U.S. Geological Survey.
Wilde, A., Ed. (1988). *Russell's Soil Conditions and Plant Growth*, 11th ed. London: Longman.
Woodward, H.P. (1936). Natural Bridge and Natural Tunnel, Virginia. *Journal of Geology*, 44(5): 604–616.

4 Sediment Properties

Sediments are the products of disintegration and decomposition of rocks. Material becomes detached and is transported to a deposition site where it may be affected by solution, cementation, consolidation, or biological action.

INTRODUCTION

The physical properties of sediments depend on a number of factors, including composition, texture, and structure of the original formation; topography; type of weathering; and sorting (Lobeck, 1939). The greatest variety of minerals and textures in sediment comes from the weathering of igneous rocks, especially from this disintegration in semiarid and arid climates. These conditions have produced great volumes of sediment containing much coarse material, including boulders, especially along mountain fronts and in intermountain valleys. These deposits commonly contain a relatively high proportion of unaltered minerals such as feldspars, amphiboles, pyroxenes, and micas. Sediments produced by erosion in more humid and deeply weathered areas generally have a finer texture and a higher proportion of minerals produced by chemical weathering. Some small grains of certain minerals, classified as detrital mineral suites, are resistant to chemical weathering. These include zircon, quartz, rutile (titanium oxide), tourmaline, topaz, and ilmenite (titanium–iron oxide); they remain in sediment relatively unchanged and may reveal the source rock type (Krumbein and Sloss, 1963). Feldspars, the most common minerals in igneous rock (Pettijohn, 1957), are much less stable and less common in sediments. In humid climates, feldspars are relatively easily decomposed to form products including clay minerals, silica, and oxides of aluminum.

ORIGIN OF SEDIMENTS

The two major natural mechanisms of sediment production are alluvial processes and earth processes (earthquakes). The powerful natural agents of wind abrasion, snow and ice, and warming and thawing on and inside rock structures occur over time, of course. The passage of time is the key factor in rock disintegration into sediment formation. Keep in mind that weathering and deterioration of rock materials are considered the primary mechanisms of sediment formation. It is also important to point out that rock deterioration and alteration and the reduction to sediment status occur at or near the Earth's surface. For illustrative purposes and for a better understanding of the material that follows, refer to Figure 4.1. It is interesting to observe that the weathering of rocks is essentially an adjustment to a new environment. Nothing in Nature is static. For example, intrusive igneous and metamorphic rocks are exposed to weathering when erosion sculpts and removes the formations covering them. The forces of weathering attack volcanic rocks and sedimentary

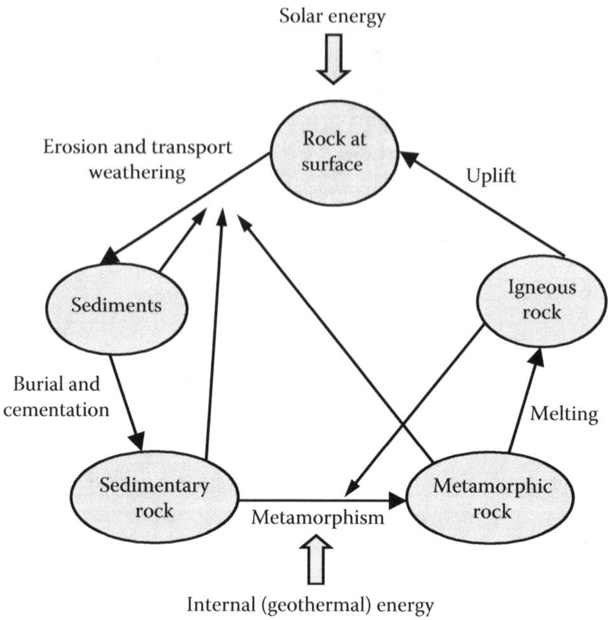

FIGURE 4.1 The rock cycle.

formations exposed at the Earth's surface. In addition to the passage of time, the rate of rock deterioration depends on many factors, including composition and structure of the formation, climate, topography (especially slope), nature of vegetal cover, and elevation.

Disintegration

Physical disruption or disintegration includes all processes by which rocks are broken into smaller pieces without much chemical change. Rocks are broken either into pieces containing all of their original minerals in a relatively unaltered state or into grains, each consisting of an original mineral. The result may be splitting of blocks from a formation or disintegration to sand or gravel. Large and rapid temperature changes can disrupt rock masses. For example, forest fires can heat exposed rock rapidly, thus fracturing and fragmenting it. Frost is a major agent in the disintegration of rocks. Water collects in voids and openings in rocks and, upon freezing, increases about 9% in volume. This transformation of water from a liquid to a solid

DID YOU KNOW?

If a rock particle is loosened, chemically or mechanically, but stays put, we call it *weathering*. When the particle begins to move, we call it *erosion*.

state can dislodge fragments of rock as large as 10 feet in maximum dimension, as found along cliffs bordering Devil's Lake, Wisconsin (Leet and Judson, 1958). Angular fragments a few inches across are a common result of frost action on rocks.

A major disruptive forces is the relief of pressure where weathering and subsequent transportation remove a load from underlying rock formations, especially on steep slopes. The same effect can be produced by landslides that remove an overlying load. *Diastrophism*, also called *tectonism* (large-scale disruption of Earth's crust), of any type disrupts rocks. It can cause new joint systems, widen preexisting joints, or produce movement along a fracture during an earthquake. Folding of rock formations over long periods of time can also be a disruptive force. Products of mechanical disintegration range from large boulders to sand. Basic igneous rocks commonly yield sand and gravel composed of calcic feldspars and relatively unaltered ferromagnesian minerals.

Absolutely nothing on Earth is safe from the heavy hand of moving water and ice formations. They are powerful disruptive forces on rock formations in several environments. These forces include wave action along shores of seas and lakes, abrasion of the banks and beds of streams, and scouring and plucking by glacial ice. The atmosphere is also a disruptive force, especially in arid regions where rocks in exposed positions are subject to attack by winds carrying abrasive mineral particles. Biological agents that have some disruptive effects on rocks include widening of crevices by root growth, pitting of rock surfaces by lichens (e.g., bare rock succession), and burrowing by some animals.

Decomposition

All rocks located at or near the surface of the Earth are subject to decomposition as well as disintegration. Decomposition is the breaking down of mineral components of rocks by chemical reaction. Most decomposition occurs above the groundwater table, but the processes of weathering extend down hundreds of feet in desert regions and in some regions of high rainfall. Igneous rocks are generally susceptible to chemical attack, as they are definitely out of equilibrium with the environment near the Earth's surface. On average, 100 g of igneous rock acquires through decomposition 5.3 g of carbon dioxide, 2 g of water, 0.7 g of carbon, and about 1 g of oxygen (Mason, 1956; Twenhofel, 1950). The resultant rocks have lower specific gravity and higher porosity than the unweathered igneous rocks. Averages of many analyses by Clarke (1924) and others indicate that the weathering of igneous rocks has produced sedimentary rocks in about the following proportions: shale, 82%; sandstone, 12%; and limestone, 6%.

Carbonation

Carbon dioxide (CO_2) is one of the most important and most common weathering agents. It comes from the atmosphere and from organic sources. It readily unites with water to form the weak acid H_2CO_3 (carbonic acid). Carbonic acid reacts with feldspars to produce clay minerals, silica, calcite, and other relatively soluble carbonates containing potassium, sodium, iron, and magnesium. The common carbonate rocks are limestone, dolomite, and marls.

Hydration

The addition of water to many of the minerals of igneous rocks results in the formation of clay minerals, which are hydrous aluminum silicates. Many minerals formed by hydration become dull earthy masses that contrast with their former hard, crystalline nature. Hydration also nearly doubles the volume of material (Lobeck, 1939). The transformation of feldspar to kaolinite is an example.

Oxidation

Through oxidation, many secondary minerals are formed from igneous rocks. The oxides of aluminum and iron are among the most stable. The oxidation of rocks in air is accelerated in the presence of moisture. Ferrous silicates in pyroxenes, amphiboles, and olivine are oxidized by the air and water to hematite (ferric oxide, Fe_2O_3). The oxidation of iron is marked by color changes from green or black to red, yellow, or brown. Oxygen combines with other elements to form sulfates, carbonates, and nitrates, most of which are relatively soluble.

Solution

Solution is important in the alteration of igneous rock. Some minerals, such as quartz and the accessory minerals, are relatively insoluble. An accumulation of quartz grains thus becomes sand or sandstone. Clays and shales contain decomposition products of the feldspars and other less common primary silicates. Some of the silica from any of the silicates may be removed in solution (see Table 4.1). The groundwater and streams contain more silica in solution in areas of igneous rock than in sedimentary terrains. This is so partly because quartz, which is more common in sediments, is less soluble than the other common silicates and partly because less stable silicates are somewhat desilicated in the earlier cycle of sedimentation. The basic igneous rocks, such as basalt and gabbro, contain much silica even if they lack free quartz. The silica in solution and the colloidal-size silica are carried away and may be redeposited in

TABLE 4.1
Chemical Weathering Products of Common Rock-Forming Silicate Minerals

Mineral	Composition	Important Decomposition Minerals
Quartz	SiO_2	Quartz grains
Orthoclase	$K(AlSi_3O_8)$	Clay, quartz (finely divided)
Albite (sodic plagioclase)	$Na(AlSi_3O_8)$	Clay, quartz (finely divided)
Anorthite (calcic plagioclase)	$Ca(Al_2Si_2O_8)$	Calcite (from Ca)
Biotite, augite, hornblende	Fe, Mg, Ca silicates of Al	Clay, calcite, limonite, hematite, quartz (finely divided)
Olivine	$(Fe, Mg)_2SiO_4$	Limonite, hematite, quartz (finely divided)

Source: Adapted from Leet, L.D. and Judson, S., *Physical Geology*, 2nd ed., Prentice Hall, Englewood Cliffs, NJ, 1958.

crevices as veins of quartz or may become a cementing material filling interstices or even a replacement mineral as in silicified wood. It has been estimated that the weight of dissolved solids carried by streams in the conterminous United States is more than 50% of the weight of the suspended sediment carried (Leifeste, 1974; USDA, 2008). The carbonates are important solution products. Some carbonates reach the ocean and become important constituents of marine deposits as a result of the chemical or biochemical action that produces limestones and marls.

PARTICLE CHARACTERISTICS

The mineral grains composing sediments have various characteristics that affect the formation and subsequent development of deposits. Size, shape, hardness, specific gravity, chemical composition, and degree of weathering of the mineral grains affect the rate and place of deposition and the nature of the deposits ultimately formed. Table 4.2 lists some of the common minerals and their hardness, specific gravity, and relative abundance.

SIZE

Size is an important particle characteristic that is readily measured. Bulk properties tend to vary with particle size in a roughly predictable manner. In fact, size alone has been found to describe sediment deposits adequately for many practical purposes. With regard to particle size, a size grade scale based on Wentworth's classification

TABLE 4.2
Common Minerals: Their Hardness, Specific Gravity, and Frequency of Occurrence in Average Sediments

Mineral	Mohs Scale of Hardness	Specific Gravity	Frequency of Occurrence (%)
Feldspars	6	2.6–2.8	15.6
Hornblende and pyroxene	5–6	2.9–3.3	—
Quartz	7	2.65	34.8
Micas	2–4	2.7–3.1	15.1
Titanium minerals	5–6	3.4–5.5	Trace
Clay minerals	—	2.0–3.0	14.5
Dolomite	3.5–4	2.8–2.9	9.1
Calcite	3	2.7	4.2
Limonite	1–5.5	3.4–4.0	4.0
Apatite	4.5–5	3.2	0.4
Gypsum	1.5–2	2.2–2.4	1.0
Other	—		1.3
Total			100.0

Source: Adapted from Clarke, F.W., *The Data of Geochemistry*, USGS Bulletin 770, U.S. Geological Survey, Reston, VA, 1924.

(Wentworth, 1922) was recommended by the American Geophysical Union (1947) and is reproduced in Table 4.3. Environmental practitioners have adopted the Wentworth scale (and other scales) for size categories of substrate rock and other mineral materials, along with the different sizes. These are listed in Table 4.3. Seven groups of sizes are presented in this table: boulders, cobbles, pebbles, gravel, sand, silt, and clay. The largest size is uncommon but is easily measured. Gravel-size particles are more important than boulders and cobbles and are transported in some streams as bedload. Gravel can be measured directly by diameter or volume or by sieving. Sand-size sediment is common and is easily sized by sieving. The finest screen, No. 200, can be used for accurate size separation of sand and silt. Silt and other fines, the clays, are best separated by measuring their rate of fall in a fluid. Silt and clay together make up most of the suspended load in streams, and they are usually distributed uniformly throughout the depth of the stream. Clay-size particles are important in their effect on density currents and on the change in volume-weight of sediment deposits during consolidation (USDA, 2008).

TABLE 4.3
Sediment Size

Class	Diameter (mm)
Boulder	>256
Cobble	64–256
Pebble	16–64
Gravel	2–16
Sand	0.0625–2
Silt	0.0039–0.0625
Clay	<0.0039

Shape

The various shapes of sediment particles are formed in numerous ways. Some shapes, such as the roundness of river and beach pebbles or the facets of wind-abraded particles, indicate the environment in which they formed. Other shapes express mineralogic characteristics; examples are the curving shards of volcanic glass and the unworn crystals of many resistant minerals. Shape is defined numerically by sphericity and roundness. Sphericity is the ratio of the surface area of a sphere having the same volume as the particle to the surface area of the particle. Sphericity is also expressed as d_n/d_s, where d_n is the nominal diameter (diameter of a sphere having the same volume as the particle) and d_s is the diameter of a circumscribing sphere. A sphere has a sphericity of 1, and all other shapes have a sphericity of less than 1 (Pettijohn, 1957). Roundness describes the sharpness of the edges and corners of a particle and is an indication of the wear the particle has received. Roundness is defined as the average radius of curvature of the edges (r_a) divided by the radius of the maximum inscribed circle (R).

Specific Gravity

The specific gravity of a mineral is the ratio of its weight to the weight of an equal volume of water. Most sediment consists of quartz or feldspar particles, which are about 2.65 times heavier than water, so a specific gravity of 2.65 is generally considered characteristic of sediment. Heavy minerals (e.g., magnetite with specific gravity of 5.18), of course, are found in many sediments, but they make up such a small percentage that their importance is minor. For many environmental practitioners, the chief value of heavy minerals in sediment deposits is that they provide a means of identifying the sediment source.

SIZE DISTRIBUTION OF SEDIMENTS

One of the most important properties of sediment deposits is the particle size distribution of the mineral grains. The distribution is important in predicting the behavior of sediment and estimating its specific weight. A number of precautions must be taken in studying deposits in the field and selecting samples for laboratory analysis. It is important to point out that laboratory studies cannot supply answers to many field problems. For example, problems such as selecting the beds or deposits to be sampled and determining the origin of deposits and the rate of deposition must be solved in the field. Field and laboratory data—the nature of the sediment and its texture, as well as its relationship to other formations, to soils, and to land use—must be interpreted. The size frequency distribution of a sediment can be measured in a number of ways. The coarsest fraction is differentiated by direct measurement of gravels or larger sizes and by sieving sands. Fine-grained sediments can be separated by elutriation (the determination of settling velocity in a sediment–liquid mixture) or by microscopic examination (USDA, 2008).

Fine-Grain Separation

Timing the settling rate of sediment particles in a column of water is one method of fine-grain separation. A suspension of sediment sample is treated with a deflocculant such as sodium carbonate, is thoroughly mixed, and is then put into a graduated cylinder containing a column of water 800 mm (31.4 in.) high. After 10 minutes, the upper part of the suspension is drawn off with a siphon. The coarse sediment containing grains 1/16 mm and larger remains at the bottom. This process is usually repeated about four times to achieve a clean separation. The coarse and fine separates can then be treated and studied separately. A popular modification of the elutriation technique involves use of a bottom withdrawal tube (Howard, 1948). The apparatus consists of a graduated glass cylinder with a constriction and a valve at the bottom through which the coarse particles are withdrawn. From the separation thus obtained, a cumulative curve showing size distribution can be plotted. Other modifications include the use of hydrometers to measure the density of the suspension at various time intervals and pipettes to withdraw fine fractions at definite time intervals.

Sieve Separation

Coarse grains (larger than 0.062 mm or 0.002 in.) are ordinarily separated by sieves having mesh openings corresponding to the grain sizes measured. The U.S. standard sieves series is based on a 200-mesh screen with a diameter of 0.074 mm. Sets of sieves with openings larger than this diameter include 0.125-mm, 0.25-mm, 0.50-mm, 2-mm, and 4-mm sizes (Twenhofel and Tyler, 1941). Grains of various sizes can be separated by this method according to the scales shown in Table 4.3. The dry sample is put in the top sieve of a stack and shaken. Usually 10 minutes in a mechanical shaker is adequate for good size separation. The material caught on each screen is weighed, and the results are expressed as a percentage of the total sample weight. For uniformity in classification of sediment texture, Table 4.3 can be used as a standard.

Fall Velocity

The settling rate of particles is influenced primarily by the size, shape, and specific gravity of the particles and by the viscosity and temperature of the medium. Of these characteristics, grain size is the most important for a given fluid. The settling rates of various minerals and aggregates vary widely.

SEDIMENT DEPOSIT SORTING

The degree of sorting in a sediment deposit is determined by the similarity or dissimilarity of the component particles. Similarity can apply to a number of characteristics, including size, shape, specific gravity, and mineral or chemical composition. In most sediment studies, the classification refers to size distribution. The engineering term "well graded" means poorly sorted and that the deposit contains a number of size grades. The following classification includes a list of environments in which sediment deposits form, arranged approximately from the most poorly sorted to the best sorted deposits. This classification, like many others, is subject to many exceptions, some of which are indicated in the following summary descriptions. Sharp distinctions between the various groups of sediment deposits cannot be made; they all grade into the adjacent groups.

Glacial and Other Ice-Action Deposits

Deposits formed by glacial action are among the most poorly sorted of all sediment deposits. Glacial till, left by melting glaciers, contains fragments of all sizes, from large boulders to finely ground fragments called *rock flour.* Moraines and glacial outwash deposits may be more uniform, but they almost always contain much gravel, as well as sand, silt, and clay. All sediment deposited as a result of glacial erosion is referred to as *glacial drift*, which consists of rock fragments that are carried by the glacier on its surface, within the ice, and at its base.

Ice-Laid Deposits

- *Till (or rock flour)*—This is nonsorted glacial drift deposited directly from ice. Consisting of a random mixture of different sized fragments of angular rocks in a matrix of fine grained, sand- to clay-sized fragments, till was produced by abrasion within the glacier. After undergoing diagenesis and turning to rock, till is called *tillite.*
- *Erratics*—An erratic is a glacially deposited rock, fragment, or boulder that rests on a surface made of different rock. Erratics are often found miles from their source, and by mapping the distribution pattern of erratics geologists can often determine the flow directions of the ice that carried them to their present locations.
- *Moraines*—These are mounds, ridges, or ground coverings of unsorted debris deposited by the melting away of a glacier. Depending on where they formed in relation to the glacier, moraines can be one of the following types:

Ground moraines—Till-covered areas deposited beneath the glacier that result in a hummocky topography with lots of enclosed small basins.

End moraines and terminal moraines—Ridges of unconsolidated debris deposited at the low elevation end of a glacier as the ice retreats due to ablation (melting); they usually reflect the shape of the glacier's terminus.

Lateral moraines—Till deposits that were deposited along the sides of mountain glaciers.

Medial moraines—The result of two glaciers meeting to form a larger glacier; the rock debris along the sides of both glaciers merge to form a medial moraine that runs down the center of a valley floor.

- *Glacial marine drift (icebergs)*—These are glaciers that reach lake shores or oceans and calve off into large icebergs which then float on the water surface until they melt. The rock debris that the icebergs contain is deposited on the lakebed or ocean floor when the iceberg melts.

Stratified Drift

Stratified drift is glacial drift that can be picked up and moved by meltwater streams which can then deposit that material as stratified drift.

- *Outwash plains*—Melt runoff at the end of a glacier is usually choked with sediment and forms braided streams, which deposit poorly sorted stratified sediment in an outwash plain; they are usually flat, interlocking alluvial fans.
- *Outwash terraces*—These river terraces form if the outwash streams cut down into their outwash deposits.
- *Kettle holes*—These are depressions sometimes filled by lakes (e.g., Minnesota, the land of 10,000 lakes) due to the melting of large blocks of stagnant ice; they are found in any typical glacial deposit.
- *Kames*—These are isolated hills of stratified material formed from debris that fell into openings in retreating or stagnant ice.
- *Eskers*—These are long, narrow, and often branching sinuous ridges of poorly sorted gravel and sand formed by deposition from former glacier streams.

Alluvial Fan Deposits

A wide range of sizes is characteristic of piedmont or alluvial fan deposits; hence, they form one of the groups of poorly sorted sediments. A lower gradient at the foot of steep slopes causes rapid deposition of most of the load of vigorous and rapid streams. Large rock blocks and boulders are commonly mixed with pebbles, sand, silt, and clay with little or no stratification.

Beach Deposits

Sorting of beach (littoral) deposits is usually poor. These deposits are primarily along shorelines and harbors along seacoasts, but they are also along the shorelines of large lakes. Locally, the sediments may be relatively well sorted and uniform in

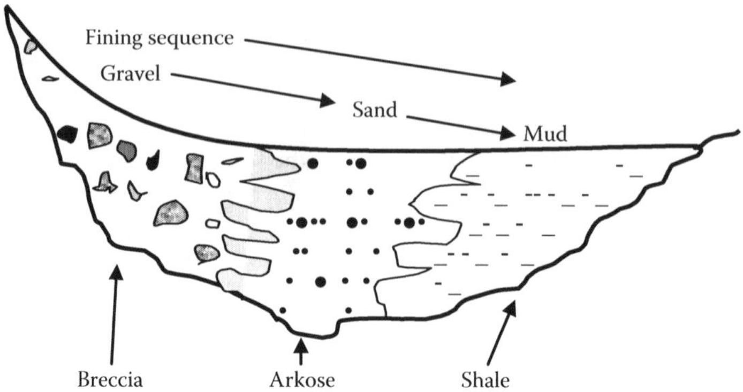

FIGURE 4.2 Alluvial fan fining process.

areas where conditions are stable, but in general the alternating rising and falling tides and the alternating dominance of tidal and river currents cause deposition of poorly sorted sediments.

Alluvial Deposits

Sediments composing alluvial deposits vary greatly in size and other characteristics. Alluvial deposits can range in area from a narrow strip in a small stream valley to a great plain such as the High Plains deposit that extends east from the Rocky Mountains. In the upstream reaches of a valley where stream action is vigorous, alluvial deposits generally are coarse and poorly sorted. In the middle reaches of most streams, the coarsest and most poorly sorted parts of the alluvial deposit form in the channel. This coarse deposit is distributed to some extent over the valley bottom as the stream meanders. The deposits that occur farther downstream usually are better sorted and contain a relatively high percentage of fine sediments—fine sands, silts, and clays. Figure 4.2 illustrates an example of the alluvial fan forming or fining process.

Colluvial Deposits

These products of upland erosion consist of heterogeneous materials of any particle size that accumulate on the lower part or base of slopes. Colluvium is transported there by gravity (talus), sheetwash, slow continuous downslope soil creep, and mudflows.

Eolian Deposits

Most sedimentary depositions of eolian origin are among the better sorted groups of terrestrial deposits. Two major types of eolian deposits are recognized: loess and dunes.

Loess

Loess is a yellowish, fine-grained, nonstratified material carried by the wind and accumulated in deposits of dust. The materials forming loess are derived from surface dust originating primarily in deserts, river flood plains, deltas, and glacial outwash deposits. Loess is cohesive and possesses the property of forming steep bluffs with vertical faces such as the deposits found in the pampas of Argentina and in the lower Mississippi River Valley. Loess is one of the best sorted and most textually uniform of the terrestrial deposits. Loess deposits can range from a featheredge to hundreds of feet thick and consist mainly of silt-size particles that have been transported by wind. Topographic irregularities such as a line of bluffs in a valley increase deposition. Loess deposits cover wide areas in the United States. Because they are mostly unconsolidated, they are subject to rapid erosion and gully development and they contribute to accelerated sediment deposition in reservoirs and stream channels and on flood plains.

Dunes

Sand dunes are windblown deposits of grains moved mostly by traction or saltation, especially in semiarid and arid areas. They form asymmetrical mounds with a gentle slope in the upwind direction and a steep slope on the downwind side (see Figure 4.3). Dunes form in areas such as lake shores, seacoasts, glacial plains, and lake beds as well as in deserts. They are generally well sorted and consist predominantly of fine- to medium-grain sands. As a result of the transporting power of the wind, the silt- and clay-size particles may be carried long distances, leaving the sand to accumulate as slow-moving dunes. Dunes vary greatly in size and shape and form when there is a ready supply of sand, a steady wind, and some kind of obstacle or barrier such as rocks, fences, or vegetation to trap some of the sand. Sand dunes form when moving air slows down on the downwind side of an obstacle (see Figure 4.3). Dunes may reach heights up to 500 m and cover large areas. Types of sand dunes include barchan, transverse, longitudinal, and parabolic:

- *Barchan dunes* are crescent shaped dunes characterized by two long, curved extensions pointing in the direction of the wind and a curved slip face on the downwind side of the dune (see Figure 4.4A). These dunes are formed in areas where winds blow steadily and from a single direction.
- *Transverse dunes* form along seacoasts and lake shores and may be 15 feet high and half a mile in length. Transverse dunes develop with their long axis at right angles to the wind (see Figure 4.4B).

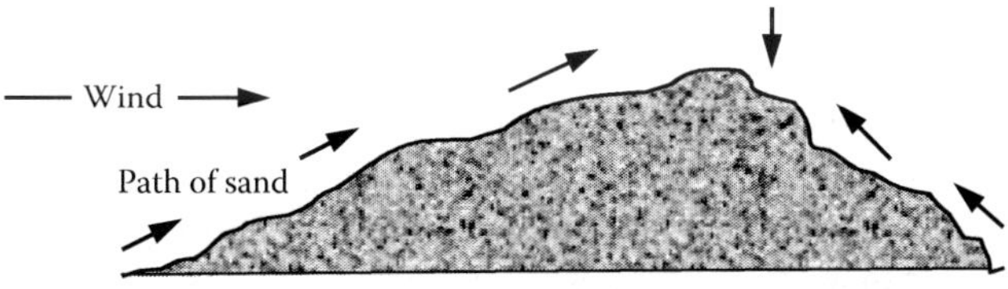

FIGURE 4.3 Profile of typical sand dune. Arrows denote paths of wind currents.

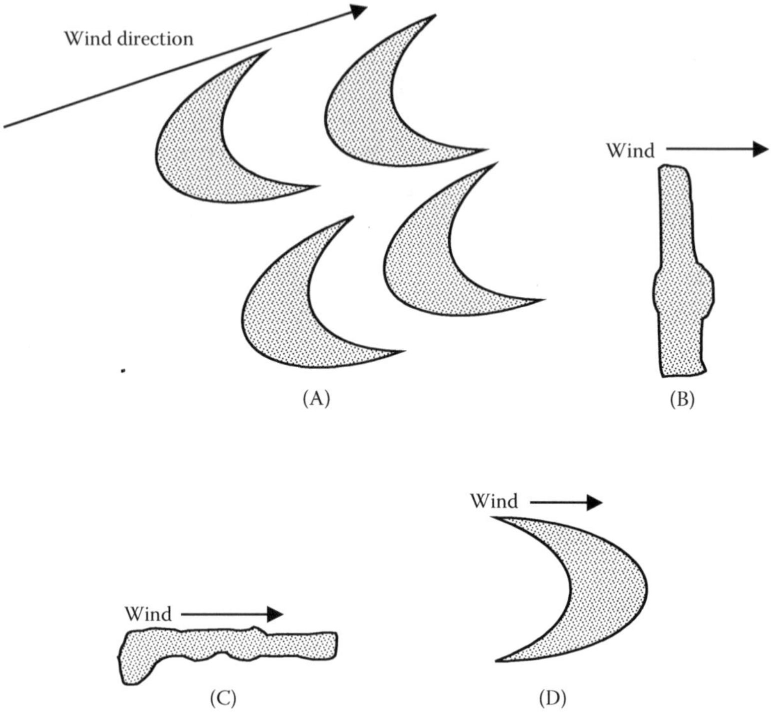

FIGURE 4.4 (A) Barchan dune; (B) transverse dune; (C) longitudinal dune; (D) parabolic dune.

- *Longitudinal dunes* are long ridge-like dunes that develop parallel to the wind (see Figure 4.4C).
- *Parabolic dunes* are U-shaped dunes with an open end facing upwind. They are usually stabilized by vegetation and occur where there is abundant vegetation, a constant wind direction, and an abundant sand supply (see Figure 4.4D).

The *mountain watershed* of the Great Sand Dunes National Park and Preserve in Colorado receives heavy snow and rain each year. Creeks flow from alpine tundra and lakes, down through subalpine and montane woodlands, and finally around the main *dunefield.* Sand that has blown from the valley floor is captured and carried back toward the valley. When creeks disappear into the valley floor, sand is again picked up and carried into the main dunefield. This recycling action of water and wind contributes to the great height of this dunefield. The 30-square mile (78-km^2) active dunefield is where the tallest (~750 ft) dunes reside. It is stabilized by opposing wind directions (southwesterly and northeasterly), creeks that recycle sand back into it, and a 7% moisture content below the dry surface. The dunefield is composed of reversing dunes, transverse dunes, start dunes, and a few barchan dunes. It is estimated to contain over 5 billion cubic meters of sand.

DID YOU KNOW?

The Great Sand Dunes tiger beetle is found nowhere else on Earth. Its specially adapted long legs and fine hairs on its underside help it survive sand temperatures of 140°F (60°C).

The *sand sheet*, the largest component of the Great Sand Dunes geological system, is made up of sandy grasslands that extend around three sides of the main dunefield. Almost 90% of the sand deposit is found here, while only about 10% is found in the main dunefield. The sand sheet is the primary source of sand for the Great Sand Dunes. Small parabolic dunes form here, then migrate into the main dunefield. Nebkha (coppice) dunes form around vegetation. The sabkha forms where sand is seasonally saturated by rising groundwater. When the water evaporates away in late summer, minerals similar to baking soda cement sand grains together into a hard, white crust. Areas of sabkha can be found throughout western portions of the sand sheet, wherever the water table meets the surface. Some wetlands in the sabkha are deeper with plentiful plants and animals, while others are shallow and salty.

Deflation, Desert Pavement, and Wind-Lag Deposits

The process of deflation (or blowing away) is the lowering of the land surface due to removal of fine-grained particles by the wind. Deflation concentrates the coarser-grained particles at the surface, eventually resulting in a relatively smooth surface composed only of the coarser-grained fragments that cannot be transported by the wind. Such a coarse-grained surface is called *desert pavement*. Some of these coarser-grained fragments may exhibit a dark, enamel-like coat of iron or manganese called *desert varnish*. Deflation may create several types of distinctive features. For example, *lag gravels* are formed when the wind blows away finer rock particles, leaving behind a residue of coarse gravel and stones. *Blowouts* may be developed where wind has scooped out soft unconsolidated rocks and soil.

Igneous Rock to Volcanic Dust

Igneous (from the Latin *ignis* for "fire") rocks are rocks that have solidified from an original molten silicate state. The occurrence and distribution of igneous rocks and igneous rock types (see Figure 4.5) can be related to the operation of plate tectonics. The molten rock material from which igneous rocks form is called *magma*. Magma, characterized by a wide range of chemical compositions and a high temperature, is a mixture of liquid rock, crystals, and gas. Magmas are large bodies of molten rock deeply buried within the Earth. Because magmas are less dense than the surrounding rocks they will move upward. As a result of this upward movement, sometimes magmatic materials are poured out upon the surface of the Earth, such as when lava flows from a volcano. Igneous rocks are volcanic or *extrusive rocks* that form when magma cools and crystallizes on the surface of the Earth. Under certain other conditions, magma does not make it to the surface and instead cools and crystallizes

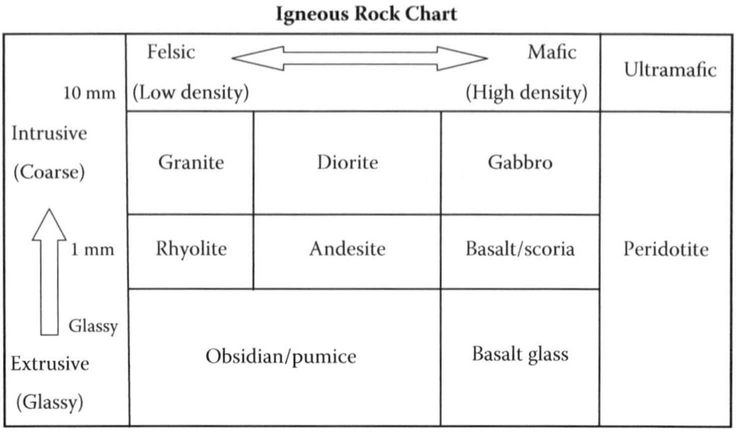

FIGURE 4.5 Igneous rock chart.

within the Earth's crust. These intruding rock materials harden and form *intrusive* or *plutonic rocks*. Magma is molten silicate material that may include already formed crystals and dissolved gases. When magma reaches the surface it is referred to as *lava*. The chemical composition of magma is controlled by the abundance of elements in the Earth. These include oxygen, silicon, aluminum, hydrogen, sodium, calcium, iron, potassium, and manganese, which combined make up 99% of the composition. Because oxygen is so abundant, chemical analyses are usually given in terms of oxides. Silicon dioxide (SiO_2), also known as silica, is the most abundant oxide. Because magma gas expands as pressure is reduced, magmas have an explosive character. The flow (or viscosity) of magma depends on temperature, composition, and gas content. Magmas with higher silicon dioxide contents and lower temperatures have higher viscosity. The three basic types of magma are basaltic, andesitic, and rhyolitic. Table 4.4 summarizes the characteristics of each type.

> ***Note:*** The magma temperature and the chemical composition of the magma determine what minerals crystallize and thus what kind of igneous rock we get.

Intrusive (or plutonic rocks) are rocks that have solidified from molten mineral mixtures beneath the surface of the Earth. Intrusive rocks that are deeply buried tend to cool slowly and develop a coarse texture; whereas, intrusive rocks near the

TABLE 4.4
Characteristics of Magma Types

Magma Type	Solidified Volcanic	Solidified Plutonic	Chemical Composition	Temperature (°C)
Basaltic	Basalt	Gabbro	45–55% silicon dioxide	1000–1200
Andesitic	Andesite	Diorite	55–65% silicon dioxide	800–1000
Rhyolitic	Rhyolite	Granite	65–75% silicon dioxide	650–800

surface that cool more quickly are finer textured. The shape, size, and arrangement of the grains comprising the igneous rock determine its texture. Because of crowded conditions under which mineral particles are formed, they are usually angular and irregular in outline. Typical intrusive rocks include the following:

- *Gabbro* is a heavy, dark-colored igneous rock consisting of coarse grains of feldspar and augite.
- *Peridotite* is a rock in which the dark minerals are predominant.
- *Granite* is the most common and best-known of the coarse-textured intrusive rocks.
- *Syenite* resembles granite but is less common in occurrence and contains little or no quartz.

Extrusive (or volcanic) rocks pour out of craters of volcanoes or from great fissures or cracks in the Earth's crust and make it to the surface in a molten state (liquid lava). Extrusive rocks tend to cool quickly and typically have small crystals (because fast cooling does not allow large crystals to grow). Some cool so rapidly that no crystallization occurs and volcanic glass is produced. Some of the more common extrusive rocks are felsite, pumice, basalt, and obsidian:

- *Felsite*—Very fine-textured igneous rocks
- *Pumice*—Frothy lava that solidifies when steam and other gases bubble out of it
- *Basalt*—World's most abundant fine-grained extrusive rock
- *Obsidian*—Volcanic glass that cools so fast that there is no formation of separate mineral crystals

Bowen's Reaction Series

Back in the early 1900s, the geologist Norman L. Bowen was able to explain why certain types of minerals tend to be found together while others are almost never associated with one another. Bowen found that minerals tend to form in specific sequences in igneous rocks, and these sequences could be assembled into a composite sequence. The idealized progression that he determined is still accepted as the general model (see Figure 4.6) for the evolution of magmas during the cooling process. In order to better understand Bowen's reaction series, it is important to define key terms:

Aphanitic—Mineral grains too small to be seen without a magnifying glass
Extrusion—Magma intruded or emplaced beneath the surface of the Earth
Feldspar—The family of minerals including microcline, orthoclase, and plagioclase
Felsic—White pumice
Mafic—A mineral containing iron and magnesium
Magma—Molten igneous rock
Phaneritic—Mineral grains large enough to be seen without a magnifying glass
Pumice—Textured form of volcanic rock; a solidified frothy lava

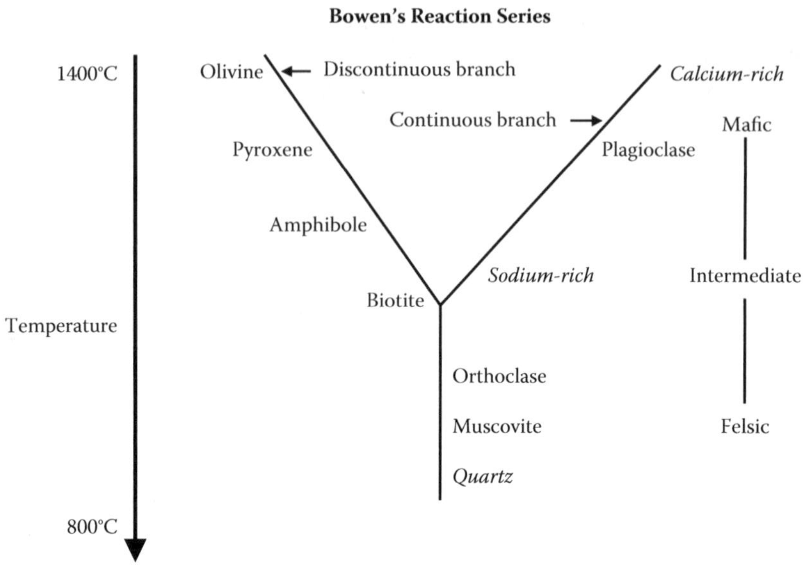

FIGURE 4.6 Bowen's reaction series.

Reaction series—A series of minerals in which a mineral reacts to change to another mineral

Rock-forming mineral—The minerals commonly found in rocks; Bowen's reaction series lists all of the common ones in igneous rocks

Specific gravity—The relative mass or weight of a material compared to the mass or weight of an equal volume of water

Some igneous rocks are named according to textural criteria:

Scoria—Porous
Pumice—Vesicular
Obsidian—Glass
Tuff—Cemented ash
Breccia—Cemented fragments
Permatite—Extremely large crystals
Aplite—Sugary texture, quartz and feldspar
Porphyry—Fine matrix, large crystals

Discontinuous Reaction Series

The left side of Figure 4.6 shows a group of mafic or iron–magnesium-bearing minerals: olivine, pyroxene, amphibole, and biotite. If the chemistry of the melt is just right, these minerals react discontinuously to form the next mineral in the series. If there is enough silica in the igneous magma melt, each mineral will change to the next mineral lower in the series as the temperature drops. Descending down Bowen's reaction series, the minerals increase in the proportions of silica in their composition. In basaltic melt, as shown in Figure 4.6, olivine will be the first mafic mineral (silicate mineral rich in magnesium and iron) to form. When the temperature is low enough to form pyroxene, all of the olivine will react with the melt to form pyroxene, and the pyroxene will crystallize out of the melt. At the crystallization temperature of amphibole, all of the pyroxene will react with the melt to form amphibole, and the amphibole will crystallize. At the crystallization temperature of biotite, all of the amphibole will react to form biotite, and the biotite will crystallize. Thus, all igneous rocks should only have biotite; however, this is not the case. In crystallizing olivine, if there is not enough silica to form pyroxene, then the reaction will not occur and olivine will remain. Additionally, in crystallizing olivine, if the temperature drops too fast for the reaction to take place (volcanic magma eruption) then the reaction will not have time to occur, the rock will solidify quickly, and the mineral will remain olivine.

Continuous Reaction Series

The right side of Figure 4.6 shows the plagioclases. Plagioclase minerals have the formula $(Ca,Na)(Al,Si)_3O_8$. The highest temperature plagioclase has only calcium (Ca). The lowest temperature plagioclase has only sodium (Na). In between, these ions mix in a continuous series from 100% Ca and 0% Na at the highest temperature to 50% Ca and 50% Na at the middle temperature to 0% Ca and 100% Na at the lowest temperature. In a basaltic melt, for example, the first plagioclase to form could be 100% Ca and 0% Na plagioclase. As the temperature drops, the crystal reacts with the melt to form 99% Ca and 1% Na plagioclase, and 99% Ca and 1% Na plagioclase crystallizes. Those then react to form 98% Ca and 2% Na and the same composition would crystallize and so forth. All of this happens continuously provided there is enough time for the reactions to take place and enough sodium, aluminum, and silica in the melt to form each new mineral. The end result will be a rock with plagioclases with the same ratio of Ca to Na as the starting magma.

> ***Note:*** On both sides of the Bowen reaction series shown in Figure 4.6, the silica content of the minerals increases as the crystallization trend heads downward. Biotite has more silica than olivine, and sodium plagioclase has more silica than calcium plagioclase.

Magma Eruptions

The volcanic processes that lead to the deposition of extrusive igneous rocks can be studied in action today and can help us to explain the textures of ancient rocks with respect to depositional processes. Some of the major features of volcanic processes and landforms are discussed in the following text (USGS, 2011). Geologists generally group volcanoes into the following four main kinds:

1. *Cinder cones* are the simplest type of volcano. They are built from particles and blobs of congealed lava ejected from a single vent. As the gas-charged lava is blown violently into the air, it breaks into small fragments that solidify and fall as cinders around the vent to form a circular or oval cone. Most cinder cones have a bowl-shaped crater at the summit and rarely rise more than 1000 feet or so above their surroundings. Cinder

cones are numerous in western North America as well as throughout other volcanic terrains of the world.

2. *Composite volcanoes*, sometimes called *stratovolcanoes*, are some of Earth's grandest mountains. They are typically steep-sided, symmetrical cones of large dimension built of alternating layers of lava flows, volcanic ash, cinders, blocks, and bombs and may rise as much as 8000 feet above their bases. Most composite volcanoes have a crater at the summit that contains a central vent or a clustered group of vents. Lavas either flow through breaks in the crater wall or issue from fissures on the flanks of the cone. Lava, solidified within the fissures, forms dikes that act as ribs that greatly strengthen the cone. The essential feature of a composite volcano is a conduit system through which magma from a reservoir deep in the Earth's crust rises to the surface. The volcano is built up by the accumulation of material erupted through the conduit and increases in size as lava, cinders, ash, etc., are added to its slopes. Probably the best known active composite or stratovolcano at the present time is Mount St. Helens (see Figures 4.7 and 4.8) in Washington State. Mount St. Helens is most notorious for its catastrophic eruption on May 18, 1980, which was the deadliest and most economically destructive volcanic event in the history of the United States. The eruption killed 57 people and destroyed 250 homes, 47 bridges, 15 miles of railways, and 185 miles of highway.
3. *Shield volcanoes* are built almost entirely of fluid lava flows. Flow after flow pours out in all directions from a central summit vent, or group of vents, building a broad, gently sloping cone of flat, domical shape, with a profile much like that of a warrior's shield. They are built up slowly by the accretion of thousands of highly fluid lava flows called *basaltic lava*, which spread widely over great distances and then cool as thin, gently dipping sheets. Lava also commonly erupts from vents along fractures (rift zones) that develop on the flanks of the cone. Some of the largest volcanoes in the world are shield volcanoes (see Figure 4.9).
4. *Lava domes* are formed by relatively small bulbous masses of lava too viscous to flow any great distance; consequently, on extrusion, the lava piles over and around its vent. A dome grows largely by expansion from within. As it grows, its outer surface cools and hardens, then shatters, spilling loose fragments down its sides. Some domes form craggy knobs or spines over the volcanic vent, whereas others form short, steep-sided lava flows known as *coulees* (from the French *couler* for "to flow"). Volcanic domes commonly occur within the craters or on the flanks of large composite volcanoes.

The type of volcanic eruption is often labeled with the name of a well-known volcano where characteristic behavior is similar—hence, the use of such terms as "Strombolian," Vulcanian," Vesuvian," "Pelean," and "Hawaiian," among others:

- *Strombolian-type eruptions* are in constant action with huge clots of molten lava bursting from the summit crater to form luminous arcs through the sky. Collecting on the flanks of the cone, lava clots combine to stream down the slopes in fiery rivulets.

FIGURE 4.7 (Top) Photograph of the north face (blast face) of Mount St. Helens taken from the start of the Truman Trail. (Bottom) Photograph of the north face of Mount St. Helens at the lahar (mudslide) runoff area in Toutle River basin. (Photographs by author.)

FIGURE 4.8 Photograph of Mount St. Helens inside the crater; notice the lava dome building and steam rising off of the upper side. (Photograph by author.)

- *Vulcanian-type eruptions* are characterized by very viscous lavas; a dense cloud of ash-laden gas explodes from the crater and rises high above the peak. Steaming ash forms a whitish cloud near the upper level of the cone.
- *Pelean-type eruptions* (or *Nuée ardente*, for "glowing cloud") are characterized by their explosiveness. These erupt from a central crater with violent explosions that eject great quantities of gas, volcanic ash, dust, incandescent lava fragments, and large rock fragments.

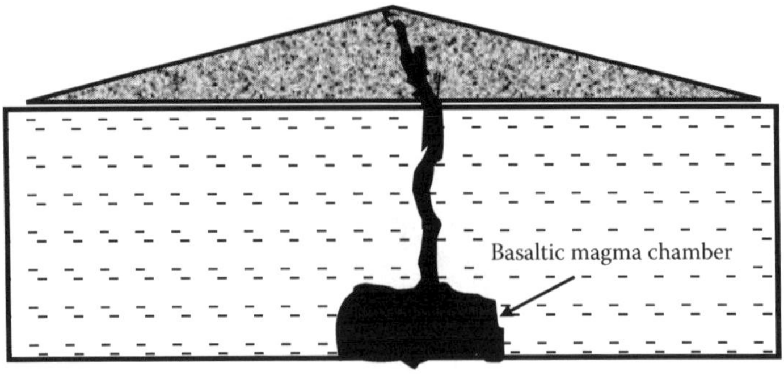

FIGURE 4.9 Cross-section of a shield volcano.

DID YOU KNOW?

Two Hawaiian words, *pahoehoe* and *a'a*, are used to describe how lava flows. Pahoehoe (pa-hoy-hoy) is smooth or ropy lava. Cooler lava hardens on the surface; hotter, more fluid lava flows under it, often leaving caves or tubes behind. A'a (ah-ah) is rough, jagged lava. The molten lava is much less fluid and usually moves slower. A crust never hardens on the surface, but chunks of cooler rock tumble along the top and sides instead. A'a can be impassable.

- *Hawaiian-type (quiet) eruptions* are characterized by less viscous lavas which permit the escape of gas with a minimum of explosive violence. In fissure-type eruptions, a fountain of fiery lava erupts to a height of several hundred feet or more. Such lava may collect in old pit craters to form lava lakes, may form cones, or may feed radiating flows.
- *Vesuvian eruptions* are characterized by great quantities of ash-laden gas that are violently discharged to form a cauliflower-shaped cloud high above the volcano.
- *Phreatic (or steam-blast) eruptions* are driven by explosive expanding steam resulting from cold ground or surface water coming into contact with hot rock or magma. The distinguishing feature of phreatic explosions is that they only blast out fragments of preexisting solid rock from the volcanic conduit; no new magma is erupted.
- *Plinian eruptions* are large explosive events that form enormous dark columns of tephra (solid material that is ejected) and gas high into the stratosphere. Such eruptions are named for Pliny the Younger, who carefully described the disastrous eruption of Vesuvius in 79 AD. This eruption generated a huge column of tephra into the sky. Many thousands of people evacuated areas around the volcano, but about 2000 were killed, including Pliny the Elder.

Lava Flow Terminology

Following are definitions of terms relevant to lava (USGS, 2016):

- *Lava cascades* are not unlike a cascade of water in a small waterfall formed as water descends over rocks. In similar fashion, a lava cascade is the rush or descent of lava over a cliff. In Hawaii, lava cascades typically occur when lava spills over the edge of a crater, a fault scarp, or a sea cliff into the ocean.
- *Lava channels* are narrow, curved, or straight open pathways through which lava moves on the surface of a volcano. The volume of lava moving down a channel fluctuates so that the channel may be full or overflowing at times and nearly empty at other times. During overflow, some of the lava congeals and cools along the banks to form natural levees that may eventually enable the lava channel to build a few meters above the surrounding ground.
- *Lava drapery* is the cooled, congealed rock on the face of a cliff, crater, or fissure formed by lava pouring or cascading over the edges.

- *Lava flows* associated with volcanoes and others are the result of fissure flow. These masses of molten rock pour onto the Earth's surface during an effusive eruption. Both moving lava and the resulting solidified deposit are referred to as lava flows. Because of the wide range in (1) viscosity of the different lava types (basalt, andesite, dacite, and rhyolite), (2) lava discharge during eruptions, and (3) characteristics of the erupting vent and topography over which lava travels, lava flows come in a great variety of shapes and sizes.
- *Lava spillways* are confined lava channels on the sides of a volcanic cone or shield that form when lava overflows the rim of the vent.
- *Lava surges* are intermittent surges or accelerations in the forward advance of lava that can occur when the supply of lava to a flow front suddenly increases or a flow front gives way. The supply of lava may increase as a consequence of a higher discharge of lava from the vent, a sudden change in the vent geometry so that a great volume of lava escapes (e.g., the collapse of a vent wall), or the escape of ponded lava from along a channel. Lava surges may be accompanied by thin, short-lived breakouts of fluid lava from the main channel and flow front.
- *Methane explosions* occur frequently near the edges of active lava flows. Methane gas is generated when vegetation is covered and heated by molten lava. The explosive gas travels beneath the ground through cracks and fills abandoned lava tubes for long distances around the margins of the flow. Methane gas explosions have occurred at least 100 m from the leading edge of a flow, blasting rocks and debris in all directions.
- *Standing waves*, in a fast-moving lava flow, appear to be stationary relative to the lava that moves over the land through them, similar to the standing waves in a water stream. In Hawaii, standing waves as high as 3 m have been observed.
- *Volcanic domes* are rounded, steep-sided mounds built by very viscous magma, usually either dacite or rhyolite. Such magmas are typically too viscous (resistant to flow) to move far from the vent before cooling and crystallizing. Domes may consist of one of more individual lava flows. Volcanic domes are also referred to as *lava domes*.

Intrusive Rocks

Intrusive (or plutonic igneous) rocks have been intruded or injected into the surrounding rocks. Some of these intrusions are invisible because they are imbedded at great depth; consequently, igneous intrusive bodies may be seen only after the underlying rocks have been removed by erosion. Intrusions are of two types:

DID YOU KNOW?

The longest historical dome-building eruption is still occurring at Santiaguito Dome, which is erupting on the southeast flank of Santa Maria volcano in Guatemala; the dome began erupting in 1922.

DID YOU KNOW?

Most obsidian is black, but red, green, and brown obsidian also occurs. Obsidian forms when magma is cooled so quickly that individual minerals cannot crystallize.

(1) concordant intrusions, which are parallel to layers of rocks, and (2) discordant intrusions, which cut across layers. Some of the more common intrusive bodies (plutons) are discussed below.

Concordant Intrusions

- *Sills* are tabular bodies of igneous rocks that spread out as essentially thin, horizontal sheets between beds or layers of rocks.
- *Laccoliths* are lens-like, mushroom-shaped, or blister-like intrusive bodies, usually near the surface, that have relatively flat under surfaces and arched or domed upper surfaces. They differ from sills in that they are thicker in the center and become thinner near their margins.
- *Lopoliths* are mega-sills, usually of gabbro or diorite, that may cover hundreds of square kilometers and be kilometers thick. They often have a concave structure and are differentiated; that is, they take so long to harden that heavy minerals have a chance to sink and light minerals can rise.

Discordant Intrusions

- *Dikes* are thin, wall-like sheets of magma intruded into fractures in the crust.
- *Stocks* or *plutons* are small irregular intrusions.
- *Batholiths*, the largest of igneous intrusions, are usually granitic and cover hundreds or thousands or square kilometers.

Volcanic Landforms

Volcanic landforms (or volcanic edifices) are controlled by the geological processes that form them and act on them after they have formed. Four principal types of volcanic landforms can be found:

- *Plateau basalts* and *lava plains* are formed when great floods of lava are released by fissure eruptions instead of central vents and spread in sheet-like layers over the Earth's surface, forming broad plateaus. Some of these plateaus are quite extensive; for example, the Columbia River Plateau of Oregon, Washington, Nevada, and Idaho is covered by 200,000 square miles of basaltic lava.
- *Volcanic mountains* are composed of the volcanic products of central eruptions and are classified as cinder cones (conical hills), composite cones (stratovolcanoes), or lava domes (shield volcanoes).
- *Volcanic craters* are circular, funnel-shaped depressions, usually less than 1 km in diameter, that form as a result of explosions that emit gases and tephra.

- *Calderas* are much larger depressions, circular to elliptical in shape, with diameters ranging from 1 km to 50 km. Calderas form as a result of the collapse of a volcanic structure resulting from evacuation of the underlying magma chamber.
- *Thermal areas* are locations where volcanic or other igneous activity takes place as is evidenced by the presence or action of volcanic gases, steam, or hot water escaping from the ground.
- *Fumaroles* are vents where gases, either from a magma body at depth or steam from heated groundwater, emerge at the surface of the Earth.
- *Hot springs* or *thermal springs* are areas where hot water comes to the surface of the Earth. Cool groundwater moves downward and is heated by a body of magma or hot rock. A hot spring results if this hot water can find its way back to the surface, usually along fault zones.
- *Geysers* result if the hot spring has a plumbing system that allows for the accumulation of steam from the boiling water. When the steam pressure builds so that it is higher than the pressure of the overlying water in the system, the steam will move rapidly toward the surface, causing eruption of the overlying water. Some geysers, such as Old Faithful in Yellowstone National Park, erupt at regular intervals, but most geysers are quite erratic in their performance. The time between eruptions is controlled by the time it takes for the steam pressure to build in the underlying plumbing system. After volcanic explosions, wind carries great quantities of volcanic dust long distances. This material is well sorted; the particles that travel the farthest are all silt and clay size.

Lacustrine Deposits

Sediment deposition in lakes and reservoirs produces some of the best sorted non-marine sedimentary deposits. The bulk of the sediment in most lakes—that found in all the larger and deeper parts of the basins, where current are not vigorous—is almost entirely silt and clay size. These deposits are, therefore well sorted and fine grained. The coarser and generally more poorly sorted lacustrine sediments are common along shore zones, where wave action is vigorous and coarse detritus is available, and in upstream segments, where inflowing streams deposit their coarse material.

Chemical Deposits and Evaporites

Sediment deposited from solution and evaporation is the best sorted of all sedimentary deposits. These deposits may consist of mineral crystals of almost uniform size. If organisms are incorporated in the deposit, the shell or skeletons add pieces of different sizes, reducing the degree of sorting.

Texture

The size, shape, and arrangement of the particles composing a sediment deposit determine its texture. These factors are influenced by the sediment volume-weight, stability of sediment deposits, porosity, and permeability. Differences in the texture of the many types of sediment deposits cause relatively large variations in the damage that results from accelerated deposition. Coarse sediments of alluvial fans consist chiefly of gravel and boulders and cause major damage if deposited on agricultural land. Overbank flood deposits produce damage that usually increases as the texture of the deposited sediment becomes coarser. Deposits of clays and silts usually have some fertility, but they may bury crops or impede drainage if thick enough. Regardless of their texture, sediment deposits occupy valuable space in reservoirs, obstruct bridge and culvert openings, decrease stream channel and ditch capacity, and cause many other types of damage (USDA, 2008).

Volume-Weight

One of the most important properties of sediment deposits is weight per unit volume, or volume-weight. Volume-weight, as it applies to measurement of eroded sediment, sediment in transport, and sediment deposits in place, is of primary importance in the sedimentary cycle. Information on the unit weight of sediment deposits for construction or other purposes reflects many variations in properties. For example, a cubic foot of quartz, which has a specific gravity of 2.65, weighs about 165 lb. Similarly, a cubic foot of solid magnetite, specific gravity 5.2, weighs 324 lb. Most sediment deposits, in contrast, weigh about 25 to 125 lb/ft^3 when water free (Dendy and Champion, 1978). The volume-weight of sediment deposits is largely determined by the proportion of voids present. If the sediment is below the water surface, the voids are filled chiefly with water. If the sediment is exposed to the atmosphere, there are fewer voids and they are filled chiefly with air or a combination of air and water, depending on rainfall, seepage, and other factors.

Stability of Sediment Deposits

A high degree of angularity of individual sediment grains of silt size or larger promotes stability. A loose aggregation of angular grains is more stable on steeper slopes than an aggregation of more rounded grains. Similarly, angular particles in earthfills increase resistance to slumping and shear. Aggregates of mostly silt- and clay-size particles usually have predominantly angular or platy pieces, but their stability in a fill is determined more by water content and overburden pressure than by the shape of the grains. Deposits of loess may be tens or hundreds of feet thick and are composed of highly angular silt-size particles that tend to stand in nearly vertical faces. Deposits of more nearly rounded grains, such as alluvial or coarse-grained eolian deposits, have lower angles of repose and are usually less stable. Deposits of platy pieces, which have an abundance of grains with two long and one short dimension, are also readily susceptible to sliding.

Porosity

Porosity has been described in connection with the volume and movement of groundwater. An arrangement of spheres providing about 49% pore space has the greatest porosity (Graton and Fraser, 1935). This arrangement, however, is unstable. The most stable arrangement of spheres of uniform diameter provides about 25% pore space, but it is not found in natural sediments. Porosity exceeding 50% has been measured in natural sedimentary deposits. Porosity in a rock or soil has been defined as the property of containing interstices or voids (Meinzer, 1923). The percentage

TABLE 4.5
Approximate Average Porosity of Various Formations

Formation	Porosity (%)	Formation	Porosity (%)
Clay	45	Shale, sandstone	18
Silt	40	Limestone	10
Sand	35	Granite, basalt	1
Gravel	25		

Source: Adapted from Leopold, L.B. et al., *Fluvial Processes in Geomorphology*, W.H. Freeman, San Francisco, CA, 1964.

of pore space is determined by the distribution of fine grains between coarse grains, the shape of the particles, and their arrangement. Grains of silt and clay size occupying spaces between sand and gravel particles can reduce porosity significantly. Both porosity and stability of sediment deposits are affected by the shape of their mineral grains. Many studies have shown that fine-grained sediments are subject to far more compaction and a decrease in volume than are deposits of sand or larger grains (USDA, 2008). Table 4.5 illustrates the range in average porosity of various materials. This table does not take into account the degree of cementation or the fact that, although a fine-grained deposit such as a clay may have high porosity, it permits little movement of water.

Permeability

Permeability of sediments varies widely. Permeability is extremely low in clay materials, even though they may have high porosity and be water saturated. The interstices between the clay particles are small enough for molecular attraction to hold water tightly. Permeability is highest for coarse, clean gravel. Table 4.6 shows the relation of permeability and porosity to grain-size distribution.

REFERENCES AND RECOMMENDED READING

Abbott, P.L. (1996). *Natural Disasters*. New York: Wm. C. Brown Co. Publishers.

American Geophysical Union. (1947). Report of the subcommittee on sediment terminology. *American Geophysical Union Transactions*, 28: 936–938.

Anderson, J.G., Bodin, P., Brune, H.N. et al. (1986). Strong ground motion from the Michoacan, Mexico, earthquake. *Science*, 233: 1043–1049.

Browning, J.M. (1973). Catastrophic rock slides: Mount Huascaran, north-central, Peru, May 32, 1970. *Bulletin American Association of Petroleum Geologists*, 57: 1335–1341.

Clarke, F.W. (1924). *The Data of Geochemistry*, USGS Bulletin 770. Reston, VA: U.S. Geological Survey.

Coch, N.K. (1995). *Geohazards, Natural and Human*. New York: Prentice Hall.

Dendy, F.E. and Champion, W.A. (1978). *Sediment Deposition in U.S. Reservoirs: Summary of Data Reported Through 1975*. Washington, DC: U.S. Department of Agriculture, Agricultural Research Service.

Eagleman, J. (1983). *Severe and Unusual Weather*. New York: Van Nostrand Reinhold.

TABLE 4.6
Permeability and Porosity Related to Grain-Size Distribution

Percent Smaller than Indicated Grain Size by Weight						
2.0 mm	**1.0 mm**	**0.5 mm**	**0.25 mm**	**0.125 mm**	**0.062 mm**	**Porosity (%)**
—	99.4	98.4	95.4	89.7	65.0	58.2
—	—	—	97.5	95.6	94.6	55.5
—	93.9	67.8	18.4	3.2	1.3	46.6
84.3	69.1	48.9	29.4	13.0	6.0	26.3
77.6	59.9	34.1	10.6	0.9	0.2 (clay)	28.9
77.7	58.1	38.4	19.3	9.6	5.3	25.0
24.3	15.7	6.3	1.1	0.4	0.2 (clay)	23.4
20.2	19.1	16.9	12.0	7.5	4.8	25.1
8.4	0.5	0.4	0.3	0.2	0.1 (clay)	38.0

Source: Adapted from Wenzel, L.K., *Methods for Determining Permeability of Water-Bearing Materials*, USGS Water Supply Paper 887, U.S. Geological Survey, Reston, VA, 1942.

Francis, P. (1993). *Volcanoes: A Planetary Perspective*. New York: Oxford University Press.

Graton, L.C. and Fraser, H.J. (1935). Systemic packing of spheres: with particular relation to porosity and permeability. *The Journal of Geology*, 43(8): 785–909.

Howard, C.S. (1948). Laboratory experiences with the bottom withdrawal tube method of size analysis. In: *Proceedings of the Federal Inter-Agency Sedimentation Conference*, pp. 235–243. Washington, DC: Bureau of Reclamation, Department of the Interior.

Keller, E.A. (1985). *Environmental Geology*, 4th ed. New York: Merrill.

Kiersh, G.A. (1964). Vaiont reservoir disaster. *Civil Engineering*, 34: 32–39.

Krumbein, W.C. and Sloss, L.L. (1963). *Stratigraphy and Sedimentation*, 2nd ed. San Francisco, CA: W.H. Freeman.

Leet, L.D. and Judson, S. (1958). *Physical Geology*, 2nd ed. Englewood Cliffs, NJ: Prentice Hall.

Leopold, L.B., Wolman, M.G., and Miller, J.P. (1964). *Fluvial Processes in Geomorphology*. San Francisco, CA: W.H. Freeman.

Lobeck, A.K. (1939). *Geomorphology*. New York: McGraw-Hill.

Mason, B. (1956). *Principles of Geochemistry*. New York: John Wiley & Sons.

Mason, M.A. (1950). Geology in shore-control problems. In: *Applied Sedimentation* (Trask, P.D., Ed.), pp. 276–290. New York: John Wiley & Sons.

Meinzer, O.E. (1923). *Outline of Groundwater Hydrology*, USGS Water Supply Paper 494. Reston, VA: U.S. Geological Survey.

Murck, B.W., Skinner, B.J., and Porter, S.C. (1997). *Dangerous Earth: An Introduction to Geologic Hazards*. New York: John Wiley & Sons.

Pettijohn, F.J. (1957). *Sedimentary Rocks*, 2nd ed. New York: Harper.

Plafker, G. and Galloway, J.P. (1989). *Lessons Learned from the Loma Prieta, California, Earthquake of October 17, 1989*, USGS Circular 1045. Reston, VA: U.S. Geological Survey.

Skinner, B.J. and Porter, S.C. (1995). *The Dynamic Earth: An Introduction to Physical Geology*, 3rd ed. New York: John Wiley & Sons.

Spellman, F.R. and Whiting, N.E. (2006). *Environmental Science and Technology*, 2nd ed. Rockville, MD: Government Institutes Press.

Stephens, J.C., Allen, L.H., and Chen, E. (1984). Organic soils subsidence. In: *Man-Induced Land Subsidence* (Holzer, T.L., Ed.), pp. 107–122. Boulder, CO: Geological Society of America.

Strickler, M. (2008). *What Is Bowen's Reaction Series?* Geoman, http://jersey.uoregon.edu/~mstrick/AskGeoMan/geoQuerry32.html.

Swanson, D.A., Wright, T.H., and Helz, R.T. (1975). Linear vent systems and estimated rates of magma production and eruption of the Yakima basalt on the Columbia Plateau. *American Journal of Science*, 275: 877–905.

Tilling, R.I. (1984). *Eruptions of Mount St. Helens: Past Present and Future*. Reston, VA: U.S. Geological Survey.

Twenhofel, W.H. (1950). *Principles of Sedimentation*, 2nd ed. New York: McGraw-Hill.

Twenhofel, W.H. and Tyler, S.A. (1941). *Methods of Study of Sediments*. New York: McGraw-Hill.

USDA. (2008). Sediment properties. In: *National Engineering Handbook*, Section 3, Chapter 2. Washington, DC: U.S. Department of Agriculture.

USGS. (2011). *Principal Types of Volcanoes*. Reston, VA: U.S. Geological Survey (http://pubs.usgs.gov/gip/volc/types.html).

USGS. (2016). *The USGS Photo Glossary of Volcanic Terms*. Reston, VA: U.S. Geological Survey (https://volcanoes.usgs.gov/vsc/glossary/).

Wentworth, C.K. (1922). *A Method of Measuring and Plotting the Shapes of Pebbles*, USGS Bulletin 730-C. Reston, VA: U.S. Geological Survey.

Wenzel, L.K. (1942). *Methods for Determining Permeability of Water-Bearing Materials*, USGS Water Supply Paper 887. Reston, VA: U.S. Geological Survey.

Williams, H. and McKinney, A.R. (1979). *Volcanology*. New York: Freeman & Copper.

5 Erosion

Acts of creation are ordinarily reserved for gods and poets, but humbler folk may circumvent this restriction if they know how. To plant a pine, for example, one need be neither god nor poet; one need only own a good shovel. By virtue of this curious loophole in the rules, any clodhopper may say: Let there be a tree—and there will be one. If his back be strong and his shovel sharp, there may eventually be ten thousand. And in the seventh year he may lean upon his shovel, and look upon his trees, and find them good. God passed on his handiwork as early as the seventh day, but I notice He has since been rather noncommittal about its merits. I gather either that He spoke too soon, or that trees stand more looking upon than do fig leaves and firmaments.

Leopold (1949)

The winds wander, the snow and rain and dew fall, the earth whirls—all but to prosper a poor lush violet.

—John Muir, naturalist and author, 1913

INTRODUCTION

When the environmental practitioner plans programs to reduce erosion and sediment yield, it is most important that the various types of erosion be thoroughly investigated as sources of sediment. Proper conservation practices and land stabilization measures can then be planned and applied. Erosion consists of a series of complex and interrelated natural processes that loosen or dissolve and move earth or rock material. The land surface is worn away through the detachment and transport of soil and rock materials by moving water, wind, or other geologic agents.

Erosion can be divided into two categories according to the conditions under which it occurs. The first category is *normal (geologic) erosion*, which has been occurring at variable rates, depending on climatic and terrestrial conditions, since the first solid materials formed on Earth. Geologic erosion is extremely slow in most places. It is, in fact, an important process in soil formation. The underlying rock is attacked by air and water, and fragments are detached, decomposed, or dissolved. This process is termed *weathering*. Generally, a rough equilibrium is reached in natural environments between geologic erosion and soil formation. The rates of normal upland erosion and soil formation are determined mainly by climate, parent rocks, soil, precipitation, topography, and vegetal cover. The second category is *accelerated erosion* caused by the activities of humans. Accelerated erosion has been defined as "erosion occurring at a rate greater than normal for the site, usually through reduction of a vegetal cover" (Roehl, 1965). Deforestation, cultivation, and destruction of vegetation accelerate erosion. Soil that normally would take 100 years to be eroded may vanish in 1 year or even a single day (United Nations, 1953). Both categories of erosion can be subdivided into two types: sheet and channel. This classification is helpful in (1) estimating the amount of erosion and sediment yield, (2) determining the relative importance of sediment sources, (3) formulating treatment measures to reduce erosion and sediment yield, and (4) evaluating the effectiveness of treatment processes.

SHEET EROSION

Sheet erosion, which includes rill erosion, is the removal of soil or earth material from the land surface by the forces of raindrop impact, overland runoff, or wind. Although it occurs on all land surfaces, sheet erosion is particularly active on cultivated areas of mild slope where the runoff is not concentrated in well-defined channels but consists largely of overland flow. The numerous small but conspicuous rills caused by minor concentrations of runoff are obliterated by normal field cultivation. This type of erosion occurs gradually over large areas as though the soil were removed in sheets (Bennett, 1939). Materials derived from sheet erosion are fine grained because overland flow, which is usually laminar, seldom exceeds a velocity of 2 or 3 ft/s. Flow of this low velocity can transport only the fine particles detached by raindrop impact. Ellison (1945) reported a grain-size diameter of less than 0.05 mm for 95% of the sediment in prechannel runoff from a silt loam soil in Ohio.

Factors Involved

The basic factors in sheet erosion are rainfall, soil properties, slope length, slope gradient, and kind and condition of cover. Several equations incorporating these factors can be used to obtain a quantitative estimate of the amount of soil material moved by sheet erosion. These equations, originally developed for the humid areas east of the Rocky Mountains, are particularly well suited for determining the effects of land treatment measures on erosion. The movement of sediment and associated pollutants over the landscape and into freshwater bodies is of increasing concern with respect to pollution control, prevention of sediment-filled floods, and environmental protection. The fate of sediment is an important issue for environmental practitioners, land managers, and decision-makers. To determine the fate of sediments, various equations and models have been used. From the late 1940s until 1972, environmental practitioners responsible for estimating sediment yield used the Musgrave equation to compute the amount of sheet and rill erosion in a watershed. Additional research on erosion resulted in the development of the Universal Soil Loss Equation (USLE) by the Agricultural Research Service (ARS). In 1972, the Musgrave equation was replaced by the USLE for computing sheet erosion for project areas.

It is important to point out that both the Musgrave equation and the USLE are empirical formulas in which sediment yield from subacre test plots is defined as "erosion" or "soil loss." The computed soil loss from large areas is usually greater than the sediment yield from the same area, and the larger the area, the greater the discrepancy between computed soil and loss and sediment yield. Neither equation allows for deposition on upland areas. Soil loss computed by these equations represents nothing that can be located or measured in the field. It therefore is an abstract figure that must not be confused with sediment yield. Computed soil loss, however, is a valuable tool for comparing the soil loss from different areas or the effects of different land treatments on a given area.

The USLE initially was used only for cropland, hayland, and pastures in rotation, because erosion factors reflecting the effect of over on uncultivated land areas were not available. Because the USLE had been used in much of the country as a tool in planning land treatment on individual operating units, use of this equation with its refined data was recommended for watersheds and other project areas. Before this could be done, however, additional plant-cover factors (*C*) had to be determined for permanent pastureland, rangeland, woodland, and idle land to estimate the effect of these types of cover on soil losses.

The complete Universal Soil Loss Equation is

$$A = RKLSCP \tag{5.1}$$

where

A = Computed annual soil loss (sheet and rill erosion) in tons per acre. A is not the sediment yield.
R = Rainfall factor, the number of erosion index units in a normal year's rain.
K = Soil erodibility factor, the erosion rate per erosion index unit for a specific soil in cultivated continuous fallow on a 9% slope 72.6 ft long.
L = Slope length factor, the ratio of soil loss from the field slope length to that from a 72.6-ft length on the same soil type and gradient.
S = Slope gradient factor, the ratio of soil loss from the field gradient to that from a 9% slope on the same soil type and slope length.
C = Cropping management factor, the ratio of soil loss from a field with specified cropping and management to that from the fallow condition from which the K factor is evaluated.
P = Erosion control practice factor, the ratio of soil loss with contouring, contour stripcropping, or contour-irrigated furrows to that with straight-row farming, upslope and downslope.

Rainfall Factor (*R*)

The energy of moving water detaches and transports soil materials. The energy intensity (EI) value is the product of the total raindrop energy of a storm and the maximum 30-minute intensity. Soil losses are linearly proportional to the number of EI units. The EI values of the storms from a 22-year (maximum) record are summed to obtain an average annual rainfall erosion index for a given location.

Soil Erodibility Factor (*K*)

The resistance of a soil surface to erosion is a function of the soil's physical and chemical properties. The soil properties most significantly affecting soil erodibility are texture, organic matter content, structure, and permeability. The *K* values assigned to named soils can be obtained from soil scientists, technical guides, or published lists.

Slope Length (*L*) and Slope Gradient (*S*)

Soil loss is affected by both length and degree of slope. For convenience in field applications, these two factors are combined into a single topographic factor, *LS*. The *LS* factor for a gradient as much as 50% and a slope length as much as 1000 ft is obtained from a slope-effect chart.

Plant Cover or Cropping Management Factor (*C*)

The erosion equation, as applied to cropland and hayland, uses established factor relationships to estimate a basic soil loss that is determined by soil properties, topographic features, certain conservation properties, topographic features, and expected rainfall patterns for a specific field. The basic soil loss is the rate at which the field would erode if it were continuously in tilled fallow. The *C* factor value indicates the percentage of this potential soil loss that would occur if the surface were partially protected by a particular combination of cover and management practices. Use of the *C* factor in other situations depends on three distinct but interrelated zones of influence: vegetal cover in direct contact with the soil surface, canopy cover, and the surface beneath it.

Erosion Control Practice Factor (*P*)

The *P* factor measures the effect of control practices that reduce the erosion potential of the runoff by their influence on drainage patterns, runoff concentration, and runoff velocity. Practices for which *P* factors have been established are contouring and contour stripcropping. The latter values are also used for contour-irrigated furrows. In contour stripcropping, strips of sod or meadow are alternated with strips of row crops or small grains. Terraces and diversions, where used, reduce the length of slope.

Water Quality and Sediment Yield

The computed soil loss for large areas is not sediment yield, and it is not directly related to water quality. Overland sediment transport is a complex process of transport and deposition. The USLE estimates the transport component and specifically excludes the deposition component. For example, only 5% of the computed soil loss may appear as sediment yield in a drainage area of 500 square miles. The remaining 95% is redistributed and deposited on uplands or flood plains and is not a net soil loss from the area. Procedures for computing sediment yield are given in Chapter 8.

CHANNEL EROSION

Channel erosion consists of the removal of soil and rock by a concentrated flow of water. Concentrated flow permits a more concerted local attack on the soil and associated materials. Channel erosion includes gully erosion, streambank erosion, streambed degradation, floodplain scour, valley trenching, and much roadbank erosion. Gullies usually follow sheet erosion. They begin in a slight surface depression into which, in time, the concentrated flow cuts a channel a foot or more deep. The shape of the channel is usually determined by the relative resistance of the soil.

Streambank erosion and streambed degradation are affected primarily by the bank materials and the resistance of the channel bottom to the character and direction of flow. Removal of the natural vegetation from streambanks increases bank erosion. The presence of coarse bed material that a stream cannot pick up during reduced flows results in an attack on the banks by the flowing water.

When estimating long-term streambank erosion, keep in mind that bank erosion is a natural process and occurs even on streams that tend to maintain a long-term constant width. On these streams, bank erosion is offset by less obvious deposition and accretion. Therefore, streams of this type are not primary sources of sediment. Streambed erosion is not a significant long-term sediment source because the material subject to this type of erosion is limited in both extent and volume. Compared with other potential sources of sediment, streambed erosion usually is minor. Floodplain scour is the removal of floodplain soil by flows sweeping across the floodplain. It may occur in the form of channelization or sheet removal of the surface soil. This form of sheet erosion cannot be computed by the USLE or similar equations.

Streams

Streams are bodies of running water that carry rock particles (sediment loads) and dissolved ions and flow downslope along a clearly defined path, called a *channel*. Thus, streams may vary in width from a few inches to several miles. Streams are important for several reasons:

- Streams are an important part of the water cycle; they carry most of the water that goes from the land to the sea.
- Streams are one of the main transporters of sediment load from higher to lower elevations.
- Streams carry dissolved ions, the products of chemical weathering, into the oceans and thus make the sea salty.
- Streams (along with weathering and mass wasting) are a major part of the erosional process.
- Most population centers are located next to streams because they provide a major source of water and transportation.

Key Terms

Evapotranspiration (plant water loss)—The process whereby plants lose water to the atmosphere during the exchange of gases necessary for photosynthesis. Water loss by evapotranspiration constitutes a major flux back to the atmosphere.

Gaining stream—Typical of humid regions, gaining streams occur where groundwater recharges the stream.

Infiltration capacity—Maximum rate at which soil can absorb rainfall.

Laminar flow—Occurs in a stream where parallel layers of water shear over one another vertically.

Losing stream—Typical of arid regions, losing streams occur where streams recharge the groundwater.

Meandering—Stream condition whereby flow follows a winding and turning course.
Perennial stream—A type of stream in which flow continues during periods of no rainfall.
Riffles—Shallow, high-velocity flow over a mixed gravel–cobble (barlike) substrate.
Sinuosity—Refers to the bending or curving shape of a stream course.
Thalweg—Line of maximum depth in a stream.
Turbulent flow—Occurs in a stream where complex mixing is the result.

Characteristics of Stream Channels

A standard rule of thumb states that flowing waters (rivers and streams) determine their own channels, and these channels exhibit relationships attesting to the operation of physical laws—laws that are not, as of yet, fully understood. The development of stream channels and entire drainage networks and the existence of various regular patterns in the shape of channels indicate that streams are in a state of dynamic equilibrium between erosion (sediment loading) and deposition (sediment deposit) and are governed by common hydraulic processes. However, because channel geometry is four dimensional with a long profile, cross-section, depth, and slope profile, and because these mutually adjust over a time scale as short as years and as long as centuries or more, cause-and-effect relationships are difficult to establish. Other variables that are presumed to interact as the stream achieves its graded state include width and depth, velocity, size of sediment load, bed roughness, and the degree of braiding (sinuosity).

Stream Profiles

Mainly because of gravity, most streams exhibit a downstream decrease in gradient along their length. Beginning at the headwaters, the steep gradient becomes less so as one proceeds downstream, resulting in a concave longitudinal profile. Though diverse geography provides for almost unlimited variation, a lengthy stream that originates in a mountainous area typically comes into existence as a series of springs and rivulets; these coalesce into a fast-flowing, turbulent mountain stream, and the addition of tributaries results in a large and smoothly flowing river that winds through the lowlands to the sea. When studying a stream system of any length, it becomes readily apparent (almost from the start of such studies) that what we are studying is a body of flowing water that varies considerably from place to place along its length. For example, a common variable—the results of which can be readily seen—is whenever discharge increases, causing corresponding changes in the stream's width, depth, and velocity. In addition to physical changes that occur from location to location along a stream's course, there are numerous biological variables that correlate with stream size and distance downstream. The most apparent and striking changes are in steepness of slope and in the transition from a shallow stream with large boulders and a stony substrate to a deep stream with a sandy substrate. The particle size of bed material at various locations is also variable along the stream's course. The particle size usually shifts from an abundance of coarser material upstream to mainly finer material in downstream areas.

Sinuosity

Unless forced by humans in the form of heavily regulated and channelized streams, straight channels are uncommon. Stream flow creates distinctive landforms composed of straight (usually in appearance only), meandering, and braided channels; channel networks; and floodplains. Simply put: Flowing water will follow a sinuous course. The most commonly used measure is the sinuosity index (SI). Sinuosity equals 1 in straight channels and more than 1 in sinuous channels.

$$\text{SI} = \text{Channel distance} \div \text{Down valley distance} \tag{5.2}$$

Meandering is the natural tendency for alluvial channels and is usually defined as an arbitrarily extreme level of sinuosity, typically an SI greater than 1.5. Many variables affect the degree of sinuosity, however, so SI values range from near unity in simple, well-defined channels to 4 in highly meandering channels (Gordon et al., 1992).

It is interesting to note that, even in many natural channel sections of a stream course that appear straight, meandering occurs in the line of maximum water or channel depth (known as the thalweg). Keep in mind that a stream has to meander; that is how streams renew themselves. By meandering, they wash plants and soil from the land into their waters, and these serve as nutrients for the plants in the rivers. If rivers are not allowed to meander, if they are channelized, the amount of life they can support will gradually decrease. That means fewer fish, ultimately—and fewer bald eagles, herons, and other fishing birds (Spellman, 1996).

Meander flow follows a predictable pattern and causes regular regions of erosion and deposition. The streamlines of maximum velocity and the deepest part of the channel lie close to the outer side of each bend and cross over near the point of inflection between the banks. A huge elevation of water at the outside of a bend causes a helical flow of water toward the opposite bank. In addition, a separation of surface flow causes a back eddy. The result is zones of erosion and deposition, explaining why point bars develop in a downstream direction in depositional zones (Morisawa, 1968).

Bars, Riffles, and Pools

Implicit in the morphology and formation of meanders are bars, riffles, and pools. Bars develop by deposition in slower, less competent flow on either side of the sinuous main stream. Onward moving water, depleted of bedload, regains competence and shears a pool in the meander—reloading the stream for the next bar. Alternating bars migrate to form riffles. As stream flow continues along its course a pool–riffle sequence is formed. Basically, the riffle is a mound or hillock and the pool is a depression.

DID YOU KNOW?

Meandering channels can be highly convoluted or merely sinuous but maintain a single thread in curves having definite geometric shape. *Straight channels* are sinuous but apparently random in the occurrence of bends. *Braided channels* have multiple streams separated by bars and islands (Leopold, 1994).

DID YOU KNOW?

Floodplain rivers are found where regular floods form lateral plains outside the normal channel which seasonally become inundated as a consequence of greatly increased rainfall or snowmelt.

Floodplains

Stream channels influence the shape of the valley floor through which they course. This self-formed, self-adjusted flat area near to the stream is the floodplain, which loosely describes the valley floor prone to periodic inundation during over-bank discharges. What is not commonly known is that valley flooding is a regular and natural behavior of the stream. Many people learn about this natural phenomenon the hard way—that is, whenever their farms, towns, streets, and homes become inundated by a river or stream that is doing nothing more than following its "natural" periodic cycle—conforming to the master plan designed by the master planner: Mother Nature.

Water Flow in a Stream

Most elementary students learn early in their education process that water on Earth flows downhill (gravity)—from land to the sea; however, they may or may not be told that water flows downhill toward the sea by various routes. For the moment, the route (channel, conduit, or pathway) we are concerned with is the surface water route taken by surface runoff. Surface runoff is dependent on various factors. For example, climate, vegetation, topography, geology, soil characteristics, and land use determine how much surface runoff occurs compared with other pathways. The primary source (input) of water for total surface runoff is, of course, precipitation. This is the case even though a substantial portion of all precipitation input returns directly to the atmosphere by evapotranspiration. Evapotranspiration is a combination process, as the name suggests, whereby water in plant tissue and in the soil evaporates and transpires to water vapor in the atmosphere.

Probably the easiest way to understand the input of precipitation to surface water runoff is to take a closer look at this precipitation input. Again, a substantial portion of precipitation input returns directly to the atmosphere by evapotranspiration. It is also important to point out that when precipitation occurs, some rainwater is intercepted or blocked or caught by vegetation where it evaporates, never reaching the ground or being absorbed by plants. A large portion of the rainwater that reaches the surface of the ground, lakes, and streams also evaporates directly back to the atmosphere. Although plants display a special adaptation to minimize transpiration, plants still lose water to the atmosphere during the exchange of gases necessary for photosynthesis. Notwithstanding the large percentage of precipitation that evaporates, rainwater or meltwater that reaches the ground surface follows several pathways to reach a stream channel or groundwater.

Soil can absorb rainfall to its infiltration capacity (i.e., to its maximum intake rate). During a rain event, this capacity decreases. Any rainfall in excess of infiltration capacity accumulates on the surface. When this surface water exceeds the depression storage capacity of the surface, it moves as an irregular sheet of overland flow. In arid areas, overland flow is likely because of the low permeability of the soil. Overland flow is also likely when the surface is frozen or when human activities have rendered the land surface less permeable. In humid areas, where infiltration capacities are high, overland flow is rare.

In rain events where the infiltration capacity of the soil is not exceeded, rain penetrates the soil and eventually reaches the groundwater—from which it discharges to a stream slowly and over a long period of time. This phenomenon helps to explain why stream flow through a dry weather region remains constant; the flow is continuously augmented by groundwater. This type of stream is known as a perennial stream, as opposed to an intermittent one, because the flow continues during periods of no rainfall.

Streams that course their way in channels through humid regions are fed water via the water table, which slopes toward the stream channel. Discharge from the water table into the stream accounts for flow during periods without precipitation and also explains why this flow increases, even without tributary input, as one proceeds downstream. Such streams are called gaining or effluent streams, as opposed to losing or influent streams, which lose water into the ground. It is interesting to note that the same stream can shift between gaining and losing conditions along its course because of changes in underlying strata and local climate.

Stream Water Discharge

The current velocity (speed) of water (driven by gravitational energy) in a channel varies considerably within a stream's cross-section due to friction with the bottom and sides and the atmosphere; the presence of sediment and obstructions (rocks, logs, etc.); and sinuosity (bending or curving). Highest velocities, obviously, are found where friction is least, generally at or near the surface and near the center of the channel. In deeper streams, current velocity is greatest just below the surface due to the friction with the atmosphere; in shallower streams, current velocity is greatest at the surface due to friction with the bed. Velocity decreases as a function of depth, approaching zero at the substrate surface. A general and convenient rule of thumb is that the deepest part of the channel occurs where the stream velocity is the highest. Additionally, both width and depth of a stream increase downstream because discharge (the amount of water passing any point in a given time) increases downstream. As discharge increases, the cross-sectional shape will change, with the stream becoming deeper and wider. Velocity is important to discharge because discharge (m^3/s) = Cross-sectional area [width × average depth] (m^2) × Average velocity (m/s):

$$Q = A \times V \tag{5.3}$$

A stream is constantly seeking balance. This can be seen whenever the amount of water in a stream increases. The stream must adjust its velocity and cross-sectional area to reach a balance. Discharge increases as more water is added through precipitation, by tributary streams, or from groundwater seeping into the stream. As discharge increases, generally width, depth, and velocity of the stream also increase.

Transport of Material (Load)

Water flowing in a channel may exhibit laminar flow (parallel layers of water shear over one another vertically) or turbulent flow (complex mixing). In streams, laminar flow is uncommon, except at boundaries where flow is very low and in groundwater. Thus, the flow in streams generally is turbulent. Turbulence exerts a shearing force that causes particles to move along the stream bed by pushing, rolling, and skipping, referred to as *bedload*. This same shear causes turbulent eddies that entrain particles in suspension (the suspended load—particles under 0.06 mm in size). Entrainment is the incorporation of particles when stream velocity exceeds the entraining velocity for a particular particle size. The entrained particles in suspension (suspended load) also include fine sediment, primarily clays, silts, and fine sands that require only low velocities and minor turbulence to remain in suspension. These are referred to as *washload* (under 0.002 mm), because this load is "washed" into the stream from banks and upland areas (Gordon et al., 1992; Spellman, 1996).

Thus, the suspended load includes the washload and coarser materials (at lower flows). Together, the suspended load and bedload constitute the *solids load*. It is important to note that in bedrock streams the bedload will be a lower fraction than in alluvial streams where channels are composed of easily transported material.

A substantial amount of material is also transported as the dissolved load. Solutes (ions) are generally derived from chemical weathering of bedrock and soils, and their contribution is greatest in subsurface flows and in regions of limestone geology. The relative amount of material transported as solute rather than solids load depends on basin characteristics, lithology (i.e., the physical character of rock), and hydrologic pathways. In areas of very high runoff, the contribution of solutes approaches or exceeds sediment load; whereas, in dry regions, sediments make up as much as 90% of the total load.

Deposition occurs when the *stream competence*—which refers to the largest particles that a stream can move, which in turn depends on the critical erosion, or competent, velocity, which is the lowest velocity at which a particle resting on the streambed will move—falls below a given velocity. Simply stated: The size of the particle that can be eroded and transported is a function of current velocity. Sand particles are the most easily eroded. The greater the mass of larger particles (e.g., coarse gravel), the higher the initial current velocities must be for movement. However, smaller particles (silts and clays) require even greater initial velocities because of their cohesiveness and because they present smaller, streamlined surfaces to the flow. Once in transport, particles will continue in motion at somewhat slower velocities than initially required to initiate movement and will settle at still lower velocities.

DID YOU KNOW?

Entrainment is a natural extension of erosion and is vital to the movement of stationary particles under changing flow conditions. Remember, all sediments ultimately derive from erosion of basin slopes, but the immediate supply usually derives from the stream channel and banks, while the bedload comes from the streambed itself and is replaced by erosion of bank regions.

Particle movement is determined by size, flow conditions, and mode of entrainment. Particles over 0.02 mm (medium to coarse sand size) tend to move by rolling or sliding along the channel bed as traction load. When sand particles fall out of the flow, they move by saltation or repeated bouncing. Particles under 0.06 mm (silt) move as suspended load, particles under 0.002 (clay) indefinitely as washload. A considerable amount of particle sorting takes place because of different styles of particle flow in different sections of a stream (Likens, 1984; Richards, 1982). Unless the supply of sediments becomes depleted, the concentration and amount of transported solids increase, but discharge is usually too low throughout most of the year to scour, shape channels, or move significant quantities of sediment in all but sand-bed streams, which can experience change more rapidly. The greatest scour occurs during extreme events, and the amount of material removed increases dramatically. Sediment inflow into streams can be increased or decreased as a result of human activities. For example, poor agricultural practices and deforestation greatly increase erosion. Manmade structures such as dams and channel diversions, on the other hand, can greatly reduce sediment inflow.

Computation Procedures

Methods of determining soil loss by the various types of channel erosion include (1) comparing aerial photographs of different dates to determine the annual growth rate of channels; (2) rerunning existing cross-sections to determine the difference in total channel cross-sectional area; (3) assembling historical data to determine the average age of channels and their average annual growth; and (4) making field studies to estimate the average annual growth rate (volume per unit length of channel). The formula for computing annual channel erosion from data obtained in these determinations is

$$S = H \times L \times R \tag{5.4}$$

where

S = Annual soil loss from streambank erosion (cubic feet).
H = Average height of bank (feet)
L = Length of bank being eroded, each side of channel (feet).
R = Annual rate of bank recession (feet).

■ EXAMPLE 5.1

Problem: If $H = 5$ ft, $L = 1800$ ft, and $R = 0.1$ ft, what is S?

Solution:

$$S = 5 \text{ ft} \times 1800 \text{ ft} \times 0.1 \text{ ft} = 900 \text{ ft}^3$$

DID YOU KNOW?

Annual recession rates of more than 0.1 ft are common on the outside of bends and meanders. This cut-bank recession is usually offset by sediment accretion on the opposite slip-off slope, which results in channel migration with no substantial change in channel width. Significant long-term changes in channel width cannot occur without equally drastic changes in discharge, slope, or depth.

■ **EXAMPLE 5.2**

Problem: If $H = 20$ ft, $L = 900$ ft, and $R = 0.2$ ft, what is S?

Solution:

$$S = 20 \text{ ft} \times 900 \text{ ft} \times 0.2 \text{ ft} = 3600 \text{ ft}^3$$

In this example, an annual degradation rate of 0.2 ft for 100 years would deepen the channel by 20 ft, but this rate is not likely to occur in a perennial stream.

WIND EROSION

During a recent research outing to several national parks in the western United States, the author stopped at several locations (as always) and photographed various natural wonders. One of the focal points of study was the weathering processes that are discussed in this chapter. The natural bridges, such as the one shown in Figure 5.1, and the natural arches or windows (eventually become hoodoos) shown in Figures 5.2 and 5.3 are all a result of some form of weathering; thus, they are highlighted here.

There was a time not that long ago when many believed that the main difference between natural bridges (see Figure 5.1) and the natural windows (preforming hoodoos) shown in Figures 5.2 and 5.3 was that the natural bridges were formed by water erosion and natural windows, or arches, were formed by wind erosion. Contrary to popular belief, however, wind is not a significant factor in the formation

FIGURE 5.1 Rainbow Bridge, located in the Lake Powell/Colorado River region in Utah, is the world's largest known natural bridge. (Photograph by author.)

FIGURE 5.2 Windows forming in future hoodoos in Bryce Canyon, Utah. (Photograph by author.)

FIGURE 5.3 Weathered-window forming in hoodoo formation in Bryce Canyon, Utah. (Photograph by author.)

DID YOU KNOW?

The velocity of the wind and the size, shape, and weight of the rock particles determine the manner in which wind carries its load. Wind-transported materials are most commonly derived from floodplains, beach sands, glacial deposits, volcanic explosions, and dried lake bottoms—places containing light ash and loose, weathered rock fragments.

The wind is capable of transporting large quantities of material for very great distances. The wind deposits sediment when its velocity decreases to the point where the particles can no longer be transported. Initially (in a strong wind), part of the sediment load rolls or slides along the ground (bedload). Some sand particles move by a series of leaping or bounding movements (saltation). And lighter dust may be transported upward (suspension) into higher, faster moving wind currents, traveling many thousands of miles.

The wind will begin to deposit its load when its velocity is decreased or when the air is washed clean by falling rain or snow. A decrease in wind velocity may also be brought about when the wind strikes some barrier-type obstacle (e.g., fences, trees, rocks, manmade structures) in its path. As the air moves over the top of the obstacle, streamlines converge and the velocity increases. After passing over the obstacle, the streamlines diverge and the velocity decreases. As the velocity decreases, some of the load in suspension can no longer be held in suspension and thus drops out to form a deposit. The major types of wind-blown or eolian deposits are dunes and loess.

of natural arches or other natural formations. Substantial studies have shown that natural arches and natural bridges are formed by many different processes of erosion that all contribute to the natural, selective removal of rock. Every process relevant to natural arch formation involves the action of water, gravity, temperature variation, or tectonic pressure on rock.

Again, wind is not a significant agent in natural arch formation; however, wind does act to disperse the loose grains that result from microscopic erosion. Moreover, sandstorms can scour and polish already existing arches. The bottom line (and an important point to remember) is that never does wind alone create arches (Barnes, 1987; Vreeland, 1994). As noted earlier, wind action or erosion is very limited in extent and effect. It is largely confined to desert regions, but even there it is limited to a height of about 18 inches above ground level. Wind does have the power, however, to transport, deposit, and erode sediment. In this section, we will discuss various aspects of the wind that are important in any study of sediments and their derivation, transport to, and eventual entry into freshwater systems.

Wind erosion is the detachment and transport of soil material by wind. The process is called *deflation*, and the resultant deposits are classified as eolian. The rate of erosion depends on the intensity and persistence of the wind, size, and availability of soil particles and the amount of protective cover. Dry soil is necessary for maximum deflation rates. In the United States, the conditions generally most favorable for wind erosion are in semiarid or arid areas west of the 100th meridian, although wind erosion does occur elsewhere. Although water erosion is dominant even in arid areas, wind erosion can approach it in amount in deserts and during periods of intensive drought in other areas.

Eolian deposits are characterized by highly sorted particles, by cross-bedded or lenticular structures, and by dunes oriented by the prevailing winds. A hummocky surface develops when wind-blown sediment lodges around isolated bushes or grass. Fence-line deposits are confined to the area alongside the fence and can be several feet thick.

Deflation areas contain scoured-out depressions or pock-marked surfaces. Such features are usually in exposed places and are not associated with water drainage rills or channels. Remnants of grass or even single pebbles may rest on small pedestals in an eroded zone. Some shrubs or bunches of grass may persist, with the root system exposed above ground. In gravity sands, selective removal of the smaller particles can produce a gravel pavement on the surface. The amount of deflation can be determined by comparing the voided area with the original ground surface. Measure enough cross-sections to delineate an average-sized depression and determine the number of depressions on recent aerial photographs or count the number per unit area. Wind-deposited materials may have come from outside a watershed. Conversely, a watershed may have lost much soil to distant areas. Windblown sediment moves progressively in the direction of the prevailing winds rather than downslope.

The most important aspect of wind erosion to be considered in studies of sediment yield is the deposition of windblown sediment in channels from which it is easily flushed and added to the sediment yield of the watershed. Channels act as natural traps for airborne sediment whether they contain water or not. If eolian deposition in channels is a factor in the watershed being studied, measure the annual volume of deposition. A sampling process will usually be adequate. Unless channel capacity is decreasing because of these deposits, add the volume of these sediments to the sediment yield. The sediment delivery ratio depends on the kind of material. Wind erosion does not occur every year in most areas. Adjust the annual sediment yield rates downward to account for years in which wind erosion does not occur.

In some areas, a significant amount of windblown soil may be deposited on snow. During snowmelt the soil is carried by water into streams or drainage ditches. This snow-caught sediment can be measured by pushing metal tubes into the snow and weighing the content after the snow in the sample melts.

Many factors affect the amount of soil moved by wind erosion. An equation has been developed (Chepil and Woodruff, 1963) to predict the average annual soil loss from wind erosion:

$$E = f(I, C, K, L, V) \tag{5.5}$$

where

E = Estimated average annual soil loss in tons per acre per year.

f = Indicates functional relationships that are not straight-line mathematical calculations.

I = Soil erodibility index.

C = Climatic factor.

K = Soil surface roughness factor.
L = Unsheltered distance.
V = Vegetative cover factor.

The I factor, expressed as the average annual soil loss in tons per acre per year from a field area, accounts for the inherent soil properties affecting erodibility. These properties include texture, organic matter, and calcium carbonate percentage. I is the potential annul wind erosion for a given soil under a given set of field conditions. The given set of field conditions for which I is referenced is that of an isolated, unsheltered, wide, bare, smooth, level, loose, and noncrusted soil surface at a location where the climatic factor (C) is equal to 100.

The K factor is a measure of the effect of ridges and cloddiness made by tillage and planting implements. It is expressed as a decimal from 0.1 to 1.0. The C factor for any given locality characterizes climate erosivity, specifically wind speed and surface soil moisture. This factor is expressed as a percentage of the C factor for Garden City, Kansas, which has a value of 100. The L factor considers the unprotected distance along the prevailing erosive wind direction across the area to be evaluated and the preponderance of the prevailing erosive winds. The V factor considers the kind, amount, and orientation of vegetation on the surface. The vegetative cover is expressed in pounds per acre of a flat, small-grain residue equivalent.

Solving the equation involves five successive steps. Steps 1, 2, and 3 can be solved by multiplying the factor values. Determining the effects of L and V (steps 4 and 5) involves more complex functional relationships.

1. $E_1 = I$. Factor I is established for the specific soil. I may be increased for knolls less than 500 ft long facing into the prevailing wind, or decreased to account for surface soil crusting, and irrigation.
2. $E_2 = IK$. Factor K adjusts E_1 for tillage-induced oriented roughness, K_{rd} (ridges), and for random roughness, K_{rr} (cloudiness). The value of K is calculated by multiplying K_{rd} by K_{rr} ($K = K_{rd} \times K_{rr}$).
3. $E_3 = IKC$. Factor C adjust E_2 for the local climatic factor.
4. $E_4 = IKCL$. Factor L adjusts E_3 for unsheltered distance.
5. $E_5 = IKCLV$. Factor V adjusts E_4 for vegetative cover.

Limitations of the Equation

When the unsheltered distance (L) is sufficiently long, the transport capacity of the wind for saltation and creek is reached. If the wind is moving all the soil it can carry across a given surface, the inflow into a downwind area of the field is equal to the outflow from the same area of the field, for saltation and creep. The net soil loss from this specific area of the field is then only the suspension component. This does not imply a reduced soil erosion problem because, theoretically, there is still the estimated amount of soil loss in creep, saltation, and suspension leaving the downwind edge of the field. The equation does not account for snow cover or seasonal changes in soil erodibility. The equation does not estimate erosion from single storm events, and surface armoring by non-erodible gravel is not usually addressed in the I factor.

MASS MOVEMENT

Mass wasting, or mass movement, takes place as earth materials (loose uncemented mixtures of soil and rock particles known as *regolith*) move downslope in response to gravity without the aid of a transporting medium such as water, ice, or wind—although these factors play a role in regolith movement. This type of erosion is apt to occur in any area with slopes steep enough to allow downward movement of rock debris. Mass movement occurs when shear stress exceeds shear strength. High shear stress can be caused by removal of lateral support; added weight of rain, snow, or talus accumulations; construction or other human activities; transitory earth stresses, such as earthquakes; regional tilling; removal of underlying support; and lateral pressure from water in cracks and caverns, freezing of water, or swelling of clay or anhydrite (NRC, 1974). Low shear strength can be caused by the following:

1. Composition; inherently weak materials such as saturated clay and silt are examples
2. Texture, such as a loose arrangement of particles or roundness of grains
3. Gross structures, including discontinuities from faults, bedding planes, or joints, or strata included toward a free face
4. Changes resulting from weathering and other physiochemical reactions
5. Changes in intergranular forces due to pore water
6. Changes in internal structure, such as fissuring in preconsolidated clays or the effect of disturbance or remolding on sensitive materials

Some of the factors that help gravity overcome resistance are discussed below.

Gravity

The heavy hand of gravity constantly pulls everything, everywhere toward Earth's surface. On a flat surface, parallel to Earth's surface, the constant force of gravity acts downward. This downward force prevents gravitational movement of any material that remains on or parallel to a flat surface. On a slope, the force of gravity can be resolved into two components: (1) a component acting perpendicular to the slope,

DID YOU KNOW?

When shear on a slope decreases, material may still be stuck to the slope and prevented from moving downward by the force of friction. It may be held in place by the frictional contact between the particles making up that material. Contact between the surfaces of the particles creates a certain amount of tension that holds the particles in place at an angle. The steepest angle at which loose material on a slope remains motionless is called the *angle of repose* (generally about 35°). Particles with angled edges that catch on each other also tend to have a higher angle of repose than those that have become rounded through weathering and that simply roll over each other.

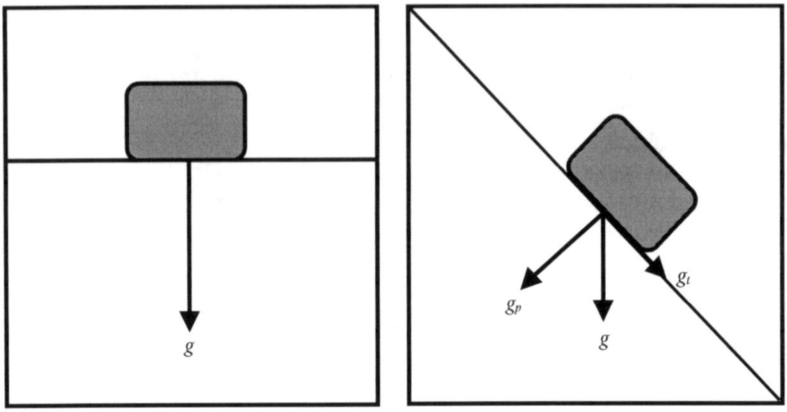

FIGURE 5.4 (A) Gravity acting perpendicular to the surface. (B) The perpendicular component (the glue) of gravity (g_p) helps to hold the material in place on the slope.

and (2) component acting tangential to the slope. Thus, material on a slope is pulled inward in a direction that is perpendicular (the glue) to the slope (see Figure 5.4A). This helps prevent material from sliding downward. However, on a slope, another component of gravity exerts a force (a constant tug) that acts to pull material down a slope, parallel to the surface of the slope. Known as *shear stress*, this force of gravity exerts stress in direct relationship to the steepness of the slope; that is, shear stress increases as the slope steepens. In response to increased shear stress, the perpendicular force (the glue) of gravity decreases (see Figure 5.4B).

Water

Even though mass wasting may occur in either wet or dry materials, water greatly facilitates downslope movements; it is an important agent in the process of mass wasting. Water will either help hold material together, acting like glue (as demonstrated by beach sandcastles built with slightly dampened sand) and increasing its angle of repose, or cause it to slide downward like a liquid (acting like a lubricant). Water may soften clays and make them slippery, add weight to the rock mass, and, in large amounts, actually force rock particles apart, thus reducing soil cohesion.

Freezing and Thawing

Frost wedging, which occurs when water contained in rock and soil expands when frozen, has significant erosive power. Mass wasting in cold climates is governed by the fact that water is frozen as ice during long periods of the year, especially in high-altitude regions. Ice, although it is solid, does have the ability to flow (glacial-movement effect), and alternate periods of freezing and thawing can also contribute to movement. In some instances, ice expansion may be great enough to force rocks downhill.

Undercutting

Undercutting occurs when streams erode their banks or when surf action along a coast undercuts a slope, making it unstable. Undercutting can also occur when man-made excavations remove support and allow overlying material to fall.

Organic Activities

Whenever animals burrow into the ground, they disturb soil materials, casting rocks out of their holes as they dig; these are commonly piled up downslope. Eventually, weather conditions and the constant force applied by gravity can put these piles into motion. Animals also contribute to mass wasting whenever they walk on soil surfaces; their motions can knock materials downhill.

Shock Waves or Vibrations

A sudden strong shock or vibration, such as an earthquake, faulting, blasting, or heavy traffic, can trigger slope instability. Minor shocks due to, for example, heavy vehicles rambling down the road, trees blowing the wind, or manmade explosions can also trigger mass-wasting events such as landslides.

Kinds of Mass Movements

A landslide is a mass movement that occurs suddenly and violently. In contrast, soil creep is mass movement that is almost imperceptible. These processes can be divided into two broad categories: rapid movements or slow movements. Rapid movements include landslides, slumps, mudflows, and earthflows. Slow movements include soil creep and solifluction.

Rapid Movements

- *Landslides* are by and far the most spectacular and violent of all mass movements. Landslides are characterized by the sudden movement of great quantities of rock and soil downslope. Such movements typically occur on steep slopes that have large accumulations of weathered material. Precipitation in the form of rain or snow may seep into the mass of steeply sloping rock debris, adding sufficient weight to start the entire mass sliding.
- *Slumps* are special landslides that occur along a curved surfaces. The upper surface of each slump block remains relatively undisturbed, as do the individual blocks. Slumps leave arcuate (curved like a bow) scars or depressions on the hill slope. Heavy rains or earthquakes usually trigger slumps. Slump is a common occurrence along the banks of streams or the walls of steep valleys.
- *Mudflows* are highly fluid, high-velocity mixtures of sediment and water that have a consistency of wet concrete. Mass wasting of this type typically occurs when certain arid or semi-arid mountainous regions are subjected to unusually heavy rains.

DID YOU KNOW?

Landslides are a major geologic hazard because they are widespread, occur in all 50 states and U.S. territories, and cause $1 to $2 billion in damages and more than 25 fatalities in an average year. Expansion of urban and recreational developments into hillside areas leads to more people being threatened by landslides each year. Landslides commonly occur in connection with other major natural disasters such as earthquakes, volcanoes, wildfires, and floods (USGS, 2016).

- *Earthflows* are usually associated with heavy rains and move at velocities between several centimeters and hundreds of meters per year. They usually remain active for long periods of time. They generally tend to be narrow tongue-like features that begin at a scarp or small cliff.

Slow Movements

- *Soil creep* is a continuous movement, usually so slow as to be imperceptible. It normally occurs on almost all slopes that are moist but not steep enough for landslides. Soil creep is usually accelerated by frost wedging, alternate thawing and freezing, and certain plant and animal activities. Evidence for creep is often seen in bent trees, offsets in roads and fences, and inclined utility poles.
- *Solifluction* is a downslope movement typical of areas where the ground is normally frozen to considerable depth—arctic, subarctic, and high mountain regions. The actual soil flowage occurs when the upper portion of the mantle rock thaws and becomes water saturated. The underlying, still frozen subsoil acts as a slide for the sodden mantle rock which will move down even the gentlest slope.

Desertification

Deserts areas are where the amount of precipitation received is less than the potential evaporation (<10 in./year); they cover roughly 30% of the Earth's land surface—areas we think of as arid. Desertification occurs in hot areas far from sources of moisture, in areas isolated from moisture by high mountains, in coastal areas with onshore winds and cold-water currents, and high-pressure areas where descending air masses produce warm, dry air. The world's great deserts were formed by natural processes interacting over long intervals of time (USGS, 1997). During most of these times, deserts have grown and shrunk independent of human activities. Desertification does not occur in linear, easily mapable patterns. Deserts advance erratically, forming patches on their borders. Scientists question whether desertification, as a process of global change, is permanent or how and when it can be halted or reversed.

Estimation Procedures

No standard procedures for calculating erosion by mass movement have been developed; it must therefore be estimated. Numerous measurements have been made in the semiarid western United States to determine the maximum angles at which slopes stand with and without vegetal cover. Nonvegetated talus material stands at gradients between 68 and 80% (angles of about 34 to 38°). Vegetated slopes underlain by fine-textured soils derived from the same parent material as the barren talus stand at gradients of as much as 173% (angle of 60°). Without vegetation, slopes of fine material would not stand, even at gradients as high as those of coarse talus (Bailey, 1941). The hazard of debris flows can be estimated on the basis of slope. These flows usually originate on slopes of more than 30%. The terminal slope of debris flows is between 7 and 10%. A procedure for calculating erosion from mass movement would require measuring the volume of materials moved. For large masses, comparing the findings of a topographical survey of the mass with the original topography (from standard quadrangle sheets if available) can provide an estimate of the volume of materials moved. For smaller masses, a grid of hand-auger borings extending into the original soil profile can provide a basis for estimating the volume.

OTHER TYPES OF EROSION

Other types of erosion not described in detail here do occur and must be evaluated by environmental professionals if found in areas under study.

WAVE EROSION

Caused by wind and water, wave erosion is an important source of sediment along the shorelines of lakes, streams, and oceans. Wave erosion can change shorelines markedly and can be measured in many places (Glymph and Jones, 1937; Jones and Rogers, 1952). The rate of erosion from wave action can be measured by comparing two sets of aerial photographs taken on different dates, as in estimating channel erosion. Historical data represent another basis for estimating wave erosion rates. Unless the shoreline was mechanically shaped during reservoir construction, wave erosion along a reservoir shore can also be determined by comparing the current shore profile with an extrapolation of the slope of the profile above the influence of the wave action (see Figure 5.5).

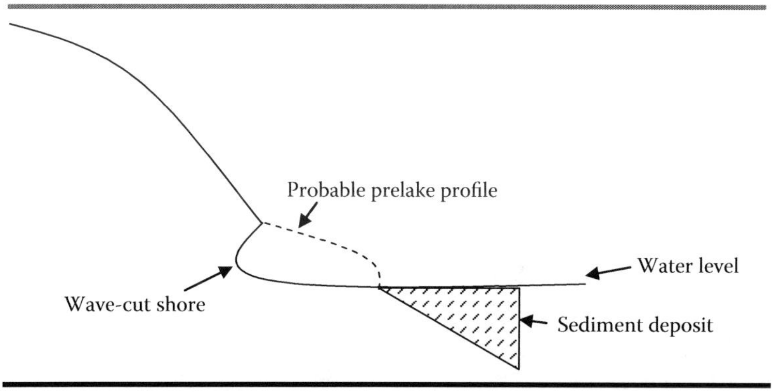

FIGURE 5.5 Projecting lines of undisturbed bank to determine the probable prelake profile.

Erosion from Stripmining and Construction

Stripmining or excavation operations and the construction of highways, industrial areas, public buildings, housing, shopping centers, and related areas have greatly accelerated the erosion of exposure sand spoil banks. Each condition must be evaluated as a separate problem. Holeman and Geiger (1959) estimated that the Lake Barcroft, Virginia, watershed yielded 25 acre-ft of sediment in 1951, when 9% of the area (13 square miles) was under construction, an increase of 21.3 acre-ft over the pre-1938 average annual rate of 3.7 acre-ft. The sediment yield was 16.3 acre-ft/mi^2 for the area under construction and 0.257 acre-ft/mi^2 for the watershed in the earlier period of agricultural use. Before 1938, 18% of the watershed was cultivated, 23.5% was pasture, 53% was woods, and 5.5% was residential. Construction activities were believed to have increased the sediment yield to more than 63 times the pre-1938 level. Wolman and Schick (1967) found that the sediment yield in construction areas averaged 72 times that in rural areas. Collier et al. (1964) found that in 1959 a watershed near Somerset, Kentucky, with 6% of its area stripmined, yielded 69 times more sediment than a similar adjacent watershed that was wooded and unmined. These findings do not mean that areas under construction always yield 70 times the sediment that they would under rural conditions, but the figures do indicate the general size of such increases. In areas undergoing urbanization, the average annual amount of soil exposed can be estimated from such factors as population curves and the number of sewer connections to determine annual trends.

The USLE is the most promising method for calculating erosion on construction sites or stripmined areas, but appropriate values for factors of the equation must be carefully selected. Keep in mind that the soil surface is probably not in the same condition as it would be under any agricultural use. The microrelief and soil surface conditions are likely to vary much more over short distances than they do in any agricultural situation. The USLE K values are indexed to "tilled continuous fallow" and a specific microrelief and surface texture that may not be common on construction sites. Topsoil K values are currently determined by use of a nomograph (Wischmeier et al., 1971). Other research (Roth et al., 1974) indicates that factors other than those considered by Wischmeier et al. may be significant in determining the erodibility of exposed cohesive subsoil. Sediment yield from construction sites and stripmined areas can be estimated from the computed erosion and a sediment delivery ratio. Projected erosion-control measures should be considered realistically when determining the sediment delivery ratio.

Ice Erosion

In watersheds likely to be studied in the Soil Conservation Service small watershed program, erosion by ice probably falls into one of three categories: (1) glacial gouging around the margin of mountain glaciers, (2) erosion by ice along river channels during spring freshets, and (3) erosion by ice shoved along the shores of northern lakes. Ice erosion usually is not an important source of sediment.

REFERENCES AND RECOMMENDED READING

ARS. (1961). *A Universal Equation for Measuring Wind Erosion: An Aid to Conservation Farming in the Great Plains*, ARS 22-69. Washington, DC: Agricultural Research Service, U.S. Department of Agriculture.

Bailey, R.W. (1941). Land erosion—normal and accelerated—in the semiarid west. *American Geophysical Union Transactions*, 22(2): 240–261.

Barnes, F.A. (1987). *Canyon Country Arches and Bridges*. Moab, UT: Canyon County Publications.

Bennett, H.H. (1939). *Soil Conservation*. New York: McGraw-Hill.

Chepil, W.S. (1962). A compact rotary sieve and the importance of dry sieving in physical soil analysis. *Soil Science Society of America Proceedings*, 26(1): 4–6.

Chepil, W.S. and Woodruff, N.P. (1963). The physics of wind erosion and its control. *Advances in Agronomy*, 15: 211–302.

Chepil, W.S., Siddoway, F.H., and Armbrust, D.V. (1962). Climatic factor for estimating wind erodibility of farm fields. *Journal of Soil and Water Conservation*, 17(4): 162–165.

Collier, C.R. et al. (1964). *Influence of Strip Mining on the Hydrologic Environment of Parts of Beaver Creek Basin, Kentucky, 1955–59*, USGS Professional Paper 427-B. Reston, VA: U.S. Geological Survey.

Eckel, E.B., Ed. (1958). *Landslides and Engineering Practice*, Highway Research Board Special Report No. 29. Washington, DC: National Academy of Sciences–National Research Council.

Ellison, W.D. (1945). Some effects of raindrops and surface flow on soil erosion and infiltration. *EOS*, 26: 415–429.

Giller, P.S. and Jalmqvist, B. (1998). *The Biology of Streams and Rivers*. Oxford: Oxford University Press.

Glymph, Jr., L.M. and Jones, V.H. (1937). *Advance Report on the Sedimentation Survey of Lake Decatur, Decatur, Illinois*, USDA SCS-SS-12. Washington, DC: U.S. Department of Agriculture.

Goodwin, P.H. (1998). *Landslides, Slumps, and Creep*. New York: Franklin Watts.

Gordon, N.D., McMahon, T.A., and Finlayson, B.L. (1992). *Stream Hydrology: An Introduction for Ecologists*. Chichester: John Wiley & Sons.

Holeman, J. N. and Geiger, A.F. (1959). *Sedimentation of Lake Barcroft, Fairfax County, Va.*, USDA SCS-TP-136. Washington, DC: U.S. Department of Agriculture.

Jennings, T. (1999). *Landslides and Avalanches*. North Mankato, MN: Thameside Press.

Jones, V.H. and Rogers, R.E. (1952). *Aspects of Sedimentation in Lake Waco, Waco, Texas*. U.S. Soil Conservation Service Technical Paper 100. Washington, DC: U.S. Department of Agriculture.

Leopold, A. (1949). *A Sand County Almanac and Sketches Here and There*. Oxford: Oxford University Press.

Leopold, L.B. (1994). *A View of the River*. Cambridge, MA: Harvard University Press.

Likens, W.M. (1984). Beyond the shoreline: a watershed ecosystem approach. *Internationale Vereinigung für Theoretische und Angewandte Limnologie*, 22: 1–22.

Morisawa, M. (1968). *Streams: Their Dynamics and Morphology*. New York: McGraw-Hill.

Musgrave, G.W. (1947). The quantitative evaluation of factors in water erosion—a first approximation. *Journal of Soil and Water Conservation*, 2(3): 133–138.

NRC. (1974). *Methodology for Delineating Mudslide Hazard Areas*. Washington, DC: National Research Council, Building Research Advisory Board.

Richards, K. (1982). *Rivers: Form and Processes in Alluvial Channels*. London: Mehuen.

Roehl, J.W. (1965). Erosion and its control on agricultural land. In: *Federal Interagency Sedimentation Conference: Proceedings*, USDA Misc. Publ. 970. Washington, DC: U.S. Department of Agriculture, pp. 18–22.

Roth, C., Nelson, D., and Romkins, M.J.M. (1974). *Alternate Criteria for Determining Subsoil "K" Values*. Washington, DC: U.S. Environmental Protection Agency.

Soil Conservation Service, U.S. Department of Agriculture. (1966). *Procedure for Determining Rates of Land Damage, Land Depreciation, and Volume of Sediment Produced by Gully Erosion: Guidelines for Watershed Management*, FAO Conservation Guide No. 1. Rome: Food and Agriculture Organization.

Spellman, F.R. (1996). *Stream Ecology and Self-Purification*. Lancaster, PA: Technomic.

Spellman, F.R. and Whiting, N. (1998). *Environmental Science and Technology*. Rockville, MD: Government Institutes.

USDA. (2008). Sediment properties. In: *National Engineering Handbook*, Section 3, Chapter 2. Washington, DC: U.S. Department of Agriculture.

United Nations. (1953). *The Sediment Problem*, Flood Control Ser. No. 5. Bangkok: Economic Commission for Asia and the Far East.

USGS. (1997). *Desertification*. Reston, VA: U.S. Geological Survey, http://pubs.usgs.gov/gip/deserts/desertification/.

USGS. (2015). *Sinkholes*. Reston, VA: U.S. Geological Survey, http://water.usgs.gov/edu/sinkholes.html.

USGS. (2016). *Landslide Hazards Program*. Reston, VA: U.S. Geological Survey, http://landslides.usgs.gov/.

Vreeland, R.H. (1994). *Nature's Bridges and Arches*. Vol. 1. *General Information*, 2nd ed. Self published.

Walker, J. (1992). *Avalanches and Landslides*. New York: Gloucester Press.

Wischmeier, W.H. (1974). New developments in estimating water erosion. In: *Proceedings of the 29th Annual Meeting of the Soil Conservation Society of America*, Syracuse, NY, August 11–14, pp. 179–186.

Wischmeier, W.H., Johnson, C.B., and Cross, B.V. (1971). A soil erodibility nomograph for farmland and construction sites. *Journal of Soil and Water Conservation*, 26: 189–193.

Wolman, M.G. and Schick, A.P. (1967). Effects of construction on fluvial sediment, urban and suburban areas of Maryland. *Water Resources Research*, 3(2): 451–464.

6 Transport of Sediment by Water

We can't fully understand freshwater sediments unless we wholly understand them.

INTRODUCTION

Understanding the principles of sediment transport by flowing water is essential to interpreting and solving many problems. The individual characteristics of water and sediment and their interaction directly affect the type and volume of material eroded and transported and the place and time of deposition. Evaluating channel instability, including erosion or aggradation, and predicting the performance of proposed channel improvements are problems that require knowledge of sediment transport and the use of procedures pertaining to it. Information derived from applying sediment transport prediction procedures can be used to determine the requirements of storage of coarse sediment in debris basins and other types of structures. This chapter includes a discussion of the characteristics of water as a medium for initiating the movement and transport of sediment. The reaction of material on the streambed to the hydraulic forces exerted and the effect of velocity and flow depth on the rate of bed material transport are described. Formulas and procedures designed to predict the rate of bed material transport are given and evaluated. Recommendations are made for applying these formulas and procedures to channel problems. The chapter concludes with a discussion of the mechanics of suspended load transport and a description of a method for computing suspended load yield from concentration and flow duration data.

FACTORS AFFECTING SEDIMENT TRANSPORT

The mechanism of entrainment and the rate at which sediment is transported depend on the characteristics of the transporting medium and on the properties and availability of particles.

Characteristics of Water as the Transporting Medium

The interrelated characteristics of water that govern its ability to entrain and move sedimentary particles are *density*, *viscosity*, and *acidity* (pH):

- Density is the ratio of mass to volume. Increasing the temperature of water increases its volume and decreases its density. With an increase in temperature from 40°C to 100°C (104°F to 212°F), water will expand to 1.04

times its original volume. When working with large volumes of moving water, the slight variations in density resulting from temperature change are usually ignored.

- Viscosity is the cohesive force between particles of a fluid that causes the fluid to resist a relative sliding motion of particles. Under ordinary pressure, viscosity varies only with temperature. A decrease in water temperature from 26.7°C to 4.4°C (80°F to 40°F) increases viscosity about 80%. Changes in viscosity affect the fall velocity of suspended sediment and thereby its vertical distribution in turbulent flow (Colby and Scott, 1965). Increasing the viscosity lowers the fall velocity of particles, particularly very fine sands and silts. A substantial decrease in water temperature and the consequent increase in viscosity smooth the bed configuration, lower the Manning *n* roughness coefficient,* and increase the velocity over a sand bed (USACE, 1968).
- The pH value is the negative logarithm (base 10) of the hydrogen ion concentration. Neutral water has a pH value of 7.0. Acid water has a pH value lower than 7.0, and alkaline water has a pH value higher than 7.0. In acid waters, sediment deposition may be promoted by the formation of colloidal masses of very fine sediments (flocculation) that settle faster than their component fine particles.

Laminar Sublayer

In turbulent flow, a thin layer forms adjacent to the bed in which the flow is laminar because the fluid particles in contact with the bed do not move. This is the laminar sublayer; the higher the velocity or the lower the viscosity, the thinner the sublayer. If the boundary is rough enough, its irregularities may project into the theoretical laminar sublayer, thereby preventing its actual development. Although laminar flow is primarily related to fluid viscosity, turbulent flow is affected by a number of factors. In laminar flow filaments of water follow parallel paths, but in turbulent flow the paths of particles crisscross and touch, mixing the liquid. A criterion defining the transition from laminar to turbulent flow is the Reynolds number (*Re*), which is the ratio of inertial force to shear force on the fluid particle. If the Reynolds number is low, shear forces are dominant, but as the Reynolds number increases they decline to little significance, thereby indicating the dominance of inertial forces. In other words, the Reynolds number is a dimensionless quantity that expresses the relative importance of inertial forces compared to viscous forces in a flow system. A small Reynolds number is associated with laminar flow; a large Reynolds number is associated with turbulent flow. The association of laminar flow with viscosity and that of turbulent flow with inertia are the same whether the fluid is moving or at rest. A small particle of sediment, such as a very fine sand, settling in still or flowing water moves slowly enough to sustain laminar flow lines in relatively viscous media. Inertial forces become increasingly important as grain size increases and are dominant when the particle size exceeds 0.5 mm.

* Manning's equation can be written as $V = (k/n)R_h^{2/3}S^{1/2}$, where V is the average flow velocity; $k = 1.49$ for English units or 1.0 for SI units; n is the Manning roughness coefficient; R_h is the hydraulic radius; and S is the slope of the water surface.

Characteristics of Transportable Materials

The characteristics of separate or distinct in form (discrete) particles were discussed in Chapter 4. The entrainment and transport of granular materials depend on the size, shape, and specific weight of the particles and their position with respect to each other. The resistance of cohesive materials depends largely on the forces of interparticle bonding. Cohesive forces can be attributed to several factors, including the amount and kind of clay minerals, the degree of consolidation or cementation, and the structure of the soil mass.

MECHANISM OF ENTRAINMENT

Forces Acting on Discrete Particles

Turbulence is a highly irregular motion characterized by the presence of eddies. The degree to which eddies form depends on the boundary roughness and geometry of the channel, and eddies are sustained by energy supplied by the flow. The eddies penetrate the laminar sublayers formed along the bed. Discrete particles resting on the bed are acted on by two components of the forces associated with the flow. One component force is exerted parallel to the flow (drag force) and the other is perpendicular to the flow (lifting force). Drag force results from the difference in pressure between the front and the back sides of a particle. Lifting force results from the difference in pressure on the upper and lower surfaces. If the lifting force exceeds the particle's immersed weight and the interference of neighboring grains, the particle goes into suspension. Because turbulence is random and irregular, discrete particles tend to move in a series of short, intermittent bursts. In each burst, particles move a short distance and many grains move simultaneously. The movement then subsides until another burst occurs. The frequency and extent of movement increases with the intensity of turbulence, and above a certain intensity some particles may be projected into the flow as suspended load (Sutherland, 1967). The coarser and rounder the particles are, however, the greater the possibility that they will begin to roll and continue rolling.

Tractive Force

Experiments to determine the forces that act on particles on a streambed were performed mainly to predict channel stability. More advanced methods are necessary to describe transport. The instantaneous interactions between turbulent flow and discrete sediment particles resting on the bed were described briefly in the preceding paragraphs. In practical application, however, it is more convenient to deal with time-average values of the force field generated by the flow near the bed. Here, the forces normal to the bed having a time average equal to zero can be eliminated, and only those forces parallel to the bed need to be considered. The time average of these forces is the tractive force. The tractive force measured over a unit surface area is the tractive stress. In a prismatic channel reach of uniform flow bounded by two end sections, the mean value of tractive stress is equal to the weight of the water prism in the reach multiplied by the energy gradient and divided by the wetted boundary surface in the reach. Shear stress or force per unit area of bed is expressed as $\tau_0 = \gamma R S_e$.

Determining Critical Tractive Stress

The most widely used and most reliable evaluation of tractive stress related to the initiation of motion is that developed by Shields (1936). The theoretical concepts, supported by experiments, resulted in a plot of $\tau_c/\{\gamma[(\gamma_s - \gamma) - 1]d_s\}$ against $(U_* d_s)/\nu$. The first expression is an entrainment function and the second is the boundary Reynolds number, indicating the intensity of flow turbulence around the particle. The Shields data are based on particles of uniform size and a flat bed. The Shields experiments indicate that beyond a certain value of the boundary Reynolds number, $(U_* d_s)/\nu$, the value of the parameter $\tau_c/\{\gamma[(\gamma_s/\gamma) - 1]d_s\}$ remains constant. Within these limits, the critical tractive stress is therefore proportional to grain size.

Data on critical tractive stresses obtained in a number of investigations were assembled by Lane (1955). These data show that the critical tractive stress in pounds per square foot is equal to $\tau_c = 0.5d_{75}$, where d_{75} is the size in inches of the bank material at which 25% by weight is larger. The limiting (allowable) tractive stress was determined from observations of canals (Land, 1955). The recommended limiting tractive stress in pounds per square foot is equal to 0.4 of the d_{75} size in inches for particles that exceed 0.25 in diameter. Results of experiments on finer particles vary considerably, probably because of variations in experimental conditions. These include differences in interpreting the initiation of sediment movement, in temperature of the water, in concentration of colloids, and in configuration of the bed. Critical conditions for initiating movement sometimes are determined by the number of particles or the frequency with which the particles start to move. For example, one observer's criterion was the time at which particles begin to move every 2 seconds at a given spot on the bed (Sutherland, 1967).

Determining Critical Velocity

Determining the critical velocity (the velocity at which particles in the bed begin to move) is another method for establishing stability criteria. Critical water velocity is a function of mean grain size. There has been less agreement on critical velocity than on critical tractive stress, probably because bottom velocity increases more slowly with increasing depth than does mean velocity. Critical conditions for initiating movement can be expressed directly in terms of tractive stress, but critical mean velocity must be related to variation in velocity with depth. Determining the correct critical value for tractive stress or velocity is important when considering stability problems in channels in which there is to be no significant movement of the boundary material. The significance of the critical value is determined by the magnitude and duration of flows that initiate sediment movement. A prolonged flow slightly exceeding the critical value may have little significance in terms of the volume of bed material transported. On the other hand, a brief flow substantially exceeding the critical value could transport a large volume of sediment.

HYDRAULIC CONSIDERATIONS

Fixed Boundaries

The relationships of velocity, stage, and discharge for stream channels with fixed boundaries have long been satisfactorily predicted by selected the appropriate *n* value in Manning's and other related formulas.

Movable Boundaries

The study of the hydraulics of movable boundaries has been directed toward two general problems. Primary interest has been in determining methods for predicting the friction coefficient and thereby the correct velocity, stage, and discharge relationships for channel design. The need for these data as a key element in predicting sediment transport has added incentive to the investigations. The changes in bed form produced on a movable bed and the consequent change in friction characteristics of the bed are among the most intensively studied flow phenomena. The literature on this subject generally describes the sequence of changes in bed configuration that can occur as the flow and transport intensity increase. Ripples, ripples on dunes, or dunes may form at a low transport rate, and antidunes or a flat bed may form at a high transport rate. These bed forms have been observed in sand-bed flumes and streams with a d_{50} size finer than 1.0 mm. The variety of bed forms in coarser material seems to be smaller.

Pioneering efforts in investigating the hydraulics of movable beds led to dividing the hydraulic radius into two parts. One part is the radius resulting from the roughness of the grain size of the individual particles (R'), and the other is the radius resulting from the roughness of the bed configuration (R'') (Einstein, 1950; Einstein and Barbarossa, 1952). From field observations, Einstein and Barbarossa developed a graph relating the dimensionless ratio V/U_*'', where $U_*'' = (gR''S_e)^{1/2}$, to Einstein's flow intensity parameter, ψ. The graph they developed indicates that for a given set of conditions it is possible to develop a unique stage–discharge relationship and thus to predict the hydraulics of a channel with movable boundaries. Vanoni and Brooks (1957) presented a graphical solution to the friction equation from which R' is determined.

Another procedure for predicting hydraulic behavior in movable channel beds is based on the division of slope (S) into two parts, S' and S'' (Meyer-Peter and Müller, 1948). In this procedure, S' is the energy gradient associated with the grain size of the bed material under a certain velocity and depth, excluding form resistance, and

DID YOU KNOW?

Whenever I present the observations of Hans Albert Einstein and his coauthors to my students or in my guest presentations here and there, I am always asked if Hans Albert Einstein is related to the genius Albert Einstein of $E = MC^2$ fame. Yes, he is. In fact, Hans Albert Einstein is the oldest son of Albert Einstein and is renowned for his doctoral thesis, "Bed Load Transport as a Probability Problem," which is considered the definitive work on sediment transport.

S'' is the additional gradient pertaining to bed form resistance. This division of slope was adopted by Alam and Kennedy (1969). A similar hydraulic consideration sometimes used as part of the preliminary procedure in sediment transport computations is the treatment of bank friction as completely distinct from bed friction. One such approach, involving the use of Manning's friction equation, is included as part of the procedure in the Einstein bedload function.

Movement of Bed Material

In this discussion the term *bed material load* is defined as that part of the total sediment load (suspended load plus bedload) that is composed of grain sizes occurring in appreciable quantities in the bed material. The part of the total load that consists of grain sizes not present in the bed material in significant quantities is the *washload.* Sand-size particles that constitute all or the major part of the bed material travel either on the bed as bedload or in suspension. Figure 6.1 illustrates how the total sediment load is classified: bedload, bed material load, or washload. Evaluation techniques are not refined enough to predict accurately what part of the bed material load moves in suspension or what part moves as bedload under specific hydraulic conditions. Establishing this separation does not seem essential to the general solution of sediment transport problems.

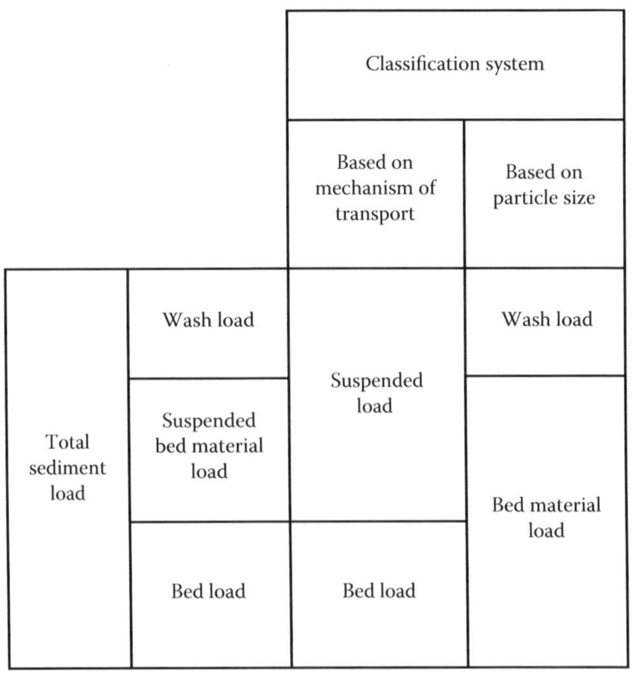

FIGURE 6.1 Sediment load classification. (Adapted from Cooper, R.H. and Peterson, A.W., *Journal of the Hydraulics Division*, 96(HY9), 1880–1886, 1970.)

Transport rates for sand and gravel have been determined by both direct measurement and computation. Measurements of the transport rate in natural streams have been few, chiefly because of the difficulty in getting representative measurements. Sampling equipment established in or on the bed tends to alter the direction of flow filaments and the sediment concentrations. The more accurate measurements have been made by using equipment installed to withdraw representative samples of the water sediment mixed during specific periods. Another method is to sample total load as the flow moves over a sill at an elevation the same as that of the slope upstream.

The existence of many procedures for predicting transport rates indicates both the difficulty of obtaining measurements and the influence of many variables on the consistency of results. Because flume studies are the most easily controlled and exclude some variables, they have become the primary means of establishing relationships between stream discharge and bed material load. The earliest bed material transport formula still in use is that of DuBoys, who published results of studies of the Rhone River in 1879. DuBoys originated a concept common to many later formulas when he assumed in his derivation that the rate of sediment transport is proportional to the tractive stress in excess of the critical value required to initiate motion:

$$q_T = \psi\tau_0(\tau_0 - \tau_c) \tag{6.1}$$

where

q_T = Rate of sediment transport per unit width of stream.
ψ = A coefficient that depends on characteristics of the sediment (not to be confused with Einstein's ψ).
τ_c = A value established by experiment (not the same as that of Shields).

Early in the 20th century, several flume studies of sand transport were started, including that of Shields. He is best known for developing criteria for the initiation of movement. Probably the most extensive early investigation of sediment transport in flumes was Gilbert's in about 1910 (Gilbert, 1914). Descriptions of a number of transport phenomena resulted from those experiments, but no general formula was derived. Of the formulas that follow, those of Schoklitsch, Meyer-Peter, Haywood, and Meyer-Peter and Müller are bedload formulas. The Einstein bedload function, the Engelund–Hansen procedure, and the Colby procedure determine the rate of bed material transport, both bedload and suspension load.

Schoklitsch Formula

Schoklitsch developed one of the more extensively used empirical formulas (Shulits, 1935; Shulits and Hill, 1968). He used his own experimental data and also data from Gilbert's flume measurements. The 1934 Schoklitsch formula in English units is

$$q_B = [86.7/(d_{50})^{1/2}]S_e^{3/2}(q - q_0) \tag{6.2}$$

where

q_B = Unit bedload discharge (pounds per second per foot of width).
d_{50} = Medium size of sediment (inches).
$q_0 = 0.00532(d_{50}/S_0^{4/3})$.

In describing the formula, Shulits recommended using a cross-section in a straight reach of river where the depth of water is as uniform as possible and the width changes as little as possible with stage. As described by Shulits, the Schoklitsch formula fits Gilbert's measurements for uniform particle sizes of about 0.3 to 7 mm and slopes ranging from 0.006 to 0.030 ft/ft for small particles and 0.004 to 0.028 ft/ft for larger particles.

Meyer-Peter Formula

In 1934, the Laboratory for Hydraulic Research at Zurich, Switzerland, published a bedload transport formula based on flume experiments with material of uniform grain size. The original analysis of the Zurich and Gilbert data for uniform particles ranging from about 3 to 28 mm in diameter was supplemented by studies of mixtures of various-size particles up to 10 mm and having various specific gravities. The Meyer-Peter formula in English units is

$$q_B = (39.25q^{2/3}S_0 - 9.95d_m)^{3/2} \tag{6.3}$$

where d_m is expressed in feet. The new term in this formula is d_m, the effective diameter of the bed material, which identifies the characteristic size of a sample. To determine this value, divide the size distribution curve of a bed material mechanical analysis into at least 10 equal-size fractions and determine the mean size and weight percentage of each fraction.

Haywood Formula

The Haywood formula is based on Gilbert's flume data and data from the U.S. Waterways Experiment Station, Vicksburg, Mississippi. In his evaluation, Haywood (1940) adjusted Gilbert's data to account for sidewall resistance. He assumed that the discharge effective in moving bedloads is midway between the discharge of walls offering no resistance and that of walls offering the same resistance as the bed. Haywood demonstrated the close relationship of his formula to the Schoklitsch formula, which is based on some of the same data. Haywood believed that his formula substantially agreed with Schoklitsch's formula for relatively large rates of bedload movement and that it was much more accurate for very small rates of movement. Haywood considered 3 mm to be the maximum particle size for application of his formula. He regarded his formula as a modification of the Meyer-Peter formula:

$$q_B = [(q^{2/3}S_0 - 1.20d^{4/3})/0.117d^{1/3}]^{3/2} \tag{6.4}$$

where d is d_{35} expressed in feet.

Meyer-Peter and Müller Formula

The Meyer-Peter and Müller formula is based on data obtained from continuing the experiments that resulted in the Meyer-Peter formula. The range of variables, particularly slope, was extended. A few tests were run with slopes as steep as 20% and sediment sizes as coarse as 30 mm. Meyer-Peter and Müller stated explicitly that their work was on bedload transport, by which they meant the movement of sediment that rolls or jumps along the bed. Transport of material in suspension was not included (Meyer-Peter and Müller, 1948). The Meyer-Peter and Müller formula as translated by Sheppard (1960) is

$$q_B = 1.606[3.306(Q_S/Q)(d_{90}{}^{1/6}/n_s)^{3/2}(DS_e - 0.627d_m]^{3/2} \tag{6.5}$$

where d_{90} and d_m are expressed in millimeters.

Nomographs are available for determined Q_S/Q (a ratio of the discharge quantity determining bedload transport to the total discharge) and n_s (a Manning n value for the streambed). The formula, a significant departure from the previously cited formulas, includes a ratio of the form roughness of the bed to the grain roughness of the bed surface.

Einstein Bedload Function

In 1950, Einstein's bedload function had a major effect on investigations of the hydraulics and sediment transport characteristics of alluvial streams. Einstein (1950) described the function as "giving rates at which flows of any magnitude in a given channel will transport as bed load the individual sediment sizes of which the channel bed is composed." It was developed on the basis of experimental data, theory of turbulent flow, field data, and intuitive concepts of sediment transport. The Einstein bedload function first computes bedload and then, by integrating the concentration at the bed layer with the normal reflection of that concentration in the remainder of the flow depth, determines the total bed material load.

Einstein introduced several new ideas into the theory of sediment transport. Included were new methods of accounting for bed friction by dividing it into two parts, one pertaining to the sand-grain surface and the other to the bed form roughness, such as ripples or dunes. An additional friction factor, that of the banks, was included in the procedure for determining hydraulic behavior before computing bed material transport.

Another idea introduced by Einstein to explain the bedload function is that the statistical properties of turbulence govern the transport of particles as bedload. This statistical character is reflected in the structure of the dimensionless parameter ϕ, defined as the intensity of bedload transport. The relationship between this factor and the dimensionless flow intensity (ψ, another dimensionless parameter reflecting the intensity of shear on the particle) is used in the procedure. The ϕ–ψ relationship has subsequently been tested by others and found to be an appropriate determinant of bedload transport.

Engelund–Hansen Procedure

Engelund and Hansen (1967) developed a procedure for predicting stage–discharge relationships and sediment transport in alluvial streams. They introduced a parameter θ (the reciprocal of Einstein's ψ) to represent the ratio of agitating forces (horizontal drag and lifting force) to the stabilizing force (immersed weight of the particle). This parameter is a dimensionless form of the bed shear, τ_0, to be divided into two parts: τ', the part acting directly as traction on the particle surface, and τ'', the residual part corresponding to bed form drag. This division is similar to that of

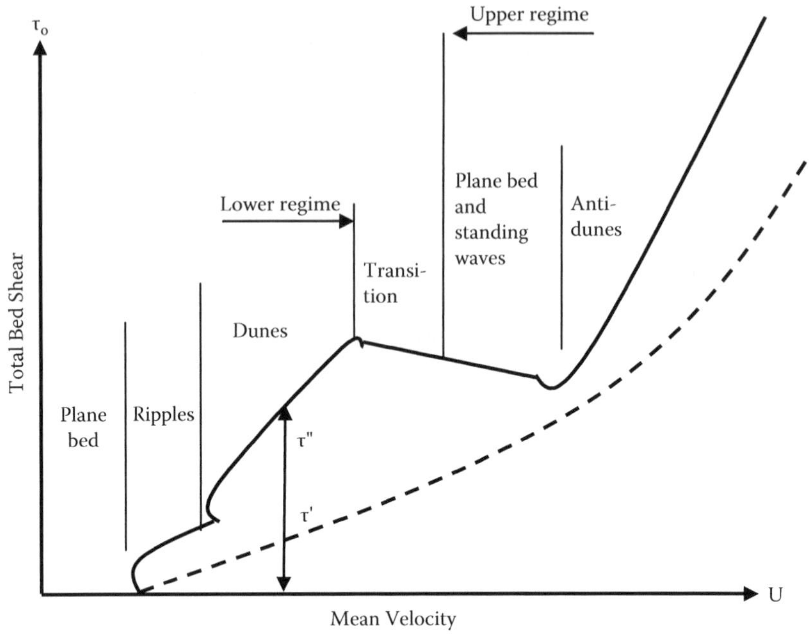

FIGURE 6.2 Relationship between grain roughness (τ') and form drag (τ''). (Adapted from Engelund, F. and Hansen, E., *A Monograph on Sediment Transport in Alluvial Streams*, Teknisk Forlag, Copenhagen, Denmark, 1967.)

the Einstein–Barbarossa R' and R''. The authors' diagram of the relationship of bed forms to the two separations of total bed shear and to velocity is shown in Figure 6.2. Principles of hydraulic similarity were used to develop a working hypothesis for describing total resistance to flow, specifically for dune-covered streambeds and bed material discharge.

The steps used in applying the Engelund–Hansen procedure are given here in some detail because the procedure demonstrates the impact of changing bed forms on bed material transport and because it was published in a foreign journal not readily available for reference. Data from flume experiments by Guy et al. (1966) were used to test the Engelund–Hansen theories. The mean sizes used in these experiments were 0.19, 0.27, 0.45, and 0.93 mm. Transport of the bed material, both in suspension and moving along the bed, was measured. It was found that the Engelund–Hansen procedure does not provide a means for determining the bed material discharge at lower flow regimes of plane beds and ripples. These regimes are not significant in terms of the volume of sediment transported.

Colby Procedure for Relating Mean Velocity to Sand Transport

The Colby procedure was developed by correlating mean velocity with sediment concentration in a sand bed stream. The procedure, partly empirical and partly derived from Einstein's bedload function, is based on measurements in flumes and channels.

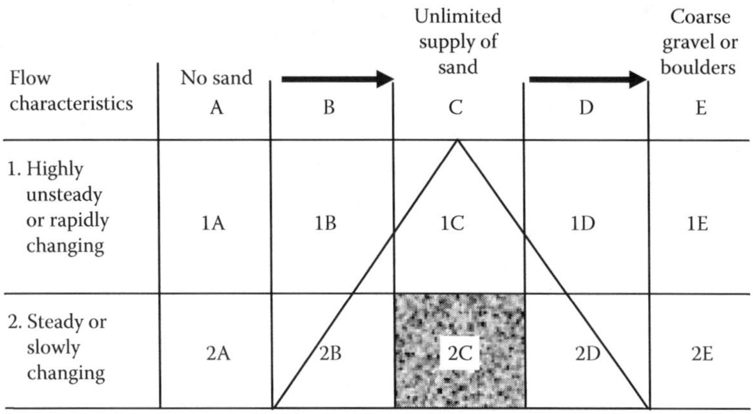

FIGURE 6.3 Characteristics of bed material.

APPLICATION AND LIMITATIONS OF FORMULAS

The lack of certainty in solving specific sediment transport problems is in part a result of the extremely limited number of situations in which predictive techniques, such as bedload or bed material transport formulas, have been substantiated by field measurement. Even for techniques that have been substantiated, little information is available about the specific hydraulic characteristics for comparison with conditions for the problem to be solved (Cooper et al., 1972). Figure 6.3 illustrates a few of the major factors that can be considered in the application and limitations of sediment transport formulas. The availability of bed material ranges from no sand (column A), to an unlimited supply of sand in sizes less than 1 mm (column C), to bed material of gravel and boulders (column E). Flow characteristics range from highly unsteady or rapidly changing to steady and slowly changing.

Of the possible conditions illustrated by this diagram, the condition in box 2C most nearly fits the flow and sediment conditions used in developing transport formulas. Box 1C pertains specifically to the smaller streams with which we are concerned, not to rivers in which deep steady flows may transport gravel as they do sand. Through limited reaches and during high flows, shallow streams may also transport gravel and boulders. Frequently there is a transition from scour to deposition over a relatively short reach. Boxes adjacent to 2C (1C, 2B, 2D) can be considered a "gray" area from which correct solutions to sediment transport problems can be obtained by including the appropriate modifiers, such as changes in slope to match variations in discharge.

The effect of rapidly changing flow (top line on the chart) on bedload transport was the subject of a flume study by DeVries (1965). The mean grain size was 2.5 mm. After an equilibrium rate of transport was attained, the tailwater was suddenly lowered while other factors were kept constant. DeVries computed the lowering of the bed level from scour and the change in rate of sediment transport during the transition to a new state of equilibrium by using several procedures, including the Meyer-Peter and Müller formula. He concluded that establishment and damping of a steady state are slow and that steady-state formulas are unreliable for predicting local, temporary transport for an unsteady state.

A subsequent flume study was made of the effect of introducing a substantial increase (65%) in bed material load into a run where equilibrium flow and transport had been established (Rathbun and Guy, 1967). The medium size of the sand used was about 0.30 mm. This increase in load increased slope, decreased depth, and increased the transport rate. In another run, the rate of sediment input was reduced to about 50% of the equilibrium rate. At first the transport rate was about the same as during equilibrium flow; then, with degradation of the upper end of the sand bed and a decrease in slope, the transport rate also decreased.

Aggradation occurs in some channels even through hydraulic computations indicate that sediment should not deposit. It is not always known whether the aggradation occurred in the rising or falling stage of the hydrograph. Some of the unpredicted changes can be explained by variable bed roughness not accounted for in conventional hydraulic computations. Variable bed roughness does not necessarily explain all of the inaccuracies in predicting the effects of hydraulic change on sediment transport, however, because some procedures do take into account the changes in bed roughness with various flows. Part of the problem may be due to unsteady flow, as steady-flow procedures fail to account for differences between stage and discharge.

When using computational procedures, it is very important that the supply of bed material just satisfies the capacity for transport under existing hydraulic conditions; that is, there can be neither a deficiency, resulting in scour, nor an excess, resulting in aggradation. A sand bed satisfies the necessary requirements for using bedload or bed material transport formulas and that of bed material availability if the bed is sand from bank to bank throughout the reach.

When he considered the availability of bed materials, Kellerhals (1966) made a distinction between channels with a sand bed and channels with a gravel bed. According to his studies, channels with a gravel bed cannot be expected to obey the same laws as channels with a sand bed. One distinction is that ripple and dune formations are less significant in channels with a gravel bed.

In terms of particle size, the scarcity of particles in the 2- to 4-mm size fraction, as described by Sundborg (1956), creates a sharp division between predominantly sand-bed streams and predominantly gravel-bed streams. This division has been substantiated by data on sizes of bed material in various parts of the United States.

The segregation of particles in a mixture of sizes, including gravel, and the depth of scour before the formation of armor were the subjects of flume studies by Harrison (1950). The purpose was to determine the most critical condition for segregation and for building an armor during degradation. Harrison used the Einstein bedload function to calculate the limiting grain diameter for equilibrium flow. He determined that a value of ψ (a dimensionless parameter of transport capability) above 27 indicates negligible transport of bed material.

Harrison (1950) found that the representative grain roughness, k_s (assumed to be d_{65} in his procedures), increases during segregation and armor formation. On the basis of data from field and laboratory studies, Kellerhals (1967) computed the k_s

values after armor formation to be the d_{90} size. On the basis of these considerations, the following treatment is suggested for sediment problems in streams as categorized in Figure 6.3:

1A, 2A—For cohesive soil, cemented gravel, and rock, initiation of movement is the important factor in channel scour or bank erosion. Critical tractive force is related to the d_{75} of bank materials. Undisturbed cohesive soil exhibits erosion resistance that may result from one or several characteristics such as structure, permeability, consolidation, cementation, or cohesion. The influence of each of these characteristics has not been identified. Their cumulative effect on erosion resistance, however, can be determined by shear strength tests on undisturbed soil that has been saturated to duplicate moisture conditions during channel flow (Flaxman, 1963).

1B, 2B—A bed only partially covered with sand and exposing different material (e.g., cohesive soil, rock) as the fixed channel boundary indicates a limited sand supply at this specific location. Sediment transport formulas applied to this condition usually yield computed rates that exceed the actual rate. Test the potential for back erosion by tractive force theory if the bank is composed of noncohesive materials; otherwise, use the procedures for cohesive soils.

1C, 2C A sand-covered bed is the condition used in sediment transport formulas if the problem to be solved requires (1) estimating the volume of bed material transport during a specific interval of time and at a specific level of discharge or (2) comparing the bed material transport in a reach with that in another reach in which changes in slope, cross-section, or discharge may influence the design of a channel. If flow is unsteady, replace the steady-state procedures with the proper unsteady flow relationships, as previously mentioned.

2D—Techniques for predicting transport rates of sand–gravel mixtures allow estimates of the potential for scour or aggradation. The probable depth of scour can be estimated by determining whether the maximum tractive force for a given flow will exceed the critical for the coarsest 5 to 10% of bed material. If the maximum tractive force exceeds the critical for the d_{90} to d_{95}, the depth of scour cannot be predicted unless still coarser material underlies the bed surface material. The amount of scour necessary to develop armor formed of the coarsest fraction can be determined from either the depth of scour or the volume of material removed in reaching this depth.

1D, 1E, 2E—For gravel and gravel–boulder mixtures, the technique used for determining depth of scour and volume of material produced by scour is similar to that for sand–gravel mixtures (2D). Do not use bedload formulas for this type of material unless confined flow, steepness of slope, and uniformity of cross-section provide relatively uniform discharge per foot of width. The highly variable velocity and discharge per foot of width in many alluvial channels is particularly conducive to deposition alternating with scour of coarse bed material.

Conditions favoring bed material transport at or near a constant and predictable rate do not include delivery in slurries or other forms that change the viscosity and natural sorting processes of flow. Alluvial fills of mountain or foothill canyons are typical of conditions favoring viscous flow. Heavy storm runoff after many years of fill accumulation may produce debris or mud flows whose volume can be predicted only by field measurement.

Example Channel Problem

Applying different procedures to determine sediment transport capacity will likely produce similarities and differences in the results obtained. In the following example, the Schoklitsch formula and the Colby procedure are used to illustrate this point. An existing channel 20 ft wide with a bed slope of 0.002 ft/ft has inadequate capacity for controlling flooding of adjacent lands. It has been proposed that the width of this channel be increased to 30 ft to provide the necessary capacity. Field investigations show that an unlimited supply of sand is available for transport in the bed of the channel and that this sand has a d_{50} size of 0.30 mm. Water temperature is 60°F, and the concentration of fine sediment does not exceed 5000 ppm. For purposes of simplification, it is assumed that the banks have no effect on depth–discharge relationships, but the roughness of the banks and differences in roughness of the banks in both unimproved and improved reaches can in fact affect depth and velocity for a given discharge and thereby affect the rate of bed material transport. The hydraulics of the flow, including distribution of shear on the banks as well as on the bed, must be determined by an established procedure before computing the bed material transport. The hydrograph used in this example is divided into segments to determine the discharge per foot of stream width as required for the computational procedures. The mean discharge and duration for each of the hydrograph segments are shown in Table 6.1.

The Schoklitsch formula requires data only for the amount of discharge per foot of width. The Colby procedure requires velocity and depth of flow. To determine velocity and depth for a given discharge (unless they are available from streamgauge records), it is necessary either to assume an n roughness coefficient for use in the Manning equation or to obtain such values empirically. For solution of the

TABLE 6.1
Discharge Data for Example Channel Problem, High Flow

	Discharge per Foot of Width	
Hydrograph Segment	**20-ft Channel (ft^3/s)**	**30-ft Channel (ft^3/s)**
Rising stage		
(a) Mean flow for 2 hours, 90 ft^3/s	4.5	3.0
(b) Mean flow for 2 hours, 280 ft^3/s	14.0	9.333
Falling stage		
(c) Mean flow for 3 hours, 240 ft^3/s	12.0	8.0
(d) Mean flow for 3 hours, 180 ft^3/s	9.0	6.0
(e) Mean flow for 3 hours, 40 ft^3/s	2.0	1.333

TABLE 6.2
Sediment Transport (lb) Computed for Various Flows

	Schoklitsch Formula		Colby Procedure: Using n = 0.020		Colby Procedure: Using Alam and Kennedy Friction Factors	
Discharge Segment	**20-ft width**	**30-ft width**	**20-ft width**	**30-ft width**	**20-ft width**	**30-ft width**
a	44,135	42,840	97,285	86,720	109,270	103,225
b	142,760	141,470	347,085	344,210	412,425	543,140
c	182,995	181,060	442,745	426,435	590,170	564,565
d	136,280	134,340	328,735	310,100	516,280	431,920
e	27,270	25,330	50,710	42,765	46,180	31,190
Total	533,440	525,040	1,226,560	1,210,230	1,674,325	1,674,040
Ratio (30-ft width/20-ft width)	525,040/533,440 = 98.43%		1,210,230/1,226,560 = 95.55%		1,674,040/1,674,325 = 99.98%	

example problem by the Colby procedure, two approaches are used. In one, a constant assumed n of 0.020 is used. In the other, the most recent and perhaps the most reliable procedure (Alam and Kennedy, 1969) for predicting friction factors (and thereby depth, velocity, and discharge relationships) is used.

The data in Table 6.2 indicate that in the stated problem the Schoklitsch formula predicts considerably less sediment transport than either of the Colby approaches. This difference may be due to the fact that the Schoklitsch formula predicts bedload and the Colby procedure accounts for suspended bed material as well as bedload. The difference between the two Colby predictions can be attributed to the different approaches for estimating the depth of flow. The first assumes $n = 0.020$ and a normal depth based on bed slope equal to friction slope; the second assumes a depth based mostly on grain roughness for friction slope. The Alam and Kennedy friction factors are never in the lower flow regime for this set of calculations; therefore, bedform changes had little effect on the results. All three results indicated a slight, but negligible, reduction (less than 5%) in sediment transport capacity for the water channel.

The next step in the analysis is to determine whether lower flows give different results. For this computation, 20% of the discharges indicated in Table 6.1 are used in Table 6.3. Table 6.4 shows the amount of sediment transported as computed by the two procedures. Table 6.4 again indicates considerable difference between the Schoklitsch and Colby predictions, but less than that shown in Table 6.2. This smaller difference can be attributed to the smaller loads in suspension for the lower flows. All three predictions, however, indicate greatly reduced sediment transport capacity for the wide (30-ft) channel (9, 17, and 32%, respectively). The most significant reduction, almost one-third, is predicted by the Colby procedure using the Alam and Kennedy friction factors. It is believed that the Colby procedure using the Alam and Kennedy factors most closely reflects the influence of variable bed forms that are more pronounced during low to moderate flows.

TABLE 6.3
Discharge Data for Example Channel Problems, Lower Flow

Hydrograph Segment	Discharge per Foot of Width: 20-ft Channel (ft^3/s)	Discharge per Foot of Width: 30-ft Channel (ft^3/s)
Rising stage		
(a) Mean flow for 2 hours, 18 ft^3/s	0.9	0.6
(b) Mean flow for 2 hours, 56 ft^3/s	2.8	1.87
Falling stage		
(c) Mean flow for 3 hours, 48 ft^3/s	2.4	1.6
(d) Mean flow for 3 hours, 36 ft^3/s	1.8	1.2
(e) Mean flow for 3 hours, 8 ft^3/s	0.4	0.267

This example clearly shows that estimates of the absolute rates of sediment transport vary according to the procedure. But, the study also shows that the relative rates can be insensitive to choice of procedure if variation in bed forms is not a factor as for channel performance at peak discharge. In many stability problems, however, the performance of the channel during one or more low to moderate flows must be considered. Formulas and procedures that determine the effect of variable bed forms on depth, velocity, and discharge relationships and, thus, on bed material discharge afford greater flexibility for all purposes.

Procedures for Evaluating Bed Material Transport Problems

Problems of bed material transport requires consideration of three elements: (1) existing conditions, (2) availability of bed material, and (3) natural or artificial changes in stream or watershed conditions. The existing conditions can be best determined

TABLE 6.4
Sediment Transport (lb) Computed for Lower Flows

	Schoklitsch Formula		Colby Procedure: Using n = 0.020		Colby Procedure: Using Alam and Kennedy Friction Factors	
Discharge Segment	20-ft width	30-ft width	20-ft width	30-ft width	20-ft width	30-ft width
a	6760	5470	9970	7195	450	700
b	26,485	25,195	53,280	46,705	61,225	41,645
c	33,500	31,560	67,580	54,615	66,255	46,245
d	24,155	22,220	43,710	36,000	39,245	24,500
e	2355	415	3315	2525	940	415
Total	93,255	84,860	177,855	147,040	168,115	113,505
Ratio (30-ft width/ 20-ft width)	84,860/93,255 = 91.00%		147,040/177,855 = 82.67%		113,505/168,115 = 67.52%	

DID YOU KNOW?

Whenever I discuss rivers and streams with students or others, invariably the term *river reach* or *reach* comes up. Many are confused as to its exact meaning—that is, its English definition. Simply, think of a river reach as an arm, which can refer to an extended portion or stretch of land or water, to a straight stretch of the stream (from one turn to another), or to a level stretch.

by field investigation and analysis. Surveys of old and new cross-sections, use of techniques for identifying depth of scour or aggradation, and comparison of series photographs all facilitate definition of the problems. Although the correct identification and analysis of existing bed material transport conditions are important, most problems require projections of what will or can occur rather than what is now occurring. The availability of bed material and the impact of change are the key elements of such projections.

Equilibrium can be achieved only if bed material is being introduced into the reach at a rate comparable to that at which bed material moves out of the reach. Problems arise when the amount introduced is greater or less than the transport capacity of the flow. In other words, equilibrium transport seldom causes problems but a change from equilibrium to nonequilibrium transport often does.

The supply of bed material can exceed transport capacity during unusually high discharges. This excess can be caused by development of new and substantial sources of bed material within or adjacent to the problem reach or by channel changes that may increase transport capacity in the upstream reach but not in the downstream reach. Determining the availability of bed material is largely a field problem. To be readily available to channel flow, sediment must be in the stream system. The coarse particles in an upland soil tend to lag behind during erosion. Gullies that feed directly into the stream system and that expose soils with a large proportion of particles of bed material size can be major contributors but do not in themselves constitute an immediate and unlimited stream channel supply.

Streambanks that have, at least in part, soil textures comparable to those in the bed, can be a ready source of supply, depending on the ease with which the flow can erode the material. A frequently used emergency flood-protection measure is to bulldoze streambed materials to each side to form banks or levees. These banks are a ready source of supply. Their erosion and the consequent deterioration of channel alignment result in overloading the flow and downstream aggradation.

Scour of bed material can result from an undersupply of sediment in an alluvial reach. Upstream changes in watershed or stream conditions that can reduce the supply of incoming bed material include the removal of supply by major flood scour and the construction of reservoirs, debris basins, or other structures.

In addition to cutting off the supply of bed material to the reach downstream, a reservoir can materially influence the stability of the channel bed and banks by modifying the flow. For example, a detention structure that controls a high flood peak can thereby extend the duration of released flows by days. The resulting bed and bank scour may be extensive because of the energetic discharge of clear water.

TABLE 6.5
Checklist of Procedures for Solving Bed Material Transport Problems

	Analysis Procedure			
Item	**Tractive Stress**[a]	**Comparative Hydraulics**[b]	**Bed Material Formulas**	**Field Evaluation**
Problem Characteristics				
Erodibility of bed	X			X
Erodibility of bed and banks	X			X
Erodibility of banks	X			X
Channel aggradation	X	X	X	
Volume of bed material			X	X
Effects of channel change		X	X	X
Channel Boundary Characteristics				
Cohesive soils	X			X
Cohesive soils or rock with intermittent deposits of sand or gravel	X			X
Sand ≤1.0 mm	X			X
Sand ≤1.0 mm with <10% gravel	X	X	X	X
Gravel, gravel mixed with sand	X	X	X	X
Gravel and boulders	X			X
Hydraulic Characteristics				
In problem reach				
Steady state or slowly changing	X	X	X	X
Rapidly changing	X	X		X
Cross-section—slope upstream vs. problem reach				
About the same	X	X	X	X
Steeper slope	X	X	X	X
Wider channel	X	X	X	X
Narrower channel	X	X	X	X

[a] For cohesive soil boundaries, analysis may include tractive power (tractive stress times mean velocity).

[b] Comparison of relationships between depth, velocity, and unit discharge in two or more reaches.

Table 6.5 provides a checklist of procedures to consider when solving problems of bed material transport. The last column in this table indicates that a field evaluation is important to the solution of any such problem. Because of the variety of factors that can influence their solution, most problems are not routine, and solving them requires the assistance of well-trained and experienced personnel. The first step should always be a field evaluation of existing or potential problems related to sediment transport. With experience, well-trained personnel frequently can find answers to questions of stability, degradation, or aggradation by relating the availability of bed material to proposed changes in the hydraulics of the flow without resorting to formulas. If formulas must be used, it should be recognized that the results are qualitative and not quantitative. Observations of similar streams having comparable drainage areas, geology, soils, topography, and runoff often provide guidance on the probable stability.

TRANSPORT OF SUSPENDED SEDIMENT

Suspended sediment load includes both the bed material load in suspension and the washload, as shown in Figure 6.2. If erosion of fine-textured soils is the chief source of sediment, the washload, not the bed material load, usually constitutes the bulk of the sediment discharge. No method exists for predicting rates of washload transport unless there is a substantial amount of data on concentrations of suspended sediment during measured discharges.

Suspension Mechanism

Bagnold (1966) explained the suspension mechanism as follows:

> Isotropic turbulence [i.e., equal physical properties along all axes] cannot by definition be capable of exerting any upward directed stress that could support a suspended load against gravity. For any suspended solid must experience over a period of time a downward flux of eddy momentum equal on the average to the upward flux. A swarm of solids would be dispersed equally in all directions by diffusing along uniform concentration gradients, but the center of gravity of the swarm would continue to fall toward a distant gravity boundary.
>
> The center of gravity of a swarm of solids suspended by shear turbulence, on the other hand, does not fall toward the gravity shear boundary. The excess weight of the solids remains in vertical equilibrium. It follows therefore that the anisotropy [property of being directionally dependent] of shear turbulence must involve as a second-order effect a small internal dynamic stress directed perpendicularly away from the shear boundary. In other words, the flux of turbulent fluid momentum away from the boundary must exceed that toward it. … The turbulence appears to be initiated and controlled by a process akin to the generation of surface waves by a strong wind. An upwelling on the part of a minor mass of less turbulent boundary fluid intrudes into an upper, faster moving layer, where its crust is progressively torn off, like spray, and mingles with the upper layer. Corresponding motions in the reverse sense are absent or inappreciable.
>
> Since there cannot be a net normal transport of fluid, the returned flow must be effected by a general sinking toward the boundary on the part of a major mass of surrounding fluid.

The settling rate for sediment particles of uniform density increases with size, but not proportionally. The settling rate for particles smaller than about 0.062 mm varies approximately as the square of the particle diameter, whereas particles of coarse sand settle at a rate that varies approximately as the square root of the diameter. The settling rate for particles of intermediate size varies at an intermediate rate. The dividing line between sediments classed as silts and those classed as sands is

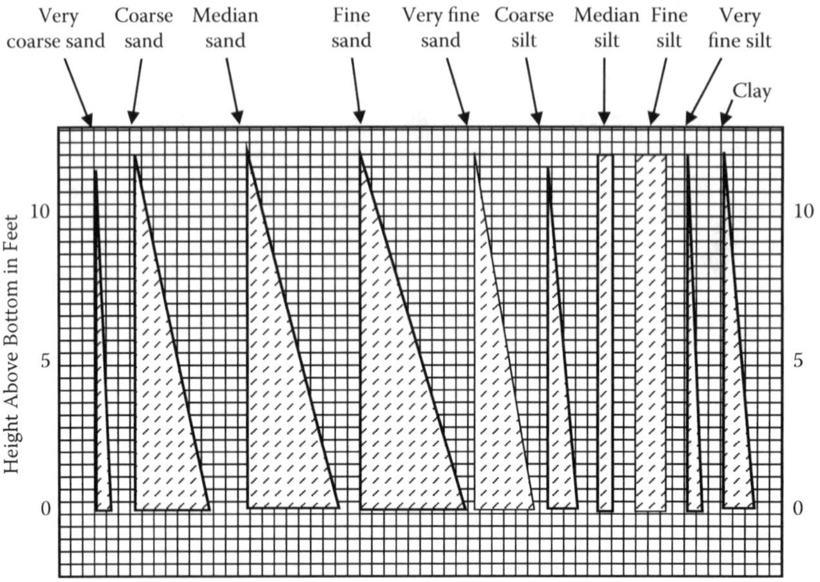

FIGURE 6.4 Approximate vertical distribution of sediment in Missouri River at Kansas City, Missouri. (Adapted from FIRBC, *A Study of Methods Used in Measurement and Analysis of Sediment Load of Streams*, Report No. 14, Federal Interagency River Basin Committee, Subcommittee on Sedimentation, Washington, DC, 1963.)

the 0.062-mm size. Clay and silt particles usually are distributed fairly uniformly in a stream, but sand particles usually are more concentrated near the bottom. The degree of variation is a function of the coarseness of the particle (see Figure 6.4).

The lateral distribution of suspended sediment across a stream is fairly uniform in both deep and shallow flows except below the junction of a tributary carrying material at a concentration substantially different from that of the main stream. The flow from the tributary tends to remain on the entrance side of the channel for some distance downstream.

REFERENCES AND RECOMMENDED READING

Alam, A.M. and Kennedy, J.F. (1969). Friction factors for flow in sand-bed channels. *Journal of the Hydraulics Division*, 95(HY6): 1973–1992.

Bagnold, R.A. (1966). *An Approach to the Sediment Transport Problem from General Physics*, USGS Professional Paper 422-I. Reston, VA: U.S. Geological Survey.

Colby, B.R. and Scott, C.H. (1965). *Effects of Water Temperature on the Discharge of Bed Material*, USGS Professional Paper 462-G. Reston, VA: U.S. Geological Survey.

Cooper, R.H. and Peterson, A.W. (1970). Discussion of "Coordination in Mobile-Bed Hydraulics." *Journal of the Hydraulics Division*, 96(HY9): 1880–1886.

Cooper, R.H., Peterson, A.W., and Blench, T. (1972). Critical review of sediment transport experiments. *Journal of the Hydraulics Division*, 98(HY5): 827–843.

DeVries, M. (1965). *Considerations About Non-Steady Bedload Transport in Open Channels*, Delft Hydraulics Laboratory Publ. No. 36. Delft, The Netherlands: Delft Hydraulics Laboratory.

Einstein, H.A. (1944). *Bed-Load Transportation in Mountain Creek*, USDA SCS-TP-55. Washington, DC: U.S. Department of Agriculture.

Einstein, H.A. (1950). *The Bed-Load Function for Sediment Transportation in Open Channel Flows*, USDA Technical Bull. No. 1026. Washington, DC: U.S. Department of Agriculture.

Einstein, H.A. and Barbarossa, N.L. (1952). River channel roughness. *Transactions of the American Society of Civil Engineers*, 117(2528): 1121–1132.

Engelund, F. and Hansen, E. (1967). *A Monograph on Sediment Transport in Alluvial Streams*. Copenhagen, Denmark: Teknisk Forlag.

FIRBC. (1963). *A Study of Methods Used in Measurement and Analysis of Sediment Load of Streams*, Report No. 14. Washington, DC: Federal Interagency River Basin Committee, Subcommittee on Sedimentation.

Flaxman, E.M. (1963). Channel stability in undisturbed cohesive soils. *Journal of the Hydraulics Division*, 89(HY2): 87–96.

Gilbert, G.K. (1914). *The Transportation of Debris by Running Water*, USGS Professional Paper 86. Reston, VA: U.S. Geological Survey.

Guy, H.P., Simons, D.B., and Richardson, E.V. (1966). *Summary of Alluvial Channel Data from Flume Experiments, 1956–61*, USGS Professional Paper 462-I. Reston, VA: U.S. Geological Survey.

Harrison, A.S. (1950). *Report on Special Investigation of Bed Sediment Segregation in a Degrading Bed*. Berkeley: University of California.

Haywood, Jr., O.G. (1940). Flume Experiments on the Transportation by Water of Sands and Light-Weight Materials, D.Sc. thesis, Massachusetts Institute of Technology, Cambridge.

Hunt, J.R. (1982). Particle dynamics in seawater: implications for predicting the fate of discharged particles. *Environmental Science and Technology*, 16(6): 303–309.

Kellerhals, R. (1967). Stable channels with gravel paved beds. *Journal of the Waterways and Harbors Division*, 93(1). 63–84.

Lane, E.W. (1955). Design of stable channels. *Transactions of the American Society of Civil Engineers*, 120(2776): 1234–1279.

Meyer-Peter, E. and Müller, R. (1948). Formulas for bed load transport. In: *Proceedings of the IAHSR 2nd Meeting*, Stockholm, June 7–9, pp. 39–64.

O'Melia, C.R. (1980). Aquasols: the behavior of small particles in aquatic systems. *Environmental Science and Technology*, 14(9): 1052–1060.

Rathbun, R.E. and Guy, H.P. (1967). Effect of non-equilibrium flow conditions on sediment transport and bed roughness in a laboratory alluvial channel. In: *Proceedings of the 12th Congress of the International Association for Hydraulic Research*, Fort Collins, CO, September 11–14, pp. 187–193.

Sheppard, J.R. (1960). *Investigation of Meyer-Peter, Müller Bedload Formulas*. Washington, DC: U.S. Bureau of Reclamation, Sedimentation Section.

Shields, A. (1936). *Application of Similarity Principles and Turbulence Research to Bed-Load Movement*. Pasadena, CA: California Institute of Technology.

Shulits, S. (1935). The Schoklitsch bed-load formula. *Engineering*, 139: 644–646.

Shulits, S. and Hill, Jr., R.D. (1968). *Bedload Formulas, Part A*. State College: Pennsylvania State University, Department of Civil Engineering.

Sundborg, Å. (1956). The River Klarälven: a study of fluvial processes. *Geografiska Annaler*, 38(2): 238–316.

Sutherland, A.J. (1967). Proposed mechanism for sediment entrainment by turbulent flows. *Journal of Geophysical Research*, 72(24): 6183–6194.

USACE. (1968). *Missouri River Channel Regime Studies*, MRD Sediment Series No. 13A. Omaha, NE: U.S. Army Corps of Engineers, Omaha District.

USBR. (1958). *Interim Report, Total Sediment Transport Program, Lower Colorado River Basin*. Denver, CO: U.S. Bureau of Reclamation, Sediment Section.

USDOT. (1975). *Highways in the River Environment: Hydraulic and Environmental Design Considerations*. Washington, DC: U.S. Department of Transportation, Federal Highway Administration.

Vanoni, V.A. and Brooks, N.H. (1957). *Laboratory Studies of the Roughness and Suspended Load of Alluvial Streams*, Report No. E-68. Pasadena: California Institute of Technology.

Vanoni, V.A., Brooks, N.H., and Kennedy, J.F. (1961). *Lecture Notes on Sediment Transportation and Channel Stability*, Report No. KH-R-1. Pasadena: California Institute of Technology.

7 Sediment Yield

My view is that any technical discussion is of doubtful value unless demonstrated to fit the facts. It is necessary, therefore, to abide, to the extent possible, by Sergeant Joe Friday's alleged mantra: "Just the facts, ma'am." Thus, in any presentation like this one, considerable space must be devoted to the assembly and discussion of as many relevant facts as possible, so that the reader may judge the presentation for himself.

INTRODUCTION

Sediment yield depends on the erosion processes at the sediment source and on the efficiency of the system that transports the sediment to the point of measurement. The sediment yield usually differs at different locations in a stream system (USDA, 2008). Many interrelated factors affect sediment yield. Knowledge of each of these factors is important when

1. Evaluating downstream sediment damages
2. Determining the location and extent of sediment sources so effective controls can be planned and installed
3. Evaluating the relative contribution of the various sources to current and future sediment yield
4. Determining the sediment storage requirement for designed proposed structural works of improvement

This chapter presents several procedures for determining sediment yields and interrelated factors such as sediment sources and delivery ratios.

Interrelated and Interdependent Processes

With regard to the interrelationship of sediment processes, keep in mind that sediment yield depends on gross erosion in the watershed and on the transport of eroded material out of the watershed. Only part of the material eroded from upland areas in a watershed is carried out of the watershed. Variation in the proportion of the eroded material deposited as colluvium at the base of slopes and in swales, as alluvium on floodplains and in channels, and as lacustrine deposits in natural or artificial lakes usually results in variation in the yield rate for different parts of a watershed. The distinctions made among sediment sources or types or yields demonstrate their interrelationship and, more accurately, their interdependence, because their characteristics are used to split into three classes something that could be defined as one thing. Nonetheless, their distinction (for this text) is important for communicating key characteristics, geomorphic setting, and inferred processes. Field determination of sediment yield may require long-term sampling and measuring procedures. A short-term procedure is to extrapolate (and adjust as appropriate) known sediment yield from a measured similar watershed in the same physiographic section.

SEDIMENT SOURCES

Sources of sediment must be delineated to plan an adequate program for reducing downstream sediment yield. Sediment sources include agricultural land, range and forest land, road banks and ditches, stream channels and banks, floodplains, spoil banks, and gullies. When planning a program to reduce sediment yield, the relative importance of the various sources and the methods for treating them must be determined before the physical and economic feasibility of the program can be determined. Sediment derived from sheet erosion can usually be reduced by land treatment measures, whereas that derived from channel-type erosion usually requires structural works. A sediment source study is made to determine: (1) the origin of the sediment; (2) the rate of erosion from each source; (3) the proportion of the sediment derived from each source; (4) with regard to program planning or structure design, the kinds of treatment that should be recommended for reducing sediment yield; and (5) the relative effect that reducing erosion from the various sources will have on reducing sediment yield and damage. The relative importance of the sediment source may differ at different locations in a watershed. Therefore, the treatment measures may also vary, depending on the location in the watershed where a reduction in sediment yield is desired.

DETERMINING THE RELATIVE IMPORTANCE OF VARIOUS SOURCES

When preparing to study and to determine the location, extent, and relative importance of the sediment source, the following items must be considered:

- *Maps and aerial photographs*—Careful review of aerial photographs often reveals where erosion is severe and which channels appear to be carrying the heaviest load of sediment. If soil surveys are available, the information on soils, slopes, land use, and erosion condition recorded on the maps is very helpful. Using all such information as fully as possible saves considerable time in locating the most obvious sources of sediment.
- *Distinctive minerals*—The presence of distinctive minerals in modern sediment deposits helps in identifying and evaluating sediment sources. Because a watershed may contain contrasting rock formations, the distinctive erosion products of these rock formations may clearly indicate the location of the sediment sources. These distinctive minerals are quartz, micas, iron oxide, feldspar, chert, and calcite; some can be easily identified and traced to their original source. Other watersheds may lack geologic variety and hence may not provide such specific clues to the location of significant erosion.
- *Colluviation*—Another aid in evaluating sediment sources is determining the extent and location of colluvial deposition. If a coarse-grained material such as sand or gravel is being actively eroded, it may produce large volumes of sediment, little of which moves very far from the site of erosion. Substantial deposits may form at the foot of the first slope. Fans and valley deposits may form in small tributary valleys or in the next lower valleys downstream.

TABLE 7.1
Sediment Yield from Various Sources

Reach	Sediment Yield from Indicated Source (%) Sheet Erosion	Gullies	Roadbanks	Streambanks	Scour	Total
1	88	5	2	3	2	100
2	64	28	3	4	1	100
3	36	64	—	—	—	100

Procedure

Any procedure requires study of the various types of erosion apparently producing sediment. Sorting the types of erosion according to the treatments that could be recommended to reduce erosion and thus sediment yield will make the effectiveness of the various treatments much easier to evaluate. Several procedures can be used to determine the relative importance of the various sediment sources. A recommended procedure is to gather information on parts of the sediment yield that can be attributed to each of the various sources. Erosion and the sediment delivery ratio should be estimated above each reach or other point of interest for the drainage area. The sediment yield at the point of interest must be allocated to the recognized sources. Analyzing the available data, studying the watershed, and considering the sediment delivery ratios and erosion estimates enable the preparation of a table such as Table 7.1, which indicates the relative important of the sediment sources.

SEDIMENT YIELD

Sediment yield is the gross (total) erosion minus the sediment deposited en route to the point of concern. Gross erosion is the sum of all the water erosion occurring in the drainage area. It includes sheet and rill erosion plus channel-type erosion (e.g., gullies, valley trenches, streambank erosion). Measurements or estimates of the sediment yield are needed to evaluate sediment damage and its reduction and to determine the sediment storage requirements for proposed structures. The yield of a given area varies with changes over time in precipitation, cover, and land use patterns. For projection into the future, the current sediment yield must be adjusted to allow for expected changes in these factors.

Climatic Factors

The effect of climatic factors such as precipitation, temperature, and wind on sediment yields varies in different parts of the country. Rainfall and runoff are the primary erosion factors throughout the country. Wind erosion is serious in some sections but is not as widespread as water erosion. The erosive power of rainfall depends on its intensity, duration, and frequency. Seasonal distribution of rainfall is of prime importance in cropland areas because of the condition of the cover at the time of erosion-producing rainfall. Prolonged low-intensity rainfalls are less erosive than brief intense storms.

Watershed Factors

Important watershed factors affecting sediment yield are size of drainage area, topography, channel density, soil and cover conditions, and land use (USDA, 2008).

Size of Drainage Area

In a given physiographic area, the larger the drainage area, the larger the sediment yield, but generally the sediment yield per unit of area (sediment yield rate) decreases as the size of the drainage area increases. In mountainous areas, however, the size of the drainage area often makes no difference in the sediment yield rate. Where active channel-type erosion increases downstream as from bank cutting on the mainstream channel, the sediment yield rate may increase as the size of the drainage area increases. The relationship between size of drainage area and the sediment yield rate must therefore be considered carefully. In a small watershed, sediment is carried shorter distances and areas of high and low sediment production are less likely to counterbalance each other than in a large watershed. There are fewer types of land use or other watershed variables in a small watershed than in a large watershed. In a small watershed, the yield rate is higher and varies more than in a large watershed.

In a small watershed in which the land is used according to the capability, both the erosion rate and the sediment yield rate are low. Conversely, a high erosion rate is sharply reflected in a high sediment yield rate. Larger watersheds tend to have lower average slopes and less efficient sediment transport than smaller watersheds. Size of the drainage area is therefore an important factor in both the total sediment yield and the sediment yield rate. The relationship between size of the drainage area and sediment yield is complicated by many other factors, such as rainfall, plant cover, texture of the sediment, and land use. All of these factors must therefore be evaluated in estimating the volume of sediment from an erosion source, the rate of deposition in a proposed reservoir, or the rate of sediment contribution to any downstream location.

Topography

Shape of the land surface is an inherent feature of the physiographic area in which a watershed is located. Many of the problems of soil and water conservation result from the topography of an individual watershed, especially the proportions of uplands, valley slopes, floodplains, or features such as escarpments, canyons, or alluvial fans. Slope is a major factor affecting the rate of onsite erosion, and topography is important in the delivery of upland erosion products to the stream system. Drainage density, amount of sloping land, and erosion rate are closely related to the stage of erosional development. Youthful areas are characterized by a relatively high proportion of high, nearly flat upland between stream valleys. Youthful watersheds at high elevations may have deep canyons along the principal streams; youthful watersheds consisting of low glacial plains or other flat areas commonly have poorly developed stream courses and relatively low slopes. Watersheds in areas of old topography also have a relatively small amount of sloping land, but most often uplands are eroded to low elevations and the greatest proportion of land consists of old, broad valley flats.

The proportion of sloping land is usually highest in mature areas, where drainage is well developed and either uplands or valley flats are limited. The average gradient and the average sediment yield tend to be higher in mature areas.

Channel Density

The efficiency of a stream system in transporting sediment out of a watershed is affected by the degree of channelization. A watershed with a high channel density (total length of channel per unit area) has the most thorough water runoff and the most rapid and complete transport sediment from the area. Channel density can be measured on aerial photographs with the aid of a stereoscope.

Soil and Cover Conditions

The kinds of soil and cover are important in sediment yield. In general, the more erodible the soil and the sparser the vegetation, the higher the sediment yield. Estimating the average annual sediment yield from a watershed having many kinds of soil and mixed cover is complex and requires a procedure such as use of a soil-loss equation to determine erosion for the various soil–slope–cover combinations in the watershed. Sediment yield tends to be similar in watersheds of similar size, topography, and cover.

Land Use

According to a 1977 National Erosion Inventory, about 28% of the 1500 million acres of non-federal land in the United States was cropland; 36% was grassland, pasture, and range; 25% was forest; 6% was in residential, industrial, transportation, and other urban and built-up areas; and 5% was in other uses. Land use is determined to some extent by the kind of soil. In turn, land use largely determines the type of cover. If a watershed is primarily agricultural and the annual precipitation is more than 20 inches, most of the sediment yield usually is from sheet erosion. In most forest and range country and in areas with less than 20 inches of annual precipitation, channel-type erosion usually produces most of the sediment (Brown, 1960).

According to the U.S. Department of Agriculture, conversion of forest land to continuous cultivation of row crops increases erosion 100- to 10,000-fold. Plowing grassland for continuous cultivation of row crops increases erosion 20- to 100-fold (Brown, 1960). In the United States, cultivated farm fields that annually lose more than 200 tons/acre from water erosion are not uncommon (Gottschalk, 1965; Gottschalk and Jones, 1955). Small, intensively cultivated watersheds in western Iowa have had annual soil losses as high as 127,000 tons/mi^2 (Gottschalk and Brune, 1950). Because it encompasses such a broad area, agricultural land produces the most sediment, but progress is being made in conserving agricultural soils. Special uses create serious local problems, such as the following:

- *Urbanization*—Construction of an industrial park near Baltimore produced at least five times more sediment than was present in the waters immediately upstream (Wolman, 1964). Areas under construction above Lake Barcroft, Virginia, and Greenbelt Lake, Maryland, yielded annual peak sediment yield rates of 25,000 and 5600 tons/mi^2, respectively (Dawdy, 1967).

- *Stripmining*—In Kentucky, a watershed with 10% of its area disturbed by active stripmining produced 57 times the sediment measured from a similar but undisturbed adjoining watershed (Collier et al., 1964).
- *Highway construction*—Sediment yield from an area in Fairfax County, Virginia, where a highway was being built was 10 times greater than that from cultivated land, 200 times greater than that from grassed areas, and 2000 times greater than that from forested area (Vice et al., 1969).

Methods of Determination

Depending on the environment and the data available, the average annual sediment yield in a watershed can be determined from (1) gross erosion and the sediment delivery ratio, (2) measured sediment accumulation, (3) suspended load records, and (4) predictive equations.

Gross Erosion and the Sediment Delivery Ratio

This method has been used extensively for many years with success, particularly in humid sections of the country. It is well suited to estimating current sediment yield and predicting the effect of land treatment and land use changes on future sediment yield. The following equation is used to estimate sediment yield:

$$Y = E(DR) \tag{7.1}$$

where

Y = Annual sediment yield (tons/unit area).
E = Annual gross erosion (tons/unit area).
DR = Sediment delivery ratio (less than 1).

The gross (total) erosion in a drainage area is the sum of all the water erosion taking place. The sediment delivery ratio is estimated from relationships discussed later in this chapter. Sediment yield is the product of gross erosion and the sediment delivery ratio.

Measured Sediment Accumulation

The measured sediment accumulation in reservoirs of known age and history is an excellent source of data for establishing sediment yield, but deposition in reservoirs and sediment yield are not synonymous. For sediment yield, the amount of accumulated sediment must be divided by the trap efficiency of the reservoir. The amount of sediment that has passed through the reservoir plus the amount deposited in the reservoir equals the sediment yield. The sediment yield of a watershed can be estimated from measured sediment yield from another watershed in the same major land resources area if the topography, soils, and land use of the two watersheds are similar. For direct extrapolation of sediment yield data, the size of the drainage area of the surveyed reservoir should be no less than one-half nor more than twice that of the watershed under consideration. Beyond these limits, the annual sediment yield can be adjusted on the basis of the ratio of the drainage area to the 0.8 power:

$$Y_e = Y_m(A_e/A_m)^{0.8} \tag{7.2}$$

where

Y_e = Sediment yield of unmeasured watershed in tones per year.

Y_m = Sediment yield of measured watershed in tons per year (measured annual sediment deposition divided by trap efficiency of surveyed reservoir).

A_e = Drainage area of unmeasured watershed.

A_m = Drainage area of measured watershed.

This relationship must be used with judgment and be confined generally to the humid areas east of the Rocky Mountains. The amount of sediment accumulated on fans and floodplains over a known period of time can sometimes be used to estimate sediment yield but generally only to verify yield determined by other methods.

Suspended Load Records

Suspended sediment can be measured by sampling, and water discharge can be determined by gauging at stream cross-sections. Sediment yield can be estimated from these data. Sediment concentration in milligrams per liter or parts per million is converted to tons per day by multiplying the average concentration by the volume of water discharged on the day of record and a conversion factor (usually 0.0027). Tons of sediment per day plotted against water discharge in cubic feet per second is a *sediment rating curve*. The data plotted on log–log paper often approximate a straight line through at least a major part of the range of discharge.

If discharge and concentration data are available, the average annual sediment yield can be estimated by using a flow–duration curve or equivalent tabulations (Anderson, 1954). Usually the length of time required to collect a range of suspended load data large enough to prepare a sediment rating curve prohibits the establishment of a suspended load station for small watersheds. If such suspended load records are available from nearby similar watersheds, however, the sediment yield rate can be derived and transposed in the same manner as reservoir sedimentation survey data. The bedload portion of the sediment load is not measured in this method; it must be estimated. It can range from practically none to 50% or more of the total load.

Predictive Equations

Predictive equations based on watershed characteristics have been developed in some areas to estimate sediment yield. These equations express sediment yield as a function of a combination of several measurable independent variables. The variables include size of the drainage area, annual runoff, watershed shape, relief/length ratio, average slope, an expression of the particle size of the surface soil, and others. Such equations are not numerous, but, where developed, they can be used with the understanding that they apply only to the specific area they represent.

SEDIMENT DELIVERY RATIO

Determining the sediment delivery ratio is of primary importance to geologists if they are to make realistic estimates of sediment yield on the basis of computed gross erosion. No characteristic relationship is known to exist between sediment yield and erosion alone. Many factors influence the sediment delivery ratio, and, because these are not uniform from watershed to watershed, the relationship between sediment yield and erosion varies considerably.

Influencing Factors

Each of the following factors can influence the sediment delivery ratio. There may be additional factors not yet identified.

Sediment Source

The sediment source affects the sediment delivery ratio. Sediment produced by channel-type erosion is immediately available to the transport system. Much of it remains in motion as suspended sediment or bedload. Materials derived from sheet erosion, however, often move only a short distance and may lodge in areas remote from the transport system. These materials may remain in the fields in which they originated or may be deposited as colluvium on more level slopes.

Proximity of Sediment Sources

Another factor that affects the sediment delivery ratio is the proximity of the source to streamflow. For example, although a large amount of material may be produced by severe erosion in an area remote from a stream, the delivery ratio and sediment yield may be less than those from a smaller amount of material produced by moderate erosion close to that stream.

Transport System

Runoff resulting from rainfall and snowmelt is the chief transport agent for eroded material. The ability to transport sediment depends on the velocity and volume of water discharge as well as on the amount and character of the material supplied to it. If the amount of sediment in transit exceeds the transport capacity of the system, sediment is deposited and the sediment delivery ratio is decreased. The frequency and duration of discharges affect the total volume of sediment delivered. The extent and condition of the transport system have considerable bearing on the amount of sediment the system can transport. A transport system with high channel density has the greatest chance of acquiring materials from the uplands and should have a high sediment delivery ratio. The condition of the channels (clogged or open, meandering or straight) affects velocity and, consequently, transport capacity. A high-gradient stream, usually associated with steep slopes and high relief, transports eroded material efficiently. The reverse is true of a low-gradient stream.

Texture of Eroded Material

The texture of the eroded material also affects the sediment delivery ratio. Transport of sand requires a relatively high velocity. Much of the sand is deposited in upstream areas wherever velocity drops significantly. Sand usually becomes part of the sediment load only if its source areas are adjacent to an efficient transport system. Eroded silt and clay are likely to stay in suspension as long as the water is moving, and most of such material is delivered downstream. Some of the coarser particles may be deposited as colluvium before they reach the transport system. The sands and larger grain-size materials are usually produced by channel erosion, and the silts and clays are common products of sheet erosion.

Depositional Area

Some sediment is deposited at the foot of upland slopes, along the edges of valleys, in valley flats, in and along main stream channels, and at the heads of and in reservoirs, lakes, and ponds. Such deposition within a watershed decreases the amount of sediment delivered to points downstream.

Watershed Characteristics

The topography of a watershed affects the sediment delivery ratio. Slope is a major factor affecting the rate of erosion. High relief often indicates both a high erosion rate and a high sediment delivery ratio. The relief/length (R/L) ratio often corresponds closely to the sediment delivery ratio. For use in the R/L ratio, relief (measured in feet) is defined as the difference between the average elevation of the watershed divide at the headwaters of the main-stem drainage and the elevation of the stream bed at the point of sediment yield. The length is defined as the maximum valley length (in feet) parallel to the main-stem drainage from the point of sediment yield to the watershed divide. The slope of a watershed can affect the sediment delivery ratio. Channel velocity also affects the sediment delivery ratio; channel density and topography are closely related. The size of the drainage area is also important. Size can be considered a composite variable that incorporates and averages out the individual effects of variability in topography, geology, and climate.

Procedure for Estimating the Sediment Delivery Ratio

Determining the sediment delivery ratio requires knowledge of the sediment yield at a given point in a watershed and the total amount of erosion. If this information is available, determining the sediment delivery ratio is simple. Values for both these required items, however, usually are not available for most small watersheds. Gross erosion in a watershed can be estimated by using standard Soil Conservation Service procedures. Sediment yield can be determined from reservoir sedimentation surveys or sediment load measurements. Many reservoirs are not located at points where measurements of sediment yield are needed, and a program of sediment load sampling may be long and expensive. But, if the ratio of known sediment yield and erosion within a homogeneous area can be analyzed in conjunction with some measurable influencing factor, these data can be used to predict or estimate the sediment delivery ratio for similar areas where measurements are lacking.

In a given physiographic area, finding measureable factors that can be definitely related to the sediment delivery ratio is the goal of any delivery ratio analysis. As already pointed out, many factors can affect the sediment delivery ratio. Some are more pronounced in their effect than others; some lend themselves to quantitative expression, and others do not. Statistical analysis is an effective means of developing information for estimating the sediment delivery ratio. The sediment delivery ratio is used as a dependent variable and the measurable watershed factors are used as the independent or controlling variables. For such an analysis, quantitative data on sediment yield, erosion, and measurable watershed factors must be available. Reservoir sedimentation surveys are a source of sediment yield data. Either maps or field surveys can be used to obtain the erosion information and determine the watershed factors. These data can be analyzed to develop a means for estimating the sediment delivery ratio for similar areas. Analyses of this type should be made in consultation with the geologist (sedimentation) of the appropriate national technical center.

Size of Drainage Area

Data obtained from past studies (Glymph, 1954; Gottschalk and Brune, 1950; Maner, 1957; Maner and Barnes, 1953; Roehl, 1962) indicate wide variation in the sediment delivery ratio of any given size of drainage area.

Relief/Length Ratio

The watershed relief/length (R/L) ratio (Maner and Barnes, 1953; Roehl, 1962) is a significant indicator of the sediment delivery ratio. Empirical equations were derived to estimate the R/L ratio for the Red Hills of Texas, Oklahoma, and Kansas and for the southern Piedmont region of the Southeast. The significance of the R/L ratio may be less pronounced in some areas than in others, but it is related to, and seems to be a reasonable expression of, several watershed factors.

Source Texture Analysis

In all the preceding discussion of methods of estimating the sediment delivery ratio, the delivery ratio is a percentage of total erosion. In many places, the individual delivery ratio of the component parts of the total erosion is of concern to geologists. Reasonable and realistic values for the delivery of component parts must be estimated from scanty data. One method of obtaining these estimates is to make certain determinations or assumptions about the source of various components of a known sediment yield.

Source Deposition

Another method of determining the sediment delivery ratio is to make a field study of a watershed and estimate the amount of deposition that can be traced to any one source. The difference in the volume of such deposition and the volume of sediment produced by the source gives an estimate of the delivery ratio from that source.

Bottom Line for Estimating Sediment Delivery Ratio

In many places, data needed for detailed analyses are insufficient or nonexistent. Using an equation to obtain sediment data outside the physiographic area for which the equation was developed is generally not recommended; yet, geologists must know the sediment delivery ratio to determine the sediment yield and the relative importance of various sediment sources and to recommend measures for reducing the sediment yield. Information about sediment yield from some watersheds is available in most areas of the country. These data can be obtained from suspended load records. Comparing sediment yield with the calculated gross erosion indicates the expected sediment delivery ratio for an area. This kind of analysis is much broader than a detailed study, and extrapolating such an estimate to other areas can cause error.

REFERENCES AND RECOMMENDED READING

Anderson, H.W. (1954). Suspended sediment discharge as related to stream flow, topography, soil, and land use. *American Geophysical Union Transactions*, 35(2): 268–281.

Brown, C.B. (1950). Effects of soil conservation. In: *Applied Sedimentation* (Trask, P.D., Ed.), pp. 380–406. New York: John Wiley & Sons.

Brown, C.B. (1960). *Effects of Land Use and Treatment on Pollution*, Public Health Services Publ. No. 819. Washington, DC: Department of Health, Education, and Welfare.

Collier, C.R. et al. (1964). *Influence of Strip Mining on the Hydrologic Environment of Parts of Beaver Creek Basin, Kentucky, 1955–59*, USGS Professional Paper 427-B. Reston, VA: U.S. Geological Survey.

Dawdy, D.R. (1967). Knowledge of sedimentation in urban environments. *Journal of the Hydraulics Division*, 93(HY6): 235–245.

Glymph, Jr., L.M. (1954). Studies of Sediment Yields from Watersheds, paper presented at the Tenth General Assembly of the International Union of Geodesy and Geophysics, Rome, Italy, September 15–19.

Gottschalk, L.C. (1948). Analysis and Use of Reservoir Sedimentation Data, paper presented at the First Federal Interagency Sedimentation Conference, Denver, CO, May 6–8.

Gottschalk, L.C. (1965). Sedimentation transportation mechanics: nature of sedimentation problems. *Journal of the Hydraulics Division*, 91(HY2): 251–266.

Gottschalk, L.C. and Brune, G.M. (1950). *Sediment Design Criteria for the Missouri Basin Loess Hills*, U.S. Soil Conservation Service Technical Paper 97. Washington, DC: U.S. Department of Agriculture.

Gottschalk, L.C. and Jones, V.H. (1955). Valleys and hills: erosion and sedimentation. In: *The Yearbook of Agriculture 1955: Water* (Stefferud, A., Ed.), pp. 135–143. Washington, DC: U.S. Department of Agriculture.

Maner, S.B. (1957). Factors affecting sediment delivery rates in the Red Hills physiographic area. *Transactions, American Geophysical Union*, 39(4): 669–675.

Maner, S.B. and Barnes, L.H. (1953). *Suggested Criteria for Estimating Gross Sheet Erosion and Sediment Delivery Rates for the Blackland Prairies Problem Area in Soil Conservation*. Ft. Worth, TX: U.S. Department of Agriculture, Soil Conservation Service, Western Gulf Region.

Roehl, J.W. (1962). *Sediment Source Areas, Delivery Ratios, and Influencing Morphological Factors*, IASH Publ. No. 59. Wallingford, U.K.: International Association of Hydrological Sciences, pp. 202–213.

USDA. (2008). Sediment properties. In: *National Engineering Handbook*, Section 3, Chapter 2. Washington, DC: U.S. Department of Agriculture.

Vice, R.B., Guy, H.P., and Ferguson, G.E. (1969). *Sediment Movement in an Area of Suburban Highway Construction, Scott Run Basin, Fairfax Co., Virginia, 1961–64*. USGS Water Supply Paper 1591-E. Reston, VA: U.S. Geological Survey.

Wolman, M.G. (1964). *Problems Posed by Sediment Derived from Construction Activities in Maryland*. Annapolis: Maryland Water Pollution Control Commission.

Woodburn, R. and Roehl, J.W. (1951). Unpublished study. Oxford, MS: U.S. Department of Agriculture, Agricultural Research Service.

Section II

Sediment Damage

8 Yellow Boy

Day: Wednesday
Date: August 5, 2015
Place: Near Silverton, Colorado
Event: Gold King Mine spill
Surface water bodies affected: Cement Creek, Animas River, San Juan River, and Colorado River
Watershed: Colorado, Utah, New Mexico, Arizona, Nevada, and California

GOLD KING MINE SPILL

On August 5, 2014, the U.S. Environmental Protection Agency (USEPA) conducted a mine site investigation of the abandoned Gold King Mine above the old adit (a mine tunnel) to

- Assess the ongoing water releases from the mine.
- Treat mine water.
- Assess the feasibility of further mine remediation.

During the excavation required for the investigation the heavy equipment disturbed loose material around a soil "plug" at the mine entrance, spilling about 3 million gallons of pressurized water stored behind the collapsed material into Cement Creek, a tributary of the Animas River (Figure 8.1).

The spill volume associated with the release on August 5 was calculated to be approximately 3 million gallons based on flow rates. Discharge rates from the mine as of November 5, 2015, averaged around 600 gallons per minute. It is important to point out, for context, that multiple mines are located along the upper Animas, and historically there have been considerable discharges at each mine site. The Red and Bonita mines, just below the Gold King Mine, currently discharge about 300 gallons per minute (see Figures 8.2 to 8.5).

One of the most striking effects of the Gold King Mine spill was the color change clearly visible in Cement Creek and Animas River and to a lesser degree downstream almost to the San Juan River. The iron from the acid mine drainage precipitated out into the water as a result of the rise in pH, turning it yellow. Old-time goldpanners and other sluice miners refer to the resulting red, orange, and yellow solids as "yellow boy." Typically, as more water is mixed in (dilution is the solution to pollution, according that mythical hero Hercules, who arguably might have been the world's first environmental engineer), the iron and other metals become even more dilute or get attached to sediments, causing them to drop out of the water, sink, and settle into river bottom sediments. The water color then returns to normal.

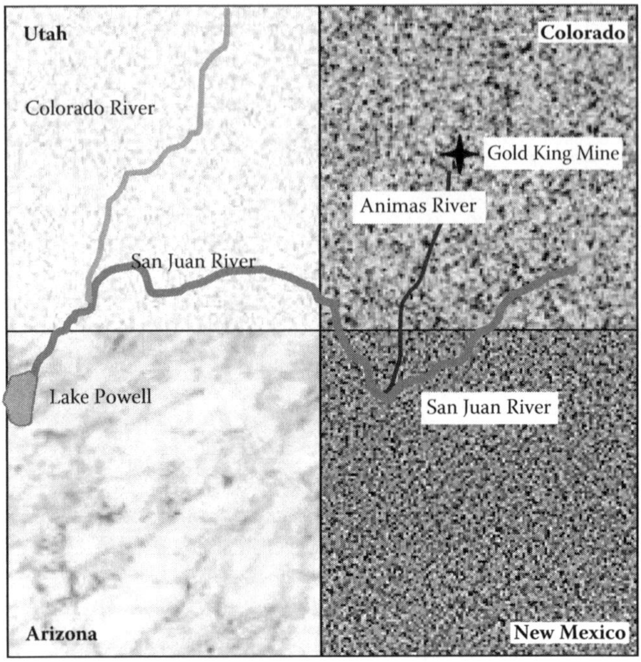

FIGURE 8.1 Illustration of the spatial relationship of Gold King Mine to the Animas, San Juan, and Colorado Rivers and to Lake Powell.

FIGURE 8.2 Photograph of the Gold King Mine entrance area, August 14, 2015. (From USEPA, *Emergency Response to August 2015 Release from Gold King Mine*, U.S. Environmental Protection Agency, Washington, DC, 2016, http://epa.gov/goldkingmine. Photograph by Eric Vance/EPA.)

FIGURE 8.3 Treatment ponds built close to the Gold King Mine. When water leaves the mines, these ponds slow it down and allow pH adjustment to be made while letting contaminants settle to bottom. (From USEPA, *Emergency Response to August 2015 Release from Gold King Mine*, U.S. Environmental Protection Agency, Washington, DC, 2016, http://epa.gov/goldkingmine.)

FIGURE 8.4 As water exits the mine, it flows into a system of four treatment ponds. The treatment ponds provide retention time to allow the pH to adjust. Here, lime is added to a settling pond to assist in the pH adjustment of the water prior to discharge to Cement Creek and the Animas River on August 14, 2015. (From USEPA, *Emergency Response to August 2015 Release from Gold King Mine*, U.S. Environmental Protection Agency, Washington, DC, 2016, http://epa.gov/goldkingmine. Photograph by Eric Vance/EPA.)

FIGURE 8.5 Settling ponds used to precipitate iron oxide and other suspended materials from nearby Red and Bonita mines drainage are shown in this August 14, 2015, photograph. (From USEPA, *Emergency Response to August 2015 Release from Gold King Mine*, U.S. Environmental Protection Agency, Washington, DC, 2016, http://epa.gov/goldkingmine. Photograph by Eric Vance/EPA.)

But, and this is the gist of this text, what appears normal in surface water bodies may not actually be normal because, as in the case of the Animas River and thousands of other polluted streams, what we are able to see at the surface does not in any way certify the quality of the water contained within the water body. To amplify this point, to demonstrate the importance of failing to consider surface water contamination beyond that which is obvious to the naked eye, refer to Case Study 8.1 below. I have used this case study, an actual study of a river pool I studied in 1988 in the Skykomish River in Washington State, often in my technical presentations because it clearly makes the point I want to make here.

Case Study 8.1. Still Waters?

Consider a river pool, isolated by fluvial processes and time from the main stream flow. We are immediately struck by one overwhelming impression: It appears so still … so very still … still enough to soothe us. The river pool provides a kind of poetic solemnity, if only at the pool's surface. No words of peace, no description of silence or motionless can convey the perfection of this place, in this moment stolen out of time.

We consider that the water is still, but does the term *still* correctly describe what we are viewing. Is there any other term we can use besides *still*—is there any other kind of still? Yes, of course, we know many ways to characterize still. *Still* can mean inaudible, noiseless, quiet, or silent. *Still* can also mean immobile, inert, motionless, or stationary—which is how this pool appears to the casual visitor on the surface. The visitor sees no more than water and rocks.

The rest of the pool? We know very well that a river pool is more than just a surface. How does the rest of the pool (the subsurface, for example) fit the descriptors we tried to use to characterize its surface? Maybe they fit, maybe they don't. In time, we will go beneath the surface, through the liquid mass, to the very bottom of the pool to find out. For now, remember that images retained from first glances are almost always incorrectly perceived, incorrectly discerned, and never fully understood.

On second look, we see that the fundamental characterization of this particular pool's surface is correct enough. Wedged in a lonely riparian corridor—formed by river bank on one side and sand bar on the other—between a youthful, vigorous river system on its lower end and a mountain source on its headwater end, almost entirely overhung by mossy old Sitka spruce, the surface of the large pool, at least at this particular location, is indeed still. In the proverbial sense, the pool's surface is as still and as flat as a flawless sheet of glass.

The glass image is a good one, because like perfect glass the pool's surface is clear, crystalline, unclouded, definitely transparent, and yet perceptively deceptive as well. The water's clarity, accentuated by its bone-chilling coldness, is apparent at close range. Further back, we see only the world reflected in the water—the depths are hidden and unknown. Quiet and reflective, the polished surface of the water perfectly reflects in mirror-image reversal the spring greens of the forest at the pond's edge, without the slightest ripple. Up close, looking straight into the bowels of the pool we are struck by the water's transparency. In the motionless depths, we do not see a deep, slow-moving reach with the muddy bottom typical of a river or stream pool; instead, we clearly see the warm variegated tapestry of blues, greens, blacks stitched together with threads of fine, warm-colored sand that carpets the bottom, at least 12 feet below.

Still waters can run deep.

No sounds emanate from the pool. The motionless, silent water does not, as we might expect, lap against its bank or bubble or gurgle over the gravel at its edge. Here, the river pool, held in temporary bondage, is patient, quiet, waiting, withholding all signs of life from its surface visitor.

Then the reality check: This stillness, like all feelings of calm and serenity, could be fleeting, momentary, temporary, you think. And you would be correct, of course, because there is nothing still about a healthy river pool. At this exact moment, true clarity is present; it just needs to be perceived … and it will be.

We toss a small stone into the river pool, and watch the concentric circles ripple outward as the stone drops through the clear depths to the pool bottom. For a brief instant, we are struck by the obvious: The stone sinks to the bottom, following the laws of gravity, just as the river flows according to those same inexorable laws—downhill in its search for the sea. As we watch, the ripples die away, leaving as little mark as the usual human lifespan creates in the waters of the world, then disappears as if it had never been. Now the river water is as before, still. At the pool's edge, we look down through the massy depth to the very bottom—the substrate.

We determine that the pool bottom is not flat or smooth, but instead is pitted and mounded occasionally with discontinuities. Gravel mounds alongside small corresponding indentations—small, shallow pits—make it apparent to us that gravel was removed from the indentations and piled into slightly higher mounds. From our topside position, as we look down through the cool, quiescent liquid, the exact height of the mounds and the depth of the indentations are difficult for us to judge; our vision is distorted through several feet of water.

However, we can detect near the low gravel mounds (where female salmon bury their eggs, and where their young grow until they are old enough to fend for themselves), and actually through the gravel mounds, movement—water flow—an upwelling of groundwater. This water movement explains our ability to see the variegated color of pebbles. The mud and silt that would normally cover these pebbles has been washed away by the water's subtle, inescapable movement. Obviously, in the depths, our still water is not as still as it first appeared. The slow, steady, inexorable flow of water in and out of the pool, along with the up-flowing of groundwater and a variety of other mechanism (hyporheic exchange) without direct groundwater movement to the surface and through the pool's substrate and through the salmon redds (nests) is only a small part of the activities occurring within the pool, including the air above it, the vegetation surrounding it, and the damp bank and sandbar forming its sides.

Let's get back to the pool itself. If we could look at a cross-sectional slice of the pool, at the water column, the surface of the pool may carry those animals that can literally walk on water. The body of the pool may carry rotifers and protozoa and bacteria—tiny microscopic animals—as well as many fish. Fish will also inhabit hidden areas beneath large rocks and ledges, to escape predators. Going down further in the water column, we come to the pool bed. This is the benthic zone, and certainly the greatest number of creatures live here, including larvae and nymphs of all sorts, worms, leeches, flatworms, clams, crayfish, dace, brook lampreys, sculpins, suckers, and water mites. And these organisms also cause internal, deep sediment water movement via the process of bioturbation, where they burrow, forage, and ingest sediment, leading to mixing of sediments and associated contaminants.

We need to go down even farther, down into the pool bed, to see the whole story. How far this goes and what lives here, beneath the water, depends on whether it is a gravelly bed or a silty or muddy one. It also depends on sediment condition—sediment health, so to speak. That is, sediment that is contaminated with certain biologicals or chemicals is only suited for housing the lowest form of organisms, mainly worms. Anyway, we assume this still water pool is healthy and thus gravel will allow water, with its oxygen and food, to reach organisms that live underneath the pool. Many of the organisms that are found in the benthic zone may also be found underneath, in the hyporheal zone.

But, to see the rest of the story, we need to look at the pool's outlet and where its flow enters the main river. These are the riffles—shallow places where water runs fast and is disturbed by rocks. Only organisms that cling very well, such as net-winged midges, caddisflies, stoneflies, some mayflies, dace, and sculpins, can spend much time here, and the plant life is restricted to diatoms and small algae. Riffles are a good place for mayflies, stoneflies, and caddisflies to live because they offer plenty of gravel to hide in. At first, we struggled to find the "proper" words to describe the river pool. Eventually, we settled on "still waters." We did this because of our initial impression and because of our lack of understanding—lack of knowledge. Things can look different from the way they really are. Thus, our still water river pool supports the idiom that appearances can be deceiving. And in Nature and in life they usually are.

TABLE 8.1
Weight of Metals Released in Mine Drainage from the Gold King Mine

Metal	Pounds	Metal	Pounds	Metal	Pounds
Iron	248,582	Copper	919	Cobalt	10
Aluminum	23,657	Sodium	586	Antimony	8
Calcium	11,365	Barium	244	Nickel	7
Magnesium	6984	Arsenic	206	Mercury	6
Potassium	5307	Vanadium	137	Cadmium	4
Lead	4481	Molybdenum	50	Beryllium	3
Manganese	1953	Silver	28	Selenium	n.d.[a]
Zinc	1101	Chromium	18	Thallium	n.d.[a]

Source: Chief, K. et al., *Understanding the Gold King Mine Spill*, The University of Arizona, Superfund Research Program, Tucson, 2015.

[a] n.d., not detected at or above the method detection limit.

Even knowing what we know now, we might still describe the river pool as still waters. However, in reality, we must call the pool what it really is: a dynamic habitat. This is true, of course, because each river pool has its own biological community, all members interwoven with one another in complex fashion, all depending on each other. Thus, our river pool habitat is part of a complex, dynamic ecosystem. On reflection, we realize, moreover, that anything dynamic certainly cannot be accurately characterized as "still"—including our river pool.

SOLIDS RELEASED

Let's channel back, so to speak, to the Animas River, which was contaminated and painted yellow by the Gold King Mine spill. Again, it was mainly the iron oxides from the acid mine waste that turned the Animas River yellow, but it is important to point out that the acid mine drainage released in the spill also contained a number of metal salts totaling some 190 tons of solids, including several forms of toxic metals such as lead, arsenic, mercury, and cadmium (Table 8.1). These solids were mixed in more than 3,000,000 gallons of water. Based on USEPA estimates, the volume of the water was approximately 9 acre-feet, or 9 football fields, spread out at 1 foot deep.

TIMELINE OF USEPA ACTIONS FROM AUGUST 12 TO SEPTEMBER 3, 2015

Following is a basic timeline account of USEPA actions in regard to the Gold King Mine Spill when the Animas River began returning to its usual color (NPS, 2015). This is an interesting and informative account of how the acid mine release was handled.

August 12, 2015

- The USEPA is leading response efforts associated with the Gold Kind Mine spill. The leading edge of the plume of contaminated water is no longer visible and appears to have been assimilated into the waters of the San Juan River.
- Based on preliminary results from water sampling conducted by the Utah Department of Environmental Quality on August 8 and 9, the Utah Department of Health recommends that recreationists on the San Juan River avoid drinking the water and wash after contact with the river as soon as possible.
- No closures are currently in effect, and additional precautions have not been identified as necessary at this time.

August 14, 2015

- The USEPA is leading response efforts associated with the Gold King Mine spill and has sent representation to the Glen Canyon area for local response efforts.
- Based on preliminary results from water sampling conducted by the Utah Department of Environmental Quality, the Utah Department of Health does not expect that recreational or agricultural use of the San Juan River will result in adverse effects in people, livestock, or crops.
- As a precaution, recreational users are still urged to carry their own drinking water and not rely on filtering or purifying water from the San Juan River.
- No closures are currently in effect, and additional precautions have not been identified as necessary at this time.

August 15–16, 2015

- The USEPA is leading response efforts associated with the Gold King Mine spill and has sent staff to the Glen Canyon area for local response efforts.
- Sampling will be conducted in locations at the confluence of the San Juan River and Lake Powell and at other locations within Lake Powell over the weekend to better understand the progress of any potentially detectable contamination. Samples will include both surface waters and sediments.
- Since August 8, the USEPA has been collecting daily surface water and sediment samples upstream of Glen Canyon between Farmington, New Mexico, and Clay Hills Crossing, Utah, and we anticipate the results next week. Results should provide information about the location and concentrations of potential contamination from the Gold King Mine release.
- As a precaution, recreational users are still urged to carry their own drinking water and not rely on filtering or purifying water from the San Juan River.
- No closures are currently in effect, and additional precautions have not been identified as necessary at this time.

August 17, 2015

- The USEPA is leading response efforts associated with the Gold King Mine spill and has sent staff to the Glen Canyon area for local response efforts.
- Sampling was conducted in locations at the confluence of the San Juan River and Lake Powell as well as other locations within Lake Powell over the weekend. Results from both surface and sediment samples will help to better understand the progress of any potentially detectable contamination in Glen Canyon.
- Since August 8, the USEPA has been collecting daily surface water and sediment samples upstream of Glen Canyon between Farmington, New Mexico, and Clay Hills Crossing, Utah, and we anticipate results this week.
- As a precaution, recreational users are still urged to carry their own drinking water and not rely on filtering or purifying water from the San Juan River.
- No closures are currently in effect, and no additional precautions have been identified as necessary at this time.

August 18, 2015

- As a precaution, recreational users are still urged to carry their own drinking water and not rely on filtering or purifying water from the San Juan River.
- No closures are currently in effect, and no precautions for the rest of Lake Powell have been identified as necessary at this time.
- The USEPA is leading response efforts, including those locally, and has sent staff to the Glen Canyon area.
- Water and sediment samples were collected in several locations on Lake Powell over the weekend, including at the confluence of the San Juan River and Lake Powell, the San Juan River arm of Lake Powell, and other locations between Dangling Rope Marina (see Figure 8.6) and Wahweap Marina (see Figure 8.7).
- Results are expected before the end of the week and should provide information about the location and concentration of any potentially detectable contamination.
- The Arizona Department of Environmental Quality (ADEQ) has also analyzed data provided by other agencies and has announced that they do not expect this spill to have short- or long-term effects on Lake Powell and the Colorado River. The National Park Service (NPS) is coordinating with other agencies to conduct long-term monitoring to assess any potential impacts.

August 25, 2015

- The USEPA is leading response efforts associated with the Gold King Mine release, and a variety of state and federal agencies are continuing to collect and analyze data to understand any potential effects of the event.

FIGURE 8.6 Lake Powell in Glen Canyon near Dangling Rope Marina, which is accessible only by boat. (Photograph by author.)

FIGURE 8.7 Lake Powell at the Arizona/Utah border, as viewed from the Wahweap Marina. (Photograph by author.)

- A previously planned mercury study is being conducted on Lake Powell by the U.S. Geological Survey (USGS) and NPS. Scientists will be on the lake over the next several days, and data collection efforts will be expanded to include additional samples that will aid in understand potential effects of the Gold King Mine spill.
- As a precaution, recreational users are still urged to carry their own drinking water and not rely on filtering or purifying water from the San Juan River.
- No closures in Glen Canyon are currently in effect, and no additional precautions have been identified as necessary at this time.
- Water and sediment samples were collected last weekend in several locations on Lake Powell, and results are expected this week that should provide information about the location and concentrations of any potentially detectable contamination.

August 27, 2015

- The USEPA is leading response efforts associated with the Gold King Mine release, and a variety of state and federal agencies are continuing to collect and analyze data to understand the long-term effects of the event.
- No closures are in effect at this time.
- Potentially detectable effects on Lake Powell are still being studied. Water samples taken from Lake Powell by the USEPA are still being analyzed, but preliminary results indicate that all samples are within drinking water standards for contaminants.
- As a precaution, recreational users are still urged to carry their own drinking water and not rely on filtering or purifying water from the San Juan River. No additional precautions have been identified as necessary at this time.

Note that during the USEPA monitoring and sampling of the Animas River, San Juan River, and Lake Powell, the Arizona Game and Fish Department was also collecting various samples, including fish tissue samples, from the Arizona portion of Lake Powell and the Colorado River at Lees Ferry (see Figure 8.8). Results were used to identify any potentially negative impacts on Lake Powell and the blue-ribbon trout fishery at Lees Ferry.

September 3, 2015

- Lake Powell water samples are declared within drinking water standards after the Gold King Mine spill.
- The USEPA tested water and sediment in Lake Powell, including the San Juan arm of Lake Powell. Testing was conducted to assess contaminant levels resulting from the Gold King Mine spill. All contaminants tested met drinking water standards, and sediment results met recreational screening levels, as well. Sampling results from Lake Powell are posted on the USEPA website (http://www2.epa.gov/goldkingmine/data-gold-king-mine-response).

FIGURE 8.8 Colorado River downstream from Lake Powell and Glen Canyon Dam near Lee's Ferry and the entrance to Grand Canyon, as viewed from the eastern landing of Navajo Bridge on the Navajo Nation. (Photograph by author.)

- No area closures or human safety advisories are currently in effect for Lake Powell. As always, visitors are encouraged to purify Lake Powell water for drinking. The Utah Department of Health continues to advise San Juan River users to carry their own drinking water and not rely on filtering or purifying river water. Local communities are open for business, including boat and paddle craft tours and rentals.
- Long-term impacts of the Gold King Mine spill in the Glen Canyon National Recreation Area are being monitored by the National Park Service, Environmental Protection Agency, U.S. Geological Survey, Bureau of Reclamation, and the states of Utah and Arizona.

Bottom Line on Gold King Mine Spill

As of November 5, 2015, there was no longer a visible leading edge of the Gold King Mine. The USEPA estimated that the water associated with the release reached Lake Powell sometime on Wednesday afternoon, August 12. Lake Powell is a large body of water, and no significant impacts on the lake, the Colorado River, or any water bodies downstream are expected (see Figures 8.2 and 8.3).

PERSISTENCE! PERSISTENCE! PERSISTENCE!

When I viewed the Animas River, San Juan River, and Lake Powell in November of 2015, no yellow boy was apparent anywhere in the watershed. This is good, of course. But we can't assume victory over the spill and its contaminating effects. Adopting an attitude of "out of sight, out of mind" can be a problem. Think again about the still waters case study. What is apparent at the surface is not always the case down below. Even though a freshwater body such as the Animas River may appear clear, healthy, and unpolluted at the surface, that is not always the case deep within the bowels of the water column of the river. Keep in mind that the color, smell, or taste of the Animas River, San Juan River, and Lake Powell is not enough to tell us whether or not the water is safe. The only way to be sure of the water quality is to sample and test it.

Why do we need to test it? Wasn't the mythical Hercules correct when he said that dilution is the solution to pollution? Don't the self-purifying characteristics of running waters naturally cleanse the water? After the Gold King Mine spill, one word best reflects the need to test and continue to test the Animas River and its downstream water bodies: *persistence*. Many compounds are resistant to environmental degradation through chemical, biological, and photolytic processes (remember Rachel Carson's *Silent Spring*). This persistence is why it is important for us to consider the long term. Although it would appear that there are no short-term environmental effects as a result of the Gold King Mine spill, we do not know what we do not know about the long-term impacts of this highly concentrated release of metals (see Table 8.1) into our environment. The simple truth is we will not know or understand the long-terms effects of this release into our environment for quite some time.

The longer term question and eventual answer come down to and are dependent on the behavior of the metals deposited in the sediments. One might think that because the metals (and other contaminants) are attached to bottom sediments that they are sequestered, so to speak, and therefore harmless. The problem is that there remains the potential for contaminated sediments to be stirred up and released during flooding, recreational activities, hyporheic exchange, bioturbation by organisms residing on or in sediments, or human disturbances related to dredging or construction activities. Moreover, the contaminants could become concentrated in fish that live in the rivers and other water bodies affected and feed on periphytons or other things that grow on the sediments. Contaminants in the sediments could seep into the groundwater, resulting in impacts to drinking and irrigation water. But that is not the end of it. If the contaminated sediments deposit on river or lake shores, they could potentially dry out and become wind-blown and airborne as dust. If this happens, they not only could contaminate the surrounding soil but could also become hazardous to anyone who breathes them in.

BOTTOM LINE ON YELLOW BOY

Only time will tell what the ultimate long-term effects are of the Gold King Mine spill and its formation of yellow boy in Cement Creek and the Animas and San Juan rivers. It can be difficult to fully imagine the potential damage to the Animas and

San Juan rivers, Lake Powell, and the Glen Canyon National Recreational Area that could result from the surrounding waters turning into a toxic stew. We can only hope that the USEPA will do all it can to prevent and quickly remediate further yellow boy occurrences and other such horrific environmental catastrophes. For the reader, it is important to keep in mind that sediment damage caused by the yellow boy described above is just one small example of sediment damage. Sediment damage can manifest itself physically, biologically, and chemically and have catastrophic effects on the environment, human life, and other forms of life. These physical, biological, and chemical damages to sediments and the repercussions of such are discussed in the chapters that follow.

REFERENCES AND RECOMMENDED READING

Chief, K., Artiola, J.F., Wilkinson, S.T., Beamer, P., and Maijer, R.M. (2015). *Understanding the Gold King Mine Spill*. Tucson: The University of Arizona, Superfund Research Program (http://www.superfund.pharmacy.arizona.edu/sites/default/files/u43/gold_king_mine_spill.pdf).

NPS. (2015). *Glen Canyon: Gold King Mine Spill Update*. Washington, DC: National Park Service (http://www.nps.gov/glca/learn/news/gold-king-mine-spill-update.htm).

USDA. (2008). Sediment properties. In: *National Engineering Handbook*, Section 3, Chapter 2. Washington, DC: U.S. Department of Agriculture.

USEPA. (2015a). *Frequent Questions Related to Gold King Mine Response*. Washington, DC: U.S. Environmental Protection Agency (https://www.epa.gov/goldkingmine/frequent-questions-related-gold-king-mine-response)

USEPA. (2015b). *Gold King Mine Incident: Preliminary Analytical Data Upper Animas River*. Washington, DC: U.S. Environmental Protection Agency (http://www2.epa.gov/sites/production/files/2015-08/documents/preliminarydata_08092015.pdf).

USEPA. (2016). *Emergency Response to August 2015 Release from Gold King Mine*. Washington, DC: U.S. Environmental Protection Agency (http://epa.gov/goldkingmine).

9 Physical Sediment Damage

Soil erosion is the most serious and prevalent disease of the land.

USDA (1954)

INTRODUCTION

Natural erosion causes about 30% of the total sediment in the United States, but human activities account for the remaining 70%. Physical sediment damage contributes to contaminated sediments in freshwater systems simply due to their presence. Earlier we stated that sediment is defined as solid material that is being transported or has been moved from its site of origin by air, water, gravity, or ice. We also defined erosion as the detachment and movement of rock or mineral materials by wind, moving water, ice, gravity, or other agents. When we discuss sediment damage of any type—physical, biological, or chemical—it is important to keep in mind the source of the material under discussion. In this book, and throughout the current discussion, for purposes of evaluation sediment and erosion damages are separated into two categories, as shown in Figure 9.1: (1) direct damages, and (2) indirect damages. Direct damages result in primary impairment of manmade properties, facilities, and utilities. Indirect damages are secondary damages related to or resulting from primary or direct damages. Computations of monetary damages are the responsibility of environmental economists. Ecology and environmental specialists are expected to supply the data on physical damages that environmental economists need for such calculations. In some instances, such as in cleanup costs, sediment damages can be estimated only in monetary terms, in which case geologists and environmental economists should jointly decide who should estimate such damages. Thorough knowledge of the use of information by environmental economists and close coordination with environmental economists are necessary to avoid collection of unnecessary data.

SEDIMENT DAMAGE

Damage to properties, facilities, and utilities as a result of sediment in transit or sediment deposition is considered to be sediment damage and/or contamination. Figure 9.2 illustrates and the following explanation describes the types of sediment damage.

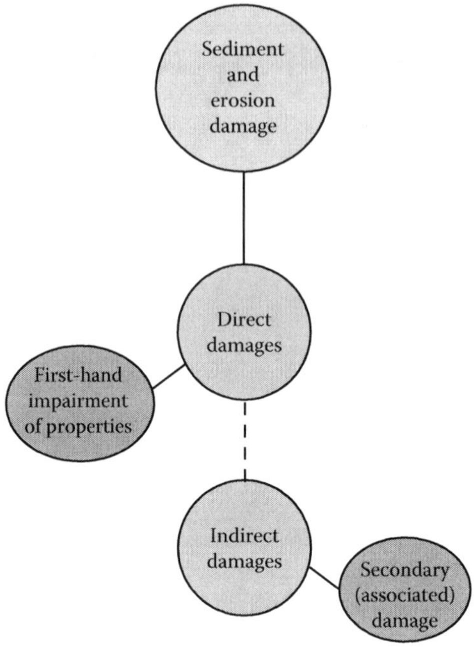

FIGURE 9.1 Direct and indirect damages resulting from sediment and erosion damages.

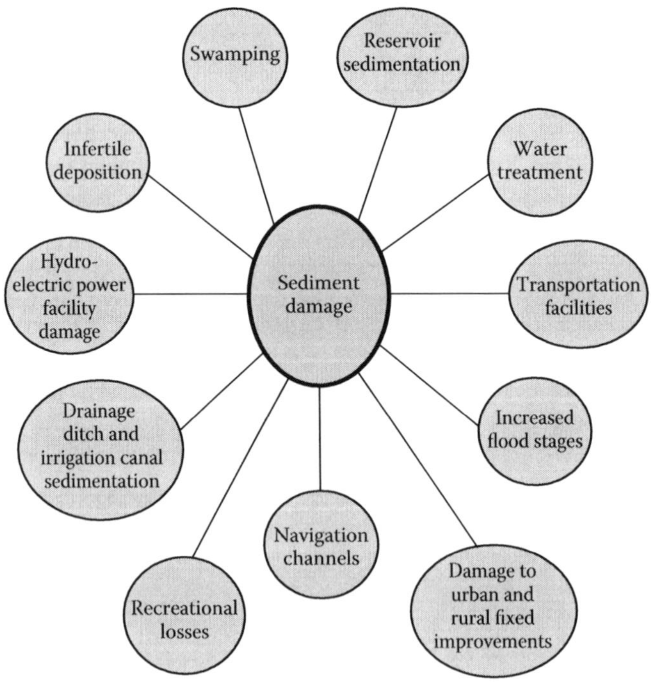

FIGURE 9.2 Types of sediment damage.

Infertile Deposition

Relatively infertile modern sediment may be deposited on floodplains or on colluvial soils. Modern sediment results from culturally accelerated erosion. It may take various forms, such as over-bank splays, fans, or vertical accretion deposits. Infertile sand or gravel deposits commonly cause this type of damage, but silts and clays derived from subsoil erosion, if low in nutrient elements, are also harmful in some areas. General principles of stream and valley sedimentation, criteria for recognition of modern valley deposits, and the relationship of stream and valley sedimentation of flood-control problems were outlined by Happ et al. (1940). The degree of damage by deposition of infertile materials depends on the type of material, the depth of deposit, and the rate of deposition, as well as on the productivity of the land in its original state. As an example, consider two areas of silt loam floodplain soils that have been damaged by sand. One has been damaged by a deposit of 16 inches of sand added gradually, at the rate of an inch each year for 16 years. It has been possible to mix the sand with the surface 8 inches (plow depth) of soil each year. Therefore, although the productive capacity of the soil has diminished to 50% of the original capacity, it is still in production. In contrast, the other area has received a deposit of only 8 inches of sand, all of which was deposited in a year. Under normal practices, there would not be much mixing of this sand with the old soil below. Unless special treatment is used, it might be taken out of production and for all practical purposes lost to cultivation.

Swamping

Swamping is any impairment of lateral or vertical drainage or floodplain soils due to sediment deposits. Swamping may be caused by the filling of stream channels with sediment, which raises the water table; by the formation of natural levees by modern sediment deposits, which prevents proper surface drainage; or by deposition of fine-grained sediment upon floodplain soils, which results in puddling or a reduction of permeability and prevention of internal drainage. Although swamping is a direct result of deposition, it is evaluated as a separate damage. Swamping often affects extensive areas of floodplains, and in its most serious form it will make formerly good cropland unfit for agricultural use.

Reservoir Sedimentation

Deposition of sediment in reservoirs results in a loss of the storage capacity required for water supply, power, recreational, irrigation, flood control, and other purposes. As a result, the services dependent upon such capacity are impaired. Damage to natural lakes is included in this category as well as artificial reservoirs.

Water Treatment

It is rather ironic that state-of-the-art water treatment processes include sand filtration as a unit process to ensure that the water is cleansed to the extent possible. Sand filtration is extremely important in ensuring the removal of *Giardia lamblia* and *Cryptosporidium* oocysts. Traditional, small-scale water treatment processes are usually good enough to ensure adequate and safe water treatment, but when *Giardia* or *Cryptosporidium* could be present traditional methods of water treatment are not adequate to ensure the removal of these disease-causing protozoans. So, what is the irony here? Simply, the irony is that water treatment systems spend considerable amounts of time and money to remove sand—which, if part of the process, is otherwise vital to filtration in the unit process—and other sediment from surface water or reservoir inlets because of blockage problems. It is a damage in this text because it is considered preventable.

Hydroelectric Power Facility Damage

After a power reservoir is completely filled with sediment, the power plant no longer has reserve, or carryover, storage and must depend on run-of-the-river flow entirely. Although this may greatly reduce the amount of power that can be generated, most power plants continue to operate under these circumstances on a reduced power output basis. At this stage, new sediment damage may occur. Coarse materials may move out of the silted-up reservoir, through the intake, and into the turbines. This causes excessive wear on the turbines, runners, and other equipment and necessitates more frequent overhaul and replacement.

Damage to Transportation Facilities

Sediment deposits damage highways and railways by collecting in ditches and culverts and on roadways and by filling and constricting channels beneath bridges. Roadway sediments become surface runoff and eventually make their way into surface water bodies.

Drainage Ditch and Irrigation Canal Sedimentation

Drainage ditches, irrigation canals, and floodways are usually vulnerable to sedimentation because of the low grades developed. As they become filled with sediment, and often with vegetative growth, they lose their capacity to transport water. This results in more frequent overflow of floodways and drainage ditches. These overflows may raise the water table adjacent to drainage ditches or impair the effectiveness of lateral outlets. Silting of irrigation canals reduces the amount of water that may be delivered to irrigated areas at critical times, resulting in a loss of crop production.

Damage to Navigation Channels

When navigation channels, pools, and harbors become shoaled because of sediment deposits, movement of vessels may be limited to high water periods or halted entirely.

Increased Flood Stages

When stream channels become clogged with sediment and floodplains are raised by sediment deposits, flood crests for the same discharges are constantly forced to higher elevations and floodwater damage increases. The damage may be caused by several conditions. If the floodplain is bounded by low terraces and the channel and floodplain are aggrading at approximately the same rate, increasing flood damage will take place on the terraces, but not on the floodplain, except in the case of buildings or other fixed installations. If the channel is aggrading at a faster rate than the floodplain, then there will be increasing floodwater damage on both the floodplain and terraces. If the channel is enlarging at a rate equal to the rate of floodplain aggradation, no increase in floodwater damage will take place on either the floodplain or terraces. Figure 9.3 illustrates these various conditions.

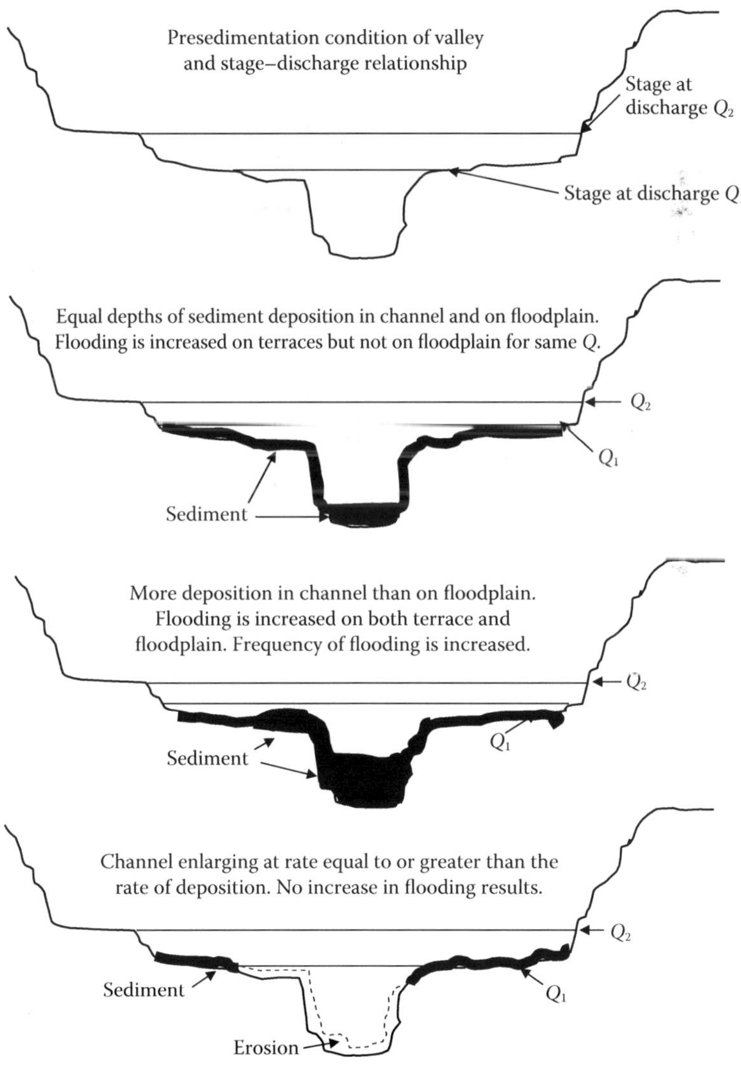

FIGURE 9.3 Interrelation between sedimentation and flood stages. (From USDA, *Geologic Investigations for Watershed Planning*, Technical Release No. 17, U.S. Department of Agriculture, Washington, DC, 1996.)

Damage to Urban and Rural Fixed Improvements

After most floods, deposits of sediment are found on streets and in homes, factories, sewers, wells, and other places where they cause damage due to the cost of removing this sediment or cleaning and replacing equipment and materials.

Recreational Losses

Sediment may cause the impairment of recreational values, such as damage to fish, wildlife, and recreational facilities (e.g., beaches, bathing facilities).

REFERENCES AND RECOMMENDED READING

Andrews, Jr., J.S. (1992). The cleanup of Kuwait: how long will it take. In: *Hydrocarbon Contaminated Soils*, Vol. II (Kostecki, P.T. et al., Eds.), pp. 43–48. Chelsea, MI: Lewis Publishers.

Happ, S.C., Rittenhouse, F., and Dobson, G.D. (1940). *Some Principles of Accelerated Stream and Valley Sedimentation*, USDA Technical Bull. No. 695. Washington, DC: U.S. Department of Agriculture.

PEDCO. (1979). *PEDCO Analysis of Eckhardt Committee Survey for Chemical Manufacturer's Association*. Washington, DC: PEDCO Environmental, Inc.

Testa, S.M. (1997). *The Reuse and Recycling of Contaminated Soil*. Boca Raton, FL: Lewis Publishers.

Tucker, R.K. (1989). Problems dealing with petroleum contaminated soils: a New Jersey perspective. In: *Petroleum Contaminated Soils*, Vol. I (Kostecki, P.T. and Calabrese, E.J., Eds.), pp. 37–53. Chelsea, MI: Lewis Publishers.

USDA. (1954). *A Manual on Conservation of Soil and Water: Handbook for Professional Agricultural Workers*, Agriculture Handbook No. 61. Washington, DC: U.S. Department of Agriculture, Soil Conservation Service.

USDA. (1966). Ge*ologic Investigations for Watershed Planning*, Technical Release No. 17. Washington, DC: U.S. Department of Agriculture.

USDA. (2008). Sediment properties. In: *National Engineering Handbook*, Section 3, Chapter 2. Washington, DC: U.S. Department of Agriculture.

10 Biologically Contaminated Sediments

"In other words, Bob, don't let the folks down there know that you are looking for sites for hog facilities or they will prevaricate and try to take us to the cleaners, they will carry on to various editors, every kind of meanness and so forth, as they have been brainwashed by the Sierra Club to think that hog facilities are bad, even the folks who love baby back ribs, even the ones hunting jobs. But I will tell you something. The panhandle region is perfect for hog operations—plenty of room, low population, nice long dry seasons, good water. There is no reason why the Texas panhandle can't produce seventy-five percent of the world's pork." … "Headaches, sore throat, dizziness. Them hogs are pumped full a antibiotics and growth hormones. Eat that pork and it gets into *you*. Bacteria and viruses adapt to the antibiotics so the day is comin' when if we get sick the antibiotics can't help."

Proulx (2002)

HERCULES' FIFTH LABOR

In mythical accounts, the great hero Hercules was assigned 12 labors. The fifth labor involved cleaning out the stables of King Augeas. For 30 years, the stables housed the single greatest number of animals (cows, bulls, goats, sheep, and horses) in the country and had never been cleaned. Every evening the cowherds, goatherds, and shepherds drove thousands of animals to the stables. Hercules told the king that he could clean out the stables in one day. Augeas couldn't believe his ears but promised to give Hercules a tenth of his cattle if he could accomplish the feat. Hercules went to work. He first tore open a gigantic hole in the wall of the cattle yard where the stables were, then he made another opening in the wall on the opposite side of the yard. Next, he dug wide trenches to reroute the Alpheus and Peneus rivers to wash out the filth from the stables. As promised, Hercules cleaned out the stables in one day. He did it by washing the huge mass of dung into the rivers he had rerouted. Hercules, however, never did get his reward because the king refused to pay him.

The mythical hero Hercules is often credited with stating that "the solution to pollution is dilution." Dilution is, indeed, a powerful natural tool that helps to disperse smoke from factories and homes into wide open spaces in the atmosphere around us. In a running water body, pollution will generally run with the stream as it flows to the sea and dissipate along the way. Remember the earlier discussion on still waters? What we see on the surface of the water does not necessarily reflect conditions in the depths. As it turns out, dilution is simply an *apparent* solution to pollution, the key word being *apparent*. Pollution does not just disappear by waving a magic wand over a contaminated surface body of water or in the air above our heads. When Hercules dumped the animal dung into the two diverted rivers, it is likely that Nature's self-purifying process purified the river water at about the 30-mile downstream marker, depending on the size and velocity of stream flow; it would have required this distance because of the huge organic load (biochemical oxygen demand, or BOD) that he dumped. The problem is that the bottom and shoreline sediments are sinks for contamination. In the lateral reaches of rivers, at sandbar areas, along the riparian corridor, and in slow-moving river pool areas, animal contaminants settle and deposit on the sediments below. Although many pure organics volatize (evaporate) with time, it can be a much slower process than self-purification by moving water.

Contaminants that reach and enter a stream or lake or other surface body are, of course, what we are concerned with in this book. More specifically, the focus of our concern is contaminants that end up in or entrapped on sediments within freshwater surface waters. Sediment damage can be classified as physical, chemical, or biological. In this chapter, we are concerned with the biological damage that may occur due to biological contaminants intermixing with bottom sediments.

BIOLOGICAL PARTICLES

Before discussing biological particles it is important to explain what a particle is, at least in the context of this book. At the submicron end of the particle size spectrum, the distinction between what is and what is not a particle becomes more unclear as the scale of observation approaches the limits between "particulate" and "dissolved." Generally speaking, 0.45 μm is the operationally defined demarcation between particulate and dissolved matter. Even so, colloids are treated as discrete particles (Edzwald et al., 1974) even when this dimension falls below the 0.45-μm mark.

Biological Organisms and the Particle Interface

Many biological organisms present in the water column influence the movement and transformation of dissolved and sorbed organic contaminants. They include bacteria, protozoans, phytoplankton, zooplankton, and other higher trophic species. Although these different biological species are not strictly considered suspended sediment, they are included here because they are suspended in the water column, participate in the transformation and transport of organic contaminants, and eventually (after they die, begin to decay, and settle) become part of what is collectively considered surficial sediment. In addition, as dead aquatic organisms, they combine with terrestrial organic decay products to form detritus, and thus they become an integral component of the material cycle of elements between the water column and surficial sediment (Sigleo et al., 1980, 1982).

The ways in which biological organisms transform organic contaminants are diverse. By exchanging water across their cellular membranes, bacteria, protozoans, plankton, and lower vertebrates can bioconcentrate dissolved contaminants within themselves (Lush and Hynes, 1973; Steen and Karickhoff, 1981). Some filter feeders, such as zooplankton, ingest lower trophic organisms and colloidal size particles and either bioaccumulate the contaminants within their bodies or excrete them with fecal pellets (Sheldon et al., 1973). The fecal pellets are denser and settle faster than the ingested colloids or organisms (Meyers et al., 1984). Consequently, as has been demonstrated for the oceanic distribution of polychlorinated biphenyls (PCBs), the

DID YOU KNOW?

Benthic macroinvertebrates are aquatic organisms without backbones that spend at least a part of their life cycle on the stream bottom. Examples include aquatic insects—such as stoneflies, mayflies, caddisflies, midges, and beetles—as well as crayfish, worms, clams, and snails. Most hatch from eggs and mature from larvae to adults. The majority of the insects spend their larval phase on the river bottom and, after a few weeks to several years, emerge as winged adults. The aquatic beetles, true bugs, and other groups remain in the water as adults. Macroinvertebrates typically collected from the stream substrate are either aquatic larvae or adults.

pelletization process can provide a significant contribution to the vertical transport of organic compounds through the water column (Elder and Fowler, 1977). Finally, some biological organisms that are resident in the surficial sediment layer (e.g., benthic macroinvertebrates) are capable of ingesting and resuspending (bioturbation) sediment particles along with their sorbed contaminants (Berner, 1980).

ANIMAL WASTE CONTAMINANTS

The tale of Hercules and his fifth labor related earlier was deliberately injected by the author because it is an excellent way to introduce an important source of biological contaminants in freshwater surface bodies: animal waste and its associated contamination effects. The contaminants most commonly associated with animal waste include nutrients (including ammonia), organic matter, solids, pathogens, and odorous compounds. Animal waste can also be a source of salts and various trace elements (including metals), as well as pesticides, antibiotics, and hormones. These pollutants can be released into the environment through discharge or runoff if manure and associated wastewater are not properly handled, treated, and managed.

Pollutants in animal waste can enter the environment through a number of pathways. These include surface runoff and erosion, overflows from lagoons, spills and other dry-weather discharges, leaching into soils and groundwater, and volatilization (evaporation) of compounds (e.g., ammonia) and subsequent redeposition on the landscape. Pollutants from animal waste can be released from an operation's animal confinement area, treatment and storage lagoons, and manure stockpiles and from cropland where manure is often land-applied. This chapter discusses the

DID YOU KNOW?

Acanthamoeba, a genus of free-living amoebae that are ubiquitous in the environment, particularly in soil, sediment in water habitats, and in drinking water, is just one example of the type of biological contaminants that can call freshwater sediment home. *Acanthamoeba* can infect a variety of mammals, including humans, and can produce severe, even fatal disease.

contaminants and pollutants associated with livestock and poultry operations, of which concentrated animal feeding operations (CAFOs) are a subset; the pathways by which the pollutants reach surface water; and the impacts of these pollutants on the environment and human health. But first, before we take a closer look at the actual contaminants and their potential impact on human and environmental health, it is important to set the CAFO stage.

Setting the Stage*

The debate over the future of agriculture and agricultural policy is not new. It can be found in the history of such agrarian protest movements as the Grange (1870s), populism (1890s), the Farm Holiday Movement (1930s), the National Farmers Organization (1950s), and the immigration and environment groups of the last two decades. There have always been farmers and rural residents who lauded progress, others who have lamented the loss of the agrarian nature of America which allowed more to provide for themselves, and many more who have been ambivalent (Schwab, 1998). A debate that is gaining steam in agriculture concerns concentrated animal feeding operations (CAFOs). The debate is fueled by the serious impacts of CAFOs on the environment and the social fabric of rural living. CAFOs are farming operations where large numbers (often in the thousands of animals) of livestock or poultry are housed inside buildings or in confined feedlots. The U.S. Environmental Protection Agency (USEPA) defines a CAFO or industrial operation as a concentrated animal feeding operation where animals are confined for more than 45 days per year. To classify as a CAFO, such an operation must also have over 1000 animal units—a standardized number based on the amount of waste each species produces, basically 1000 pounds of animal weight. Under this system, dairy cattle count as 1.4 animal units each. A CAFO could house more than 750 mature dairy cattle (milking and or dry cows) or 500 horses and discharge into navigable water through a manmade ditch or a similarly manmade device. CAFO classification sets numbers for various species per 1000 animal units:

- 2500 hogs
- 700 dairy cattle
- 1000 beef cattle
- 125,000 broiler chickens
- 82,000 layer hens

It can be difficult to grasp the scope of the problem. As an example, how do the amounts of CAFO-generated animal manure compare to human waste production? Let's take a look. Here is a small-scale number: one hog per day excretes 2.5 times more waste than an adult human—nearly 3 gallons (Cantrell et al., 2004). Here is a medium-scale number: 10,000 hogs produce as much waste in a single day as a town of 25,000 people (Sierra Club, 2004), but the difference is that the town has a treatment plant. Here is a bigger picture (Ohio EPA, 2008; USEPA, 2003):

* Adapted from Spellman, F.R., *Environmental Management of Concentrated Animal Feeding Operations (CAFOs)*, CRC Press, Boca Raton, FL, 2007.

- The USEPA estimates that human uses generate about *150 million tons* (wet weight) of human sanitary waste annually in the United States, assuming a U.S. population of 285 million and an average waste generation of about 0.518 tons per person per year.
- The U.S. Department of Agriculture (USDA) estimates that operations that confine livestock and poultry animals generate about *500 million tons* of excreted manure annually.
- The USEPA estimates that there are over 257,000 animal feeding operations (AFOs) in the United States today, of which 15,500 meet the more narrow criteria for CAFOs, producing *575 billion pounds* (or over 280 million tons) of manure annually in the United States today.

Here's the bottom line: By these estimates, all confined animals generate well over *three times* more raw waste than is generated by humans in the United States. Much of this waste undergoes no—or very little—waste treatment. Waste handling for any CAFO is a major business concern and expense. Unless regulation and legislation support sound environmental practices for these operations, CAFO owners have little incentive to improve their waste handling practices.

Water Supply, Use, and Wastewater Treatment

Wastewater treatment is considered a water use because it is so interconnected with the other uses of water. Much of the water used by homes, industries, and businesses is treated prior to its release back into the environment, where it becomes part of the endless water cycle. The scope of treatment processes involved in wastewater treatment is unknown to most people, who generally think of it solely in terms of "sewage treatment." Nature has an amazing ability to cope with small amounts of water wastes and pollution through its self-purification process. However, Nature would be overwhelmed if the billions of gallons of wastewater and sewage produced every day were not treated before being released back into the environment. Wastewater treatment plants reduce pollutants in wastewater to a level Nature can handle.

Used water is treated for many reasons. Principle among them is the matter of caring for our environment and our own public health. For example, we treat used water because clean water is critical to our water supply, as well as to plants and animals that live in water. Human health, environmental health, and many commercial interests (e.g., fishing industry, sport fishing enthusiasts) depend on biota that can only survive in clean, healthy water systems—and we, of course, as today's responsible adults, hold our water system in its entirety for future generations.

Our rivers and ocean waters teem with life that depends on healthy shorelines, beaches, and marshes, which provide critical habitats for hundreds of species of aquatic and semiaquatic life. Migratory water birds use the areas for feeding and resting. Species of both flora and fauna are adapted to live in the zones that lie between or on the verge of water. These areas are extremely vulnerable to certain types of pollution. Water is one of our most used playgrounds. The scenic and recreational values of our waters are serious factors for many people in deciding where to live. Tourists are drawn to water activities: swimming, fishing, boating, hunting, and picnicking. Improper treatment of wastes, of course, impacts these activities as well.

In short, if used water is not properly cleaned, it carries waterborne disease. Because we live, work, and play so close to water, harmful pathogenic organisms must be removed or made harmless to make water safe, regardless of the aesthetic factors involved in untreated wastes in the water system. So we treat our wastewater or used water before releasing it to the environment. The major aim of used water treatment is to remove as much of the suspended solids and other contaminants as possible before the remaining water, called *effluent*, is discharged to the environment.

Treatment requires several interrelated steps. *Primary treatment* removes about 60% of suspended solids from used water and involves aerating (mixing up) the used water to put oxygen back in. This step is essential because as solid material biodegrades it uses up oxygen needed by the plants and animals living in the water. *Secondary treatment* removes more than 90% of suspended solids. In some cases, *tertiary treatment* takes the waste removal further and addresses the removal of specific waste elements (e.g., nutrients such as nitrogen and phosphorus) not removed by other means. After treatment, used water is returned to the water cycle as treated effluent. Whether from consumer or industrial sources, the returned treated water should be returned at least as clean as—if not cleaner than—the receiving body of water.

What does wastewater contain and where does it come from? Used water (wastewater) carries substances that include human waste, food scraps, oils, soaps, and chemicals. Use by consumers in their homes includes water from showers, sinks, toilets, bathtubs, washing machines, and dishwashers. Businesses and industries also contribute their share of used water that must be cleaned and thus recycled. The treatment of both household wastewaters and industrial wastewaters is regulated and monitored. These are point source pollutants; the sources are identifiable (end of the pipe) and limited in scope—we know what is in there, where the sources are, and, in a general way, how much will be produced (the quantity can be predicted to fall within the capacity of the system's treatment capability).

Other wastes that enter our water supply are more difficult to define, quantify, and control. Stormwater or storm runoff, for example, is also a major contributor to the endless wastewater stream. The average person might assume that rain that runs off their homes, into their yards, and then down the streets during a storm event is fairly clean. It is not. Harmful substances wash off roads, fields, lawns, parking lots, and rooftops and can harm our still waters (e.g., lakes, ponds) and running waters (e.g., rivers, streams). Stormwater is incident related. If it does not rain, then stormwater obviously will not enter the system. Stormwater provides nonpoint source pollutants to the wastewater stream. We know in a general way what stormwaters will carry, and we put systems in place to channel and control the stream, but the use of complex programs and modeling is essential to evaluate how to handle stormwater to avoid serious problems with treatment and control, specifically to handle many different levels of force in storm incidents of differing durations.

Agricultural sources also contribute several problems to wastewater treatment that have been difficult to evaluate, identify, and control. Historically, especially before the chemical industry provided manufactured fertilizers and pesticides for crop farm production and growth hormones and antibiotics for livestock production, the techniques individual farmers used were reasonably environmentally friendly—because the size of the farm dictated the limits of production: Any individual farmer has limits, including the farmer's physical and financial ability to work, as well as the effective natural limits on how many bushels of grain or animals per acre the land itself can support, because overloading those capacities generally provides negative results. As modern practices have evolved, our water systems have suffered. Fertilizers, pesticides, hormones, and antibiotics send nonpoint source pollution directly into local water systems, creating downstream problems in the water system. A number of both crop and livestock farming practices contribute to soil erosion, which, of course, affects both soil and water quality. However, changes in farming practices over the last several decades (perhaps best characterized as switching from the small farmer to agribusiness, or factory farms) and the increasing demand for inexpensive meat products have created a new set of problems to address. Factory farming of crops presents its own set of issues, but this discussion addresses only the problems created by the factory farming of livestock, which creates agricultural point source pollution of extreme scope.

Animal Feeding Operations and Animal Waste Treatment

The fact is that other animal waste problems and solutions are not new; for example, field spreading of human and animal wastes is accomplished naturally under nomadic and pasture social systems. Intentional manure conservation and reuse were practiced by early Chinese cultures. In Iceland, slotted floors that allowed waste material to drop below the floor surface date back at least 200 years. From at least the 19th century, dairy operations generating wastes from scores of animals were contained in one building. Huge poultry centers with wastes concentrated in a small area have been around for decades, as have some very large swine and beef producers. Figuratively speaking, and in general, animal manure deposited by animals managed by standard grazing livestock methods does not create serious environmental problems, especially if the farmer limits herd size to numbers the acreage can support without environmental damage, restricts livestock access to stream beds, and applies practices that include soil erosion prevention methods—greenbelts for waterways and shoreline planting. Accidentally stepping into such deposits is an occupational hazard, of course.

Small-farm animal manure waste is not the problem we are addressing here, though. It is important to understand that the manure deposited by a large herd of animals that is not assimilated through the soil surface and is instead carried off by storm runoff into local streams or other water bodies is a problem. Agribusiness and large-scale factory farming practices have created a different farm category, the livestock version of factory crop farming: concentrated animal feeding operations (CAFOs), which produce massive quantities of manure. In the 1920s, no one was capable of spilling millions of gallons of manure into a local stream in a single event. Such an event is possible today, though, because of the piling up of too much manure in one place (Ikerd, 1998). Simply, such piling up is the result of greater concentration and reduced diversity in today's farm operations.

Agribusinesses do not use traditional pastures and feeding practices. Typically, the manure is removed from the livestock buildings or feedlots and stored—in stockpiles or lagoon/pond systems—until it can be spread on farm fields, sold to other farmers as fertilizer, or composted. When properly designed, constructed, and managed, CAFO-produced manure is an agronomically important and environmentally safe source of nutrients and organic matter necessary for the production of food, fiber, and good soil health. Experience has demonstrated that when properly applied to land, at proper levels, manure will not cause water quality problems. When properly stored or deposited in holding lagoons or ponds, properly conveyed to the disposal outlet, and properly applied to the appropriate end use, potential CAFO waste environmental problems can be mitigated.

As mentioned earlier, CAFOs inherently are potential sources of contaminants for the three environmental mediums of air, water, and soil. Let's take a closer look at manure handling and the storage practices recommended by the USEPA and USDA (USDA/USEPA, 1998) that should be employed to prevent water pollution from CAFOs. In addition to water pollution prevention, it should be noted that manure and wastewater handling, storage, and subsequent application or treatment practices should also consider odor issues and other environmental and public health problems.

- *Divert clean water*—Siting and management practices should divert clean water from contact with feed lots and holding pens, animal manure, or manure storage systems. Clean water can include rainfall falling on roofs of facilities, runoff from adjacent lands, or other sources.
- *Prevent leakage*—Construction and maintenance of buildings, collection systems, conveyance systems, and permanent and temporary storage facilities should prevent leakage of organic matter, nutrients, and pathogens to ground or surface water.
- *Provide adequate storage*—Liquid manure storage systems should safely store the quantity and contents of animal manure and wastewater produced, contaminated runoff from the facility, and rainfall. Dry manure, such as that produced in certain poultry and beef operations, should be stored in production buildings or storage facilities or otherwise stored in such a way so as to prevent polluted runoff. The location of manure storage systems should consider proximity to water bodies, floodplains, and other environmentally sensitive areas.
- *Manure treatments*—Manure should be handled and treated to reduce the loss of nutrients to the atmosphere during storage, to make the material a more stable fertilizer when land-applied, or to reduce pathogens, vector attraction, and odors, as appropriate.
- *Management of dead animals*—Dead animals should be disposed of in a way that does not adversely affect ground or surface water or create public health concerns. Composting, rendering, and other practices are common methods used to dispose of dead animals.

Animal Waste Treatment and Lagoons

> Advanced technologies are being developed for the biological, physical, and chemical treatment of manure and wastewaters. Some of these greatly reduce constituents in the treated solids and liquids that must be managed on the farm. Byproduct recovery processes are being developed that transform waste into value-added products that can be marketed off the farm.
>
> **Sutton and Humenik (2003)**

Primarily because it is an economical means of treating highly concentrated wastes from confined livestock operations, the most widespread and common treatment technique for managing animal waste is the use of lagoons. In the late 1960s, considerable attention was paid to the impact of lagoons on surface water quality; since the 1970s, that attention has shifted to the potential impacts on groundwater quality. Unfortunately, these lagoons are prone to leaks and breakage. Groundwater has been contaminated with bacteria from them. The lagoons can also be overrun by floods that push the wastes into streams, lakes, and oceans. North Carolina, with its concentration of factory farms, has been the site of massive water contamination due to its waste lagoons. The storage lagoons for factory farms are often stinking manure lakes the size of several football fields, containing millions of gallons of liquefied manure. A single animal factory can generate the waste equivalent of a small town.

In the past 30 years, several studies on the effectiveness of factory farm lagoons, specifically on lagoon liners, in preventing environmental damage have been conducted. Consider the following review of studies on effective lagoon construction vs. defective construction. Sewell et al. (1975) studied anaerobic dairy lagoons and found that the lagoon bottom sealed within 2 months of start up; few or no pollutants were found in the groundwater after that time. Ritter et al. (1984) studied a two-stage anaerobic swine lagoon for 4 years and determined that the contaminant concentration increased in wells 50 m from the lagoon the first year and then steadily decreased afterwards. Their data led them to speculate that biological sealing takes place over a period of time depending on the loading rate to a lagoon. Collins et al. (1975) studied three swine lagoons, each within a high water table area. They found that there was no significant effect on groundwater beyond 3 m from the lagoon edge. Miller et al. (1985) studied the performance of beef lagoons in sandy soil and found that the lagoons were effectively sealed to infiltration within 12 weeks of the addition of manure. Humenik et al. (1980) summarized research conducted by others on the subject on lagoon sealing and concluded that the studies indicated that lagoon sealing may be expected within about 6 months, after which the area of seepage impact becomes restricted to approximately 10 m.

On the other hand, Hegg et al. (1978, 1981) collected data from a dairy lagoon and from newly established swine lagoons and found that some of the monitoring wells became contaminated but others did not. This led them to conclude that seepage does not occur uniformly over the entire wetted perimeter of the lagoon, but at specific unpredictable sites where sealing has not taken place. Similarly, Ritter et al. (1980) monitored an anaerobic two-stage swine lagoon for 2 years and found that one of the wells showed contamination that indicated localized seepage, while the other monitoring well indicated that the lagoon system had minimal impact on groundwater quality and that sealing had gradually taken place.

In many states, notwithstanding the USEPA and USDA's manure handling, storage, and treatment recommendations, lawsuits against CAFOs for unsound environmental practices demonstrate that CAFO operations are still creating problems. In short, regulations and legislation have fallen behind CAFO creation and operation, enforcement of existing regulations is spotty, and problems associated with CAFOs are still being identified—although you can be sure that those who neighbor CAFOs can identify some big issues, both environmental and social.

ANIMAL WASTE POLLUTANTS OF CONCERN

The primary pollutants associated with animal waste are nutrients (particularly nitrogen and phosphorus), ammonia, pathogens, and organic matter. Animal waste is also a source of salts and trace elements and, to a lesser extent, antibiotics, pesticides, and hormones. Each of these types of CAFO pollutants is discussed in the sections that follow.* The actual composition of manure depends on the animal species, size, maturity, and health, as well as on the composition (e.g., protein content) of animal feed (Phillips et al., 1992). After waste has been excreted, it may be altered further by the bedding and waste feed and may be diluted with water (Loehr, 1972; USDA, 1992).

> ***Note:*** Ammonia is also a nutrient but is listed separately here because it exhibits additional environmental effects, such as aquatic toxicity and direct dissolved oxygen demand.

NUTRIENTS

The three primary nutrients in manure are nitrogen, phosphorus, and potassium. Much of the past research on animal manure has focused on these constituents, given their importance as cropland fertilizers. The following discussions provide more detail on nitrogen and phosphorus characteristics and concentrations in manure. Scientific literature and policy statements commonly cite these two nutrients as key sources of water quality impairments. It was estimated that, in the central United States in 1995, 37% of all nitrogen and 65% of all phosphorus inputs to watersheds came from manure (USFWS, 2000). Actual or anticipated levels of potassium in groundwater and surface water are unlikely to pose hazards to human health or aquatic life (Wetzel, 1983). Potassium does contribute to salinity, however, and applications of high-salinity manure are likely to decrease the fertility of the soil.

* The estimates of manure pollutant production are based on average values reported in the scientific literature and compiled by the American Society of Agricultural Engineers (ASAE, 1999), USDA/NRCS (1996), and USDA/ARS (1998).

TABLE 10.1
Primary Nutrients in Both Livestock and Human Manures

	Animal Group						
	Swine	**Layer**	**Broiler**	**Turkey**	**Beef**	**Dairy**	**Human**
	Mass of Animal (lb)						
	135	4	2	15	800	1400	150
	Pounds per 1000 Pounds Live Animal Weight per Day						
Nitrogen (total Kjeldahl)	0.52	0.84	1.10	0.62	0.34	0.45	0.20
Phosphorus (total)	0.18	0.30	0.30	0.23	0.092	0.094	0.02
Orthophosphorus	0.12	0.09	n/a	n/a	0.03	0.061	n/a
Potassium	0.29	0.30	0.40	0.24	0.21	0.29	0.07

Note: Livestock data are "as excreted" and are from ASAE (1999); human waste data are "as excreted" and are from USDA/NRCS (1996). Values are rounded to two significant figures; n/a = not available.

Table 10.1 presents the amounts of total Kjeldahl nitrogen, total phosphorus, orthophosphorus, and potassium generated per 1000 pounds live animal weight per day (ASAE, 1995). For comparison, Table 10.1 also presents similar information for humans. The figures illustrate that per-pound nutrient output varies among animal types and is much higher for animals than humans.

> ***Note:*** Total Kjeldahl nitrogen is the sum of organic nitrogen in the tri-negative oxidation state and ammonia.

Nitrogen

Nitrogen (N) is an essential nutrient required by all living organisms. It is ubiquitous in the environment, accounting for 78% of the atmosphere as elemental nitrogen (N_2). This form of nitrogen is inert and does not impact environmental quality. It is also not bioavailable to most organisms and therefore has no fertilizer value. Nitrogen also forms other compounds that are bioavailable, mobile, and potentially harmful to the environment. The nitrogen cycle (Figure 10.1) shows the various forms of nitrogen and the processes by which they are transformed and lost to the environment.

> ***Note:*** Nitrogen occurs in the environment in gaseous forms (elemental nitrogen, N_2; nitrogen oxide compounds, N_2O and NO_x; and ammonia, NH_3); water-soluble forms (ammonia, NH_3; ammonium, NH_4^+; nitrite, NO_2^-; and nitrate, NO_3^-); and as organic nitrogen, bound up in the proteins of living organisms and decaying organic matter (Brady, 1990). The transformation of the different forms of nitrogen among land, water, air, and living organisms is shown in Figure 10.1.

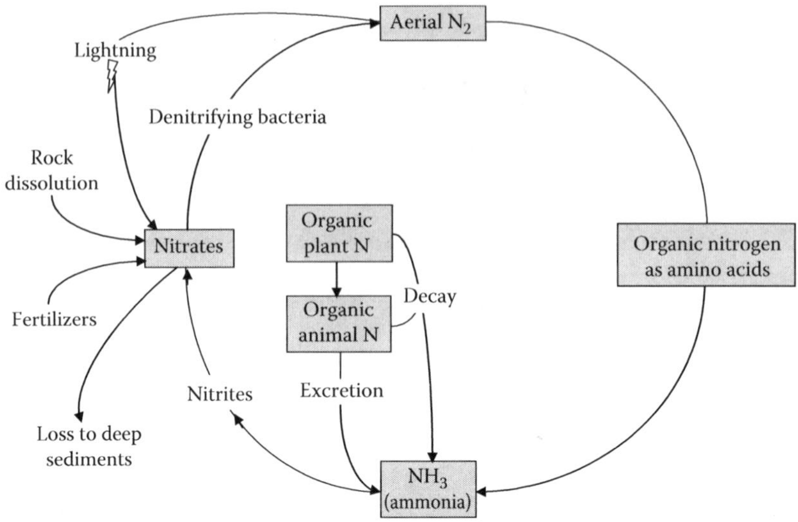

FIGURE 10.1 Nitrogen cycle.

Manure nitrogen is primarily in its organic form (organic nitrogen and ammonia nitrogen compounds) (NCAES, 1982). Organic nitrogen in the solid content of animal feces is mostly in the form of complex molecules associated with digested food, and organic nitrogen in urine is mostly in the form of urea ($(NH_2)_2CO$) (USDA, 1992). In organic form, nitrogen is unavailable to plants; however, via microbial processes, organic nitrogen is transformed to ammonium (NH_4^+) and nitrate (NO_3^-) forms, which are bioavailable and therefore have fertilizer value. These forms can also produce negative environmental impacts when they are transported to the environment.

Note: In an anaerobic lagoon, the nitrogen organic fraction is about 20 to 30% of total nitrogen (USDA, 1992).

Under aerobic conditions, ammonia can oxidize to nitrites and nitrates. Subsequent anaerobic conditions can result in denitrification (transformation of nitrates/nitrites to gaseous nitrogen forms). Overall, depending on the animal type and specific waste management practices, between 30 and 90% of nitrogen excreted in manure can be lost before its use as a fertilizer (Vanderholm, 1975).

Phosphorus

Phosphorus exists in solid and dissolved phases, in both organic and inorganic forms. Over 70% of the phosphorus in animal manure is in the organic form. Like nitrogen, the various forms of phosphorus are subject to transformation (Figure 10.2). Dissolved phosphorus in the soil environment consists of orthophosphates (PO_4^{3-}, HPO_4^{2-}, or $H_2PO_4^-$), inorganic polyphosphates, and organic phosphorus. Solid phosphorus exists as organic phosphorus in dead and living materials; mineral phosphorus in soil components; adsorbed phosphorus on soil particles; and precipitate phosphorus, which forms upon reaction with soil cations such as iron, aluminum,

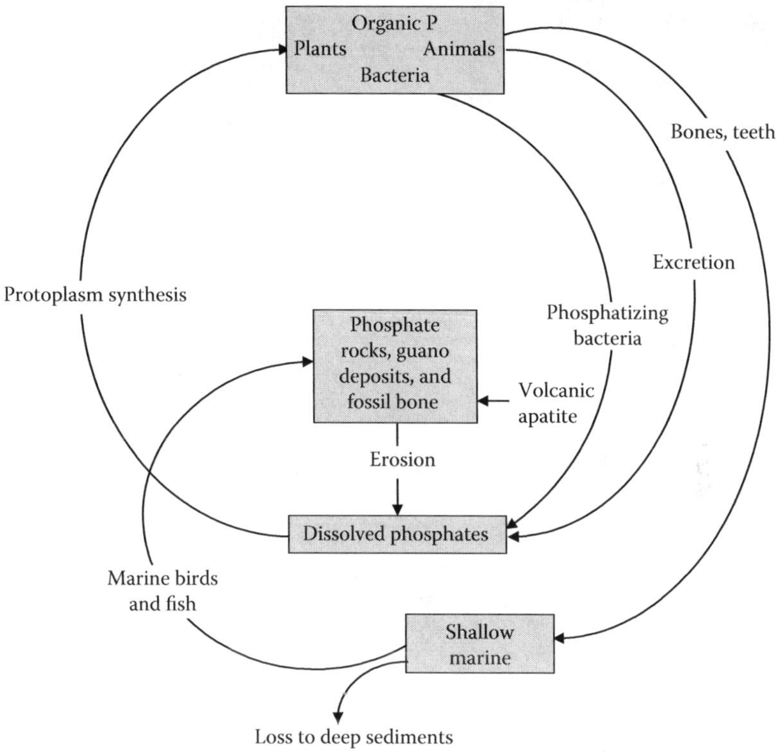

FIGURE 10.2 Phosphorus cycle.

and calcium (Poultry Water Quality Consortium, 1998). Orthophosphate species, both soluble and attached, are the predominant forms of phosphorus in the natural environment (Bodek et al., 1988). Soluble (available or dissolved) phosphorus generally accounts for a small percentage of total soil phosphorus; however, soils saturated with phosphorus can have significant occurrences of phosphorus leaching. Soluble phosphorus is the form used by plants and is subject to leaching. About 73% of the phosphorus in most types of fresh livestock waste is in the organic form (USDA, 1992). As animal waste ages, the organic phosphorus mineralizes to inorganic phosphate compounds and becomes available to plants.

Note: Inorganic phosphorus tends to adhere to soils and is less likely to leach into groundwater.

Note: Soil test data in the United States confirm that many soils in areas dominated by animal-based agriculture have elevated levels of phosphorus.

Ammonia

Ammonia-nitrogen includes the ionized form (ammonium, NH_4^+) and the un-ionized form (ammonia, NH_3). Ammonium is produced when microorganisms break down organic nitrogen products such as urea and proteins in manure. This decomposition

DID YOU KNOW?

Fish kills due to ammonia toxicity are a potential consequence of the direct discharge of animal wastes to surface waters. This is effectively illustrated by a May 1997 incident in Wabasha County, Minnesota, when ammonia in a dairy manure release killed 16,500 minnows and white suckers (CWAA, 1998).

can occur in either aerobic or anaerobic environments. In solution, ammonium enters into an equilibrium reaction with ammonia, as shown in the following equation:

$$NH_4^+ \Leftrightarrow NH_3 + H^+$$

As the equation indicates, higher pH levels (lower H^+ concentrations) favor the formation of ammonia, while lower pH levels (higher H^+ concentrations) favor the formation of ammonium. Both forms are toxic to aquatic life, although the un-ionized form (ammonia) is much more toxic. Up to 50% or more of the nitrogen in fresh manure may be in the ammonia form or converted to ammonia relatively quickly once manure is excreted (Vanderholm, 1975). Ammonia is very volatile, and much of it is emitted as a gas, although it may also be absorbed by or react with other substances.

Higher pH levels (lower H^+ concentrations) favor the formation of ammonia, and lower pH levels (higher H^+ concentrations) favor the formation of ammonium. The ammonia form is subject to volatilization. The ammonia content of fresh manure varies among animal species and changes as the manure ages. Ammonia content may increase as organic matter breaks down; it may decrease when volatilization occurs or when nitrate oxidizes to nitrite under aerobic conditions.

Pathogens

Pathogens are disease-causing organisms including bacteria, viruses, protozoa, fungi, and algae. Both manure and animal carcasses can be sources of pathogens in the environment (Juranek, 1995). Livestock manure may contain bacteria, viruses, fungi, helminthes, protozoa, and parasites, many of which are pathogenic (Jackson et al., 1987; USDA/ARS, 1998). For example, researchers have isolated pathogenic bacteria and viruses from feedlot wastes (Derbyshire and Brown, 1978; Derbyshire et al., 1966; Hrubant, 1973). In addition, USFWS (2000) has shown that fields receiving animal waste applications have elevated levels of fecal coliforms and fecal streptococci. Specifically, bacteria such as *Escherichia coli* O157:H7, *Salmonella* species, *Campylobacter jejuni*, *Listeria monocytogenes*, and *Leptospira* species are often

DID YOU KNOW?

Multiple species of pathogens may be transmitted directly from a host animal's manure to surface water, and pathogens already in surface water may increase in number due to loadings of animal manure nutrients and organic matter.

DID YOU KNOW?

Over 150 pathogens found in livestock manure are associated with risks to humans.

found in livestock manure and have also been associated with waterborne disease. A study by the USDA revealed that about half of the beef cattle presented for slaughter during July and August 1999 carried *Escherichia coli* O157:H7 (Elder et al., 2000). Also, protozoa, including *Cryptosporidium parvum* and *Giardia* species (e.g., *Giardia lamblia*), may occur in animal waste. *Cryptosporidium parvum* is associated with cows in particular; newborn dairy calves are especially vulnerable to infection and excrete large numbers of infectious oocysts (USDA/ARS, 1998). Most pathogens are shed from host animals with active infections.

The presence of bacteria (and other pathogens) is often measured by the level of fecal coliforms, *Escherichia coli*, or enterococci in manure (Bouzaher et al., 1993). The use of indicator organisms such as these has limitations; specifically, there are no established relationships between fecal coliform and pathogen contamination. However, indicators are still used because specific pathogen testing protocols are too time consuming, expensive, or insensitive to be used for monitoring purposes (Shelton, 2000). Table 10.2 lists the number of total coliform bacteria, fecal coliform bacteria, and fecal *Streptococcus* bacteria per cubic foot of manure for swine, poultry, beef, and dairy animals (ASAE, 2005).

Organic Matter

Livestock manures contain many carbon-based, biodegradable compounds. These compounds are of concern in surface water because dissolved oxygen is consumed as aquatic bacteria and other microorganisms decompose these compounds. This process reduces the amount of oxygen available for aquatic animals.

TABLE 10.2
Coliform Bacteria in Manure (Colonies per Cubic Foot of Manure, as Excreted)

Animal Group	Total Coliform Bacteria	Fecal Coliform Bacteria	Fecal *Streptococcus* Bacteria
Swine	1.6×10^{11}	5.9×10^{10}	18×10^{11}
Poultry (layers)	4.7×10^{11}	3.2×10^{10}	0.69×10^{11}
Beef	3.2×10^{11}	14×10^{10}	1.5×10^{11}
Dairy	36×10^{11}	5.2×10^{10}	3.0×10^{11}

Source: ASAE, *Manure Production and Characteristics*, ASAE D384.1, American Society for Agricultural Engineers, St. Joseph, MI, 1999.

Note: Values are rounded to two significant figures.

DID YOU KNOW?

An investigation in Iowa of chemical and microbial contamination near large-scale swine operations demonstrated the presence of pathogens not only in manure lagoons used to store swine waste before it was land applied but also in drainage ditches, agricultural drainage wells, tile line inlets and outlets, and an adjacent river (CDC, 1998).

Table 10.3 lists BOD and COD estimates for manure generated by swine, poultry, beef, and dairy animals and, for comparison, provides values for domestic sewage. Reported BOD values for various untreated animal manures range from 24,000 to 33,000 mg/L. COD values range from 25,000 to 260,000 mg/L. Dairy and beef cattle manure have BOD and COD values of similar magnitude. By comparison, the BOD value for raw domestic sewage ranges from 100 to 300 mg/L. Even after biological treatment in anaerobic lagoons, animal waste BOD concentrations (200 to 3800 mg/L) are much higher than those of municipal wastewater treated to the secondary level (about 20 mg/L) (USDA, 1992).

Salts and Trace Elements

The salinity of animal manure is directly related to the presence of the nutrient potassium and dissolved mineral salts that pass through the animal. In particular, significant concentrations of soluble salts containing the cations sodium and potassium remain from undigested feed that passes unabsorbed through the animals (NCAES, 1982). Other major cations contributing to salinity are calcium and magnesium, and the major anions are chloride, sulfate, bicarbonate, carbonate, and nitrate (NRC, 1993). Salinity tends to increase as the volume of manure decreases during decomposition and evaporation (Gresham et al., 1990). Salt buildup deteriorates soil structure, reduces permeability, contaminates groundwater, and reduces crop yields.

DID YOU KNOW?

Oxygen-depleting substances are the second leading stressor in estuaries. They are the fourth greatest stressor in impaired rivers and streams and in impaired lakes, ponds, and reservoirs (Spellman, 1996). Biochemical oxygen demand (BOD) is an indirect measure of the concentration of biodegradable substances present in an aqueous solution. Alternatively, the chemical oxygen demand (COD) test uses a chemical oxidant. This test provides an approximation of the ultimate BOD and can be estimated more quickly than the 5 days required for the BOD test. If the waste contains only readily available organic bacterial food and no toxic matter, the COD values correlate with BOD values obtained from the same wastes (Dunne and Leopold, 1978).

TABLE 10.3
Reported BOD and COD Concentrations for Manures and Domestic Sewage

Waste	BOD (mg/L)	COD (mg/L)
Swine manure		
Untreated	27,000 to 33,000	25,000 to 180,000
Anaerobic lagoon influent	13,000	n/a
Anaerobic lagoon effluent	300 to 3600	n/a
Poultry manure		
Untreated (chicken)	24,000	100,000 to 260,000
Anaerobic lagoon influent (poultry)	9800	n/a
Anaerobic lagoon effluent (poultry)	600 to 3800	n/a
Dairy cattle manure		
Untreated	26,000	68,000 to 170,000
Anaerobic lagoon influent	6000	n/a
Anaerobic lagoon effluent	200 to 1200	n/a
Beef cattle manure		
Untreated	28,000	73,000 to 260,000
Anaerobic lagoon influent	6700	n/a
Anaerobic lagoon effluent	200 to 2500	n/a
Domestic sewage		
Untreated	100 to 300	400 to 600
After secondary treatment	20	n/a

Note: Untreated values, except for beef manure BOD, are from NCAES (1982). The BOD value for beef manure is from ASAE (1999). Lagoon influent and effluent concentrations are from USDA/NRCS (1996). Values are rounded to two significant figures; n/a = not available.

Trace elements in manure that are of environmental concern include arsenic, copper, selenium, zinc, cadmium, molybdenum, nickel, lead, iron, manganese, aluminum, and boron. Arsenic, copper, selenium, and zinc are often added to animal feed as growth stimulants or biocides (Sims, 1995). Trace elements may also end up in manure through the use of pesticides that farmers apply to livestock to suppress houseflies and other pests (USDA/ARS, 1998). Trace elements have been found in

DID YOU KNOW?

In freshwaters, increasing salinity can disrupt the balance of the ecosystem, making it difficult for resident species to remain. In laboratory settings, drinking water high in salt content has inhibited the growth and slowed the molting of mallard ducklings. Salts also contribute to degradation of drinking water supplies.

manure lagoons used to store swine waste before being land applied and in drainage ditches, agricultural drainage wells, and tile line inlets and outlets. They have also been found in rivers adjacent to hog and cattle operations.

It is useful to compare trace element concentrations in manure to those in municipal sewage sludge, which is regulated by the USEPA's Standards for the Use or Disposal of Sewage Sludge promulgated under the Clean Water Act and published in 40 CFR Part 503 (USEPA, 1993c). Regulated trace elements in sewage biosolids include arsenic, cadmium, chromium, copper, lead, mercury, molybdenum, nickel, selenium, and zinc. Total concentrations of trace elements in animal manures have been found to be comparable to those in some municipal biosolids, with typical values well below the maximum concentrations allowed by 40 CFR Part 503 for land-applied sewage biosolids (Sims, 1995).

Antibiotics

Antibiotics are used in animal feeding operations and can be expected to appear in animal wastes. The practice of feeding antibiotics to poultry, swine, and cattle evolved from the 1949 discovery that very low levels usually improved growth. Antibiotics are used to treat illness and as feed additives to promote growth or improve feed conversion efficiency. In 1991, farmers used an estimated 19 million pounds of antibiotics for disease prevention and growth promotion in animals. From 60 to 80% of animals receive antibiotics during their productive life span (Tetra Tech, 2000a). Use as feed additives accounts for most of the mass of antibiotics used in both the swine and poultry industries and accounts for the presence of antibiotics in the resulting manure. Although antibiotic residues in beef and dairy manure are also a concern, little information can be found in the literature regarding the levels of antibiotics in manure. Estimated concentrations of the antibiotic chlortetracycline in the lagoon systems of a pork producer in Nebraska ranged from 150 to 300 mg/L; that producer was using 16 different antibiotics as feed and drinking water additives (USFWS, 2000).

Pesticides and Hormones

Pesticides and hormones are compounds used in animal feeding operations that can be expected to appear in animal wastes. Both of these types of pollutants have been linked with endocrine disruption. Pesticides are used in animal feeding operations

DID YOU KNOW?

Of greater concern than the presence of antibiotics in animal manure is the development of antibiotic-resistant pathogens. Use of antibiotics in raising animals, especially broad-spectrum antibiotics, is increasing. As a result, more strains of antibiotic-resistant pathogens are emerging, along with strains that are growing more resistant. Normally, about 2% of a bacterial population is resistant to a given antibiotic; however, up to 10% of bacterial populations from animals regularly exposed to antibiotics have been found to be resistant.

DID YOU KNOW?

Pesticides are applied to livestock to suppress houseflies and other pests. There has been very little research on losses of pesticides in runoff from manured lands, although it has been shown that losses of cyromazine (used to control flies in poultry litter) in runoff increased with the rate of poultry manure applied and the intensity of rainfall.

and can appear in animal wastes. Farmers may use pesticides on crops grown for animal consumption or directly in animal housing areas to control parasites (among other reasons). However, little information is available regarding the concentrations of pesticides in animal wastes or on their bioavailability in waste-amended soils.

Hormones are the chemical messengers that carry instructions to target cells throughout the body and are normally produced by the body's endocrine glands. The target cells read and follow the hormones' instructions, sometimes building a protein or releasing another hormone. These actions lead to many bodily responses such as a faster heart beat or bone growth. Hormones include steroids (estrogen, progesterone, testosterone), peptides (antidiuretic hormone), polypeptides (insulin), amino acid derivatives (melatonin), and proteins (prolactin, growth hormone). Natural hormones are potent; only very small amounts are needed to have an effect.

Specific hormones are administered to cattle to increase productivity in the beef and dairy industries, and several studies have shown that hormones are present in animal manures (Mulla, 1999). For example, poultry manure has been shown to contain about 30 ng/g of estrogen and about the same levels of testosterone, and estrogen was found in concentrations up to 20 ng/L in runoff from fields fertilized with chicken manure (Shore et al., 1995).

Other Pollutants of Concern

In addition, CAFOs can be a source of gas emissions and particulates:

- *Gas emissions*—The degradation of animal wastes by microorganisms produces a variety of gases. Sources of odor include animal confinement buildings, waste lagoons, and land application sites. In addition to ammonia, which was discussed earlier, three main gases generated from manure are carbon dioxide, methane, and hydrogen sulfide. Aerobic conditions yield mainly carbon dioxide, while anaerobic conditions generate both methane and carbon dioxide. Anaerobic conditions, which dominate in typical,

DID YOU KNOW?

In 1995, an irrigation pond and three streams in the Conestoga River watershed near the Chesapeake Bay had both estrogen and testosterone present. All of these sites were affected by fields receiving poultry litter.

DID YOU KNOW?

Animal agriculture is a common source of pollutants in watersheds, but it is never the only source. Indeed, the diverse and ubiquitous nature of pollutants in the environment introduces significant complexity to the increasingly important task of managing pollutants in watersheds.

unaerated animal waste lagoons, also generate hydrogen sulfide and over 150 other odorous compounds, including volatile fatty acids, phenols, mercaptans, aromatics, sulfides, and various esters, carbonyls, and amines (Bouzaher et al., 1993; O'Neill and Phillips, 1992; USDA, 1992).

- *Particulates*—Sources of particulate emissions from CAFOs may include dried manure, feed, epithelial cells, hair, and feathers. The airborne particles make up an organic dust that includes endotoxins (the toxic protoplasm liberated when a microorganism dies and disintegrates), adsorbed gases, and possibly steroids. At least 50% of dust emissions from swine operations may be respirable (Thu, 1995).

SURFACE WATER CONTAMINATION

Earlier, the types of major animal-generated contaminants were briefly described. This section provides a discussion of how these animal-generated contaminants are conveyed to surface freshwater bodies and their fate within. It is important to point out that these contaminants (found mostly in animal manure) can reach surface water by several mechanisms. These can be characterized as either surface discharges or other discharges. Surface discharges can result from runoff, erosion, spills, and dry-weather discharges. In surface discharges, the pollutant travels overland or through drain tiles with surface inlets to a nearby stream, river, or lake. Direct contact between confined animals and surface waters is another means of surface discharge. For other types of discharges, the pollutant travels via another environmental medium (groundwater or air) to surface water.

Surface Discharges

It is appropriate near the outset of this section to attempt a systematic quantification of pollutant sources in surface waters as a means of exploring the relative importance of the influence of animal agriculture on pollutant control in aquatic ecosystems under different conditions.

Runoff

Runoff occurs when water that falls on manmade surfaces or soil fails to be absorbed and flows across the surface. Surface discharges of manure pollutants can originate from feedlots and from overland runoff at land applications. Runoff is especially likely at open-air feedlots, when rainfall occurs soon after application and when farmers over-apply or incorrectly apply manure. It has been found that for all animal wastes, the application rate has a significant effect on the runoff concentration (Daniel et al., 1995). Other factors that promote runoff to surface waters are steep land slope, high rainfall, low soil porosity or permeability, and close proximity to surface waters. In addition, manure applied to saturated or frozen soils is more likely to run off the soil surface (Mulla et al., 1999). Runoff of pollutants dissolved in rainwater is a significant transport mechanism for water-soluble pollutants, including nitrate, nitrite, and organic forms of phosphorus. Runoff of manure pollutants has been identified as a factor in a number of documented impacts for CAFOs. For example, in 1994, an environmental advocacy group noted multiple runoff problems for a swine operation in Minnesota (CWAA, 1998), and, in 1996, the State of Ohio identified runoff from manure spread on land at several Ohio operations that were feeding swine and chicken (ODNR, 1997). More discussion of runoff and its impacts on the environment and human health appears later in this section.

Erosion

In addition to runoff, surface discharges can occur by erosion, in which the soil surface is worn away by the action of water or wind. Erosion is a significant transport mechanism for land-applied pollutants, such as phosphorus, that are strongly sorbed to soils, of which phosphorus is just one example (Gerritse and Zugec, 1977). In 1999, the Agricultural Research Service (ARS) noted that phosphorus bound to eroded sediment particles makes up 60 to 90% of phosphorus transported in surface runoff from cultivated land. For this reason, most agricultural phosphorus control measures have focused on soil erosion control to limit transport of particulate phosphorus. However, soils do not have infinite adsorption capacity for phosphate or any other adsorbing pollutant, and dissolved pollutants including phosphate can still enter waterways via runoff and leachate even if soil erosion is controlled. The Natural Resources Conservation Service (NRCS) reviewed the manure production in a watershed in South Carolina. Agricultural activities in the project area are a major influence on the streams and ponds in the watershed and contribute to nutrient related water quality problems in the headwaters of Lake Murray. NRCS found that bacteria, nutrients, and sediment from soil erosion are the primary contaminants affecting the waters in this watershed. The NRCS calculated that soil erosion, occurring on over 13,000 acres of cropland in the watershed, ranges from 9.6 to 41.5 tons per acre per year (USEPA, 1997).

Spills and Dry-Weather Discharges

Surface discharges can occur through spills or other discharges from lagoons. Catastrophic spills from large manure storage facilities can occur primarily through overflow following large storms or by intentional releases (Mulla et al., 1999). Other causes of spills include pump failures, malfunctions of manure irrigation guns, and breakage of pipes or retaining walls. Manure entering tile drains has a direct route to surface water. (Tile drains are a network of pipes buried in fields below the root zone of plants to remove subsurface drainage water from the root zone to a stream, drainage ditch, or evaporation pond.) In addition, spills can occur as a result of washouts from floodwaters when lagoons are sited on floodplains. There are also indications that discharges from siphoning lagoons occur deliberately as a means to reduce the volume in overfull lagoons (CWAA, 1998). An independent review of the Indiana Department of Environmental Management records indicated that two common causes of waste releases in that state were intentional discharges and accidental discharges resulting from lack of operator knowledge (Hoosier Environmental Council, 1997).

Numerous such dry-weather discharges have been identified; for example, the Ohio Department of Natural Resources (ODNR) documented chicken manure traveling through tile drains into a nearby stream in several instances occurring in 1994, 1995, and 1996 (ODNR, 1997). In 1995, a discharge of 25 million gallons of manure from swine farms in North Carolina was documented (Meadows, 1995; Warrick, 1995). Subsequent discharges of hundreds of thousands of gallons of manure were documented from swine operations in Iowa (1996), Illinois (1997), and Minnesota (1997) (Anon., 1999; CWAA, 1998; IDNR, 1998; ISA, 1997). Between 1994 and 1996, half a dozen discharges from poultry operations in Ohio resulted when manure entered drain tiles (ODNR, 1997). In 1996, more than 40 animal waste spills occurred in Iowa, Minnesota, and Missouri alone (U.S. Senate, 1997). In 1998, a dairy feedlot in Minnesota discharged 125,000 gallons of manure (CWAA, 1998). Acute discharges of this kind frequently result in dramatic fish kills; for example, fish kills were reported as a result of the North Caroline, Iowa, Minnesota, and Missouri discharges mentioned above.

Direct Contact between Confined Animals and Surface Water

Finally, surface discharges can occur as a result of direct contact between confined animals and the rivers, streams, or ponds that are located within their reach. Historically, people located their farms near waterways for both water access by animals and discharge of wastes. Certain animals, particularly cattle, wade into the waterbody, linger to drink, and often urinate and defecate in the water. This practice is now restricted for CAFOs; however, enforcement actions are the primary means for reducing direct access as described below (McFall, 2000).

In the more traditional farm production regions of the Midwest and Northeast, dairy barns and feedlots are often in close proximity to streams or other water sources. This close proximity to streams was formerly necessary in order to provide drinking water for the dairy cattle, to cool the animals in hot weather via direct access, and to cool the milk prior to the widespread use of refrigeration. For CAFO-size facilities, this practice has now been replaced with more efficient means of providing drinking water for the dairy herd. In addition, the use of freestall barns and modern milking centers minimizes the exposure of dairy cattle to the environment. For example, in New York, direct access of animals to surface water is more of a problem for the smaller, traditional dairy farms that use older methods of housing animals. However, at these smaller facilities, direct access to surface water has a relatively lower impact on surface water compared with impacts associated with silage leachate and milk house waste (Dimura, 2000).

In the arid west, feedlots are typically located near water bodies to allow for inexpensive and easy stock watering. Many existing lots were configured to allow the animals direct access to the water. The direct deposition of manure and urine contributes greatly to water quality problems. Environmental problems associated with allowing farm animals access to waters that are adjacent to the production area are well documented in the literature. Dramatically elevated levels of *Escherichia coli* have been documented in rivers downstream of CAFOs with direct access to surface water. Recent enforcement actions against direct access facilities have resulted in the assessment of tens of thousands of dollars in civil penalties (McFall, 2000).

Other Discharges to Surface Water

Leaching to Groundwater

Leaching of land-applied pollutants is a significant transport mechanism for water-soluble pollutants. In addition, leaking lagoons are a source of manure pollutants in groundwater. Although manure solids purportedly "self-seal" lagoons to prevent groundwater contamination, some studies have shown otherwise. A study for the Iowa legislature published in 1999 indicates that leaking is part of lagoon design standards and that all lagoons should be expected to leak (Simpkins et al., 1999). A survey of swine and poultry lagoons in the Carolinas found that nearly two-thirds of the 36 lagoons sampled had leaked into the groundwater (Meadows, 1995). Even clay-lined lagoons have the potential to leak because they can crack or break as they age and can be susceptible to burrowing worms. In a 3-year study of clay-lined swine lagoons on the Delmarva Peninsula, researchers found that leachate from lagoons located in well-drained loamy sand had a severe impact on groundwater quality (Ritter and Chirnside, 1990). Pollutant transport to groundwater is also greater in areas with high soil permeability and shallow water tables. Percolating water can transport pollutants to groundwater, as well as to surface waters via interflow. Contaminated groundwater can deliver pollutants to surface waters through hydrologic connections. Nationally, about 40% of the average annual stream flow is from groundwater (USEPA, 1993b). In the Chesapeake Bay watershed, the USGS estimates that about half of the nitrogen loads from all sources to nontidal streams and rivers originate from groundwater (ASCE, 1998).

Discharge to the Air and Subsequent Deposition

Atmospheric deposition can be a significant mechanism of transport to surface waters, as nitrogen emissions to air can return to terrestrial or aquatic environments in dry form or dissolved in precipitation (Agricultural Animal Waste Task Force, 1996). Discharges to air can occur as a result of volatilization of pollutants already present in the manure, and of pollutants generated as the manure decomposes.

DID YOU KNOW?

Understanding the connection between groundwater and surface water is important when developing surface water protection strategies, because groundwater moves much more slowly than surface water. For example, groundwater in the Chesapeake Bay region takes an average of 10 to 20 years to reach the bay; thus, it may take several decades to realize the full effect of pollutant additions or reductions.

Ammonia is very volatile and can have significant impacts on water quality through atmospheric deposition (Aneja et al., 2001). Ammonia losses from animal feeding operations can be considerable, arising from manure piles, storage lagoons, and land application fields. Other ways that manure pollutants can enter the air are from spray application methods for land applying manure and from particulates windborne in dust. The degree of volatilization of manure pollutants is dependent on the manure management system. For example, losses are greater when manure remains on the land surface rather than being incorporated into the soil and are particularly high when farmers perform spray applications. Environmental conditions such as soil acidity and moisture content also affect the extent of volatilization; ammonia also readily volatizes from lagoons. Losses are reduced by the presence of growing plants (Follet, 1995). Once airborne, pollutants can find their way into nearby streams, rivers, and lakes. One study found that atmospheric deposition was the third largest cause of water quality impairment for estuaries and the fifth largest cause of water quality impairment for lakes, ponds, and reservoirs (USEPA, 2000).

Pollutant-Specific Transport

Nitrogen Compounds

Livestock waste can contribute up to 37% of total nitrogen loads to surface water (Mulla et al., 1999). Nitrogen compounds and nitrates in manure can reach surface water through several pathways. As suggested by Follet (1995), agricultural nitrate contributions to surface water are primarily from groundwater connections and other subsurface flows. Although potentially less significant, overland runoff can also carry nitrate to surface waters. An investigation in Iowa of chemical and microbial contamination near large-scale swine operations demonstrated the presence of pathogens not only in manure lagoons used to store swine waste before it was land applied but also in

DID YOU KNOW?

A 1994 USGS report analyzed nitrogen sources (manure, fertilizers, point sources, and atmospheric deposition) in 107 watersheds. The "manure" source estimates included waste from both confined and unconfined animals. As may be expected, the USGS found that the proportions of nitrogen originating from the various sources differed according to climate, hydrologic conditions, land use, population, and physical geography. Results of the analysis for selected watersheds for the 1987 base year showed that in some instances manure nitrogen was a large portion of the total nitrogen added to the watershed. The study showed that, for the following nine watersheds, more than 25% of nitrogen originated from manure: Trinity River, Texas; White River, Arkansas; Apalachicola River, Florida; Altamaha River, Georgia; Potomac River, Washington, DC; Susquehanna River, Pennsylvania; Platte River, Nebraska; Snake River, Idaho; and San Joaquin River, California. Of these, California, Texas, Florida, Arkansas, and Idaho had large populations of confined animals.

DID YOU KNOW?

Nitrate-nitrogen in streams originates from a variety of sources. Agricultural sources include nitrogen fertilizer, animal manure, mineralization of soil nitrogen, and nitrogen-fixing crops. Other sources include human waste from sewage treatment plants, septic systems, and landfills and nitrogen produced as a waste or byproduct of some industrial processes (Rodecap, 2002).

drainage ditches, agricultural drainage wells, tile line inlets and outlets, and an adjacent river (CDC, 1998). Studies of small geographical areas have revealed evidence of nitrate contamination in groundwater. In 1988, 40% of wells in the Chino Basin, California, had nitrate levels in excess of the maximum contaminant level (MCL); dairy operations were identified as the major source of the contamination (USEPA, 1993b). Such findings could have potentially widespread impacts, as water from the Chino Basin is used to recharge the primary source of drinking water for residents of heavily populated Orange County. On the Delmarva Peninsula, in Maryland, where poultry production is dominant, over 15% of wells were found to have nitrate levels exceeding the MCL. Wells located close to chicken houses contained the highest median nitrate concentrations; measured nitrate levels in groundwater beneath Delaware poultry houses were as high as 100 mg/L (Ritter et al., 1989).

Elevated nitrate levels can also exist in surface waters, although these impacts are typically less severe than groundwater impacts. In a historical assessment, USGS (1997) found that nitrate levels in streams in agricultural areas were elevated compared to undeveloped areas. Nevertheless, the in-stream nitrate concentrations were generally less than those for groundwater in similar locations, and the drinking water MCL was rarely exceeded. The primary exception to this pattern was in the Midwest, where poorly drained soils restrict water percolation and artificial drainage provides a quick path for nutrient-rich runoff to reach streams (USGS, 1997).

When farmers apply manure to land as fertilizer, the risk of nitrate pollution generally increases at higher rates of nitrogen application. Even when farmers land apply manure at agronomic rates, nitrogen transport to surface water and groundwater can still occur for the following reasons: (1) nitrate is extremely mobile and may move below the plant root zone before being taken up; (2) ammonia may volatize and be redeposited in surface water; (3) the waste may be unevenly distributed, resulting in local "hot spots"; (4) it may be difficult to obtain a representative sample of the waste to determine the amount of mineralized (plant-available) nitrogen; (5) there are uncertainties about the estimated rate of nitrogen mineralization in the applied waste; (6) transport is affected by the manure application method (e.g., drip irrigation, spray irrigation, knifing); and (7) transport is affected by uncontrollable environmental factors such as rainfall and other local conditions (Follett, 1995).

Phosphorus Compounds

Phosphorus can reach surface waters via discharges directly into surface water and runoff of manure to surface water from feedlots, and via runoff and erosion from land application sites. The organic phosphorus compounds in manure are generally water soluble and subject to leaching and dissolution in runoff (Gerritse and Zugec, 1977). Once in receiving waters, these compounds can undergo transformation and become available to aquatic plants. Overall, land-applied phosphorus is less mobile than nitrogen, because the mineralized (inorganic phosphate) form is easily adsorbed to soil particles. The Agricultural Research Service (USDA/ARS, 1999) reported that phosphorus bound to eroded sediment particles made up 60 to 90% of phosphorus transported in surface runoff from cultivated land. For this reason, most agricultural phosphorus control measures have focused on soil erosion control to limit transport of particulate phosphorus. However, soils do not have infinite phosphate adsorption capacity, and dissolved inorganic phosphates can still enter waterways via runoff even if soil erosion is controlled (National Research Council, 1993).

In the field of water quality chemistry, phosphorus is described by several terms. Some of these terms are chemistry based (referring to chemically based compounds), and others are methods based (they describe what is measured by a particular method). The term *orthophosphate* is a chemistry-based term that refers to the phosphate molecule all by itself. *Reactive phosphorus* is a corresponding method-based term that describes what is actually being measured when the test for orthophosphate is being performed. Because the lab procedure is not quite perfect, mostly orthophosphate is obtained along with a small fraction of some other forms. More complex inorganic phosphate compounds are referred to as *condensed phosphates* or *polyphosphates*. The method-based term for these forms is *acid hydrolyzable*. Livestock waste can contribute up to 65% of total phosphorus loads in surface waters (Mulla et al., 1999). Animal wastes typically have lower N:P ratios than crop N:P requirements, such that application of manure at a nitrogen-based agronomic rate can result in application of phosphorus at several times the agronomic rate (Sims, 1995). Summaries of soil test data in the United States confirm that many soils in areas dominated by animal-based agriculture have excessive levels of phosphorus (Sims, 1995). Research also indicates that there is a potential for phosphorus to leach into groundwater through sandy soils with already high phosphorus content (Citizens *Pfiesteria* Action Commission, 1997).

Ammonia

Ammonia can reach surface waters in a number of ways, including discharge directly to surface waters, leaching, dissolution in surface runoff, erosion, and atmospheric deposition. Leaching and runoff are generally not significant transport mechanisms for ammonia compounds in land-applied manure, because ammonium can be sorbed to soils (particularly those with high cation exchange capacity), incorporated (fixed) into clay or other soil complexes, or transformed into organic form by soils microbes (Follet, 1995). However, in these forms, erosion can transport nitrogen to surface waters. An investigation in Iowa of chemical and microbial contamination near large-scale swine operations demonstrated the presence of ammonia not only in manure lagoons used to store swine waste before it was land applied but also in drainage ditches, agricultural drainage wells, tile line inlets and outlets, and an adjacent river (CDC, 1998). Ammonia losses from animal feeding operations to the air and subsequent deposition to surface waters can be considerable, arising from sources such as manure piles, storage lagoons, and land application fields. For example, in North Carolina, animal agriculture is responsible for over 90% of all ammonia emissions (Aneja et al., 2001). Ammonia composes more than 40% of the total estimated nitrogen emissions from all sources (Aneja et al., 2001). Data from Sampson County, North Carolina, indicate that ammonia levels in rain have increased with increases in the size of the pork industry. Levels more than doubled between 1985 and 1995 (Aneja et al., 2001). Based on 1995 USEPA estimates, swine operations in eastern North Carolina were responsible for emissions of 135 million pounds of nitrogen per year. If deposited in a single basin, this would have resulted in nitrogen loadings of almost 2.1 million pounds of nitrogen per year (Nowlin et al., 1997).

Pathogens

Sources of pathogen contamination from CAFOs include surface discharges and lagoon leachate. Surface runoff from land application fields can be a source of pathogen contamination, particularly if a rainfall event occurs soon after application or if the land is frozen or snow covered (Mulla et al., 1999). Researchers have reported concentrations of bacteria in runoff water from fields treated with poultry litter at several orders of magnitude above contact standards (Coyne and Blevins, 1995; Giddens and Barnett, 1980). An investigation in Iowa of chemical and microbial contamination near large-scale swine operations demonstrated the presence of pathogens not only in manure lagoons used to store swine waste before it was land applied but also in drainage ditches, agricultural drainage wells, tile line inlets and outlets, and an adjacent river (CDC, 1998). Also, studies have reported that lands receiving fresh manure applications can be the source of up to 80% of the fecal bacteria in surface waters (Mulla et al., 1999). Similarly, both *Cryptosporidium parvum* and *Giardia* species were found in over 80% of 66 surface water sites tested (LeChevallier et al., 1991). Because these protozoa do not multiply outside of the host, livestock animals are one potential source of this contamination. The bacterium *Erysipelothrix* spp., primarily a swine pathogen, has been isolated from many fish and avian species (USFWS, 2000).

High levels of indicator bacteria in surface water near CAFOs have been documented. For example, Zirbser (1998) documented a report of fecal coliform counts of 3000/100 mL and fecal streptococci counts over 30,000/100 mL downstream from a

DID YOU KNOW?

Waterborne disease outbreaks caused by microbial agents can be divided into three categories: "(1) Those associated with intestinal infection and feces from multiple species including humans such as *Cryptosporidium parvum*, *Giardia* species (sp.), *Escherichia coli* O157:H7, *Campylobacter jejuni*, and *Salmonella* sp.; (2) those associated with human intestinal infection and feces such as *Shigella* sp., *Salmonella typhi*, and human intestinal viruses; and (3) those which live in the environment such as *Pseudomonas* and *Legionella* that are associated with a variety of human illnesses including skin infections (dermatitis) and Legionnaires' disease. Intestinal infections are the most common type of waterborne infection and affect the most number of people" (Stehman, 2000).

swine waste lagoon site. (No sampling was performed upstream of the lagoon site.) Fecal coliform pollution from treated and partially treated sewage and stormwater runoff is often cited in beach closures and shellfish restrictions.

The natural filtering and adsorption action of soils typically causes a majority of the microorganisms in land-applied manure to be stranded at the soil surface (Crane et al., 1980). This phenomenon helps protect underlying groundwater but increases the likelihood of runoff losses to surface waters. Pathogens discharged to the water column can subsequently adsorb to sediments, presenting long-term health hazards. Benthic sediments harbor significantly higher concentrations of bacteria than the overlying water column (Mulla et al., 1999).

Although surface waters are typically more prone to pathogen contamination than groundwaters, subsurface flows may also be a mechanism for pathogen transport depending on weather, site, and operating conditions. Groundwaters in areas of sandy soils, limestone formations, or sinkholes are particularly vulnerable. For example, the bacteria *Clostridium perfringens* was detected in the groundwater below plots of land treated with swine manure, and fecal coliform has been detected in groundwater beneath soils amended with poultry manure (Mulla et al., 1999). *Campylobacter jejuni* was isolated from groundwater, and some of the strains were the same type as those from a dairy farm in the same hydrologic area (Stanley et al., 1998).

There are other accounts of high levels of microorganisms in groundwater near feedlots. In cow pasture areas of Door County, Wisconsin, where a thin topsoil layer is underlain by fractured limestone bedrock, groundwater wells have commonly been shut down due to high bacteria levels; for example, a well at one rural household produced brown, manure-laden water (Behm, 1989). Private wells are more prone to contamination than public wells, as they tend to be shallower and therefore more susceptible to contaminants leaching from the surface. In a survey of drinking water standard violations in six states over a 4-year period, the U.S. General Accounting Office (USGAO, 1997) found that bacterial standard violations occurred in 3 to 6% of community water systems each year; however, some bacterial contamination occurred in 15 to 42% of private wells, according to statistically representative assessments performed by others.

Several factors affect the likelihood of disease transmission by pathogens in animal manure, including pathogen survivability in the environment. For example, *Salmonella* can survive in the environment for 9 months or more, providing for increased dissemination potential (USFWS, 2000), and *Campylobacter* can remain dormant, making water an important vehicle for campylobacteriosis (Altekruse, 1998). Some studies have better characterized the survivability and transport of pathogens in manure once it has been land applied. Several researchers (Dazzo et al., 1973; Himathongkham et al., 1999; Kudva et al., 1998; Maule, 2000; Van Donsel et al., 1967) found that soil type, manure application rate, temperature, moisture level, aeration, soil pH, and the amount of time that manure is held before it is applied to pastureland are dominating factors in bacteria survival.

Experiments on land-applied poultry manure (Crane et al., 1980) indicated that the population of fecal organisms decreases rapidly as manure is heated, dried, and exposed to sunlight on the soil surface. However, regrowth of fecal organisms also occurred in these experiments. More recent research indicated that pathogens can

DID YOU KNOW?

In a series of experiments, Edwards and Daniel (1992, 1993a,b) measured runoff from fescue grass plots treated with poultry litter, poultry manure slurry, and swine manure slurry to determine how runoff quality is impacted by application rate and rain intensity. They found that for all wastes, the application rate had a significant effect on the runoff concentration and mass loss of COD (as well as other constituents).

survive in manure for 30 days or more (Himathongkham et al., 1999; Kudva et al., 1998; Maule, 2000). Kudva et al. (1998) found that *Escherichia coli* survived for 47 days in aerated cattle manure piles that were exposed to outdoor weather; drying the manure reduced the number of viable pathogens. Stehman (2000) observed that *Escherichia coli* O157:H7, *Cryptosporidium parvum*, and *Giardia* can survive and remain infectious in surface waters for a month or more. The continued application of waste on a particular area could lead to extended pathogen survival and buildup (Dazzo et al., 1973). Additionally repeated applications and/or high application rates increase the likelihood of runoff to surface water and transport to groundwater.

Organic Matter

Discharge and runoff of manure from feedlots cause large loadings of organic matter to surface waters. There have been numerous incidents of discharges from CAFOs nationwide directly to surface waters. Discharges can also originate from land application sites when farmers over-apply or incorrectly apply manure. Even if farmers apply manure such that there is not a concentrated discharge, organic matter will be present in runoff from land application sites. Runoff or organic matter has been shown to increase with application rate; for example, Daniel et al. (1995) reported that when the swine manure slurry application rate increased from 193 lb N per acre to 387 lb N per acre, COD levels in runoff (generated from a rainfall intensity of 2 inches/hour) increased from 282 mg/L to 504 mg/L. By comparison, runoff from a control plot yielded 78 mg/L COD.

Salts and Trace Elements

Salts can reach surface waters via discharges from feedlots and runoff from land application sites. Salts can also leach into groundwater and subsequently reach surface water. Trace elements can also be transported by these mechanisms. An investigation in Iowa of chemical and microbial contamination near large-scale

DID YOU KNOW?

The USEPA assumes that 175 lb N per acre is a typical requirement for a fescue crop in Arkansas, based on information from U.S. Department of Agriculture extension agents (Tetra Tech, 2000b).

swine operations demonstrated the presence of trace elements not only in manure lagoons used to store swine waste before it was land applied but also in drainage ditches, agricultural drainage wells, tile line inlets and outlets, and an adjacent river (CDC, 1998). Selenium concentrations have been detected in swine manure lagoons at levels up to 6 µg/L, copper has been detected in liquid swine manure prior to land application at 15 mg/L, and zinc has been detected in soils that receive applications of cattle manure at levels up to 9.5 mg/kg in the upper 60 cm of soil (USFWS, 2000).

Antibiotics

Little information is available regarding the fate and transport properties of antibiotics, or the potential releases from animal waste compared to other sources such as municipal and industrial wastewaters, septic tank leachate, runoff from land-applied sewage biosolids, crop runoff, and urban runoff. However, it is known that the primary mechanisms of eliminating antibiotics from livestock are through urine and bile. Also, essentially all of an antibiotic administered to an animal is eventually excreted, whether unchanged or in metabolite form (Tetra Tech, 2000a). Although the presence of excreted antibiotics themselves may be of concern, the development of antibiotic-resistant pathogens due to exposure to environmental levels of antibiotics is generally of greater concern. The risk for development of antibiotic-resistant pathogens from this exposure is unknown.

Hormones

Hormones can reach surface waters through the same route as other manure pollutants, including runoff and erosion as well as direct contact of animals with the water. Considering specific hormones used, however, estrogen is more likely to be lost by runoff than leaching, while testosterone is lost mainly through leaching (Shore et al., 1995). Several sites have documented the presence of hormones in runoff and surface waters. For example, runoff from a field receiving poultry litter was found to contain estrogen. Also, an irrigation pond and three streams in the Conestoga River watershed near the Chesapeake Bay had both estrogen and testosterone. Each of these sites was affected by fields receiving poultry litter (Shore et al., 1995). Runoff from fields with land-applied manure has been reported to contain estrogens, estradiol, progesterone, and testosterone, as well as their synthetic counterparts. Estrogens have also been found in runoff from heavily grazed land (Addis et al., 1999).

Other Pollutants

There has been almost no research on losses of pesticides in runoff from manured lands. A 1999 literature review by the University of Minnesota discussed a 1994 study showing that losses of cyromazine (used to control flies in poultry litter) in runoff increased with the rate of poultry manure application and the intensity of rainfall. The 1999 literature review also included a 1995 study documenting that about 1% of all pesticides enter surface water. However, the magnitude of the impacts of these losses on surface water are unknown (Mulla, 1999). In general, little information is available regarding the fate and transport of pesticides or their bioavailability in waste-amended soils. Furthermore, there is little information comparing potential releases of these compounds from animal waste to other sources such as municipal and industrial wastewaters, septic tank leachate, runoff from land-applied biosolids, crop runoff, and urban runoff.

POTENTIAL HAZARDS FROM CAFO POLLUTANTS

As described in the previous section, animal feeding operations are associated with a variety of pollutants, including nutrients (specifically nitrogen and phosphorus), ammonia, pathogens, organic matter, salts, trace elements, solids, antibiotics, hormones, gas and particulate emissions, and pesticides. These CAFO pollutants can produce multimedia impacts, such as the following:

- *Surface water*—Organisms that live in, on, or under rocks, sediments, or small spaces occupy what is known as a *microhabitat.* Some organisms make their own microhabitats; many of the caddisflies build a case around themselves and use it for their shelter. Rocks and sediments are not the only physical features of streams where aquatic organisms can be found. For example, fallen logs and branches, commonly referred to as large woody debris (LWD), provide an excellent place for some aquatic organisms to burrow into and surfaces for others to attach themselves to, as they might to a rock. They also create areas where small detritus such as leaf litter can pile up underwater. These piles of leaf litter are excellent shelters for many organisms, including large, fiercely predaceous larvae of dobsonflies. Another important aquatic organism habitat is found in the matter, or *drift*, that floats along downstream. Drift is important because it is the main source of food for many fish. It may include insects such as mayflies (*Ephemeroptera*), some true flies (*Diptera*), and some stoneflies (*Plecoptera*) and caddisflies (*Trichoptera*). In addition, dead or dying insects and other small organisms, terrestrial insects that fall from the trees, leaves, and other matter are common components of drift. Among the crustaceans, amphipods (small crustaceans) and isopods (small crustaceans including sow bugs and gribbles) also have been reported in the drift. Impacts have been associated with surface discharges of waste, as well as leaching to groundwater and subsurface flow to surface water. Generally, states with high concentrations of feedlots experience 20 to 30 serious water quality pollution problems per year involving manure lagoon spills and feedlot runoff (Mulla et al., 1999). The oxygen demand of the waste and its ammonia content can result in fish kills and reduced biodiversity. Solids can increase turbidity and impact benthic organisms. Nutrients contribute to eutrophication and associated algal blooms. Algal decay and nighttime respiration can depress dissolved oxygen levels, potentially leading to fish kills and reduced biodiversity. Eutrophication is also a factor in blooms of toxic algae and other toxic microorganisms, such as *Pfiesteria piscicida.* Human and animal health impacts are primarily associated with drinking contaminated water (pathogens and nitrates), coming into contact with contaminated water (pathogens such as toxic algae and *Pfiesteria*), and consuming contaminated shellfish (pathogens such as toxic algae). Trace elements (e.g., arsenic, copper, selenium, zinc) may also present human health and ecological risks. Salts contribute to salinization and disruption of ecosystem balance, as well as degradation of drinking water supplies. Antibiotics, pesticides, and hormones may have low-level, long-term ecosystem effects.

- *Groundwater*—Impacts have been associated with pollutants leaching to groundwater. Human and animal health impacts are associated with pathogens and nitrates in drinking water. Leaching salts can increase health risks to salt-sensitive individuals and can make the water unpalatable. Trace elements, antibiotics, pesticides, and hormones may also present human health and ecological risks through groundwater pathways.
- *Air*—Air impacts include human health effects from ammonia, hydrogen sulfide, other odor-causing compounds, particulates, and the contribution to global climate change due to methane emissions. In addition, volatilized ammonia can be redeposited on the Earth and contribute to eutrophication.
- *Soil*—Trace elements and salts in animal manure can accumulate in soil and become toxic to plants. Salts also deteriorate soil quality by leading to reduced permeability and overall poor physical condition. Crops may provide a human and animal exposure pathway for trace elements and pathogens.

This section describes in greater detail the known or potential adverse human health and ecological effects of CAFO pollutants that become intermixed with, housed by, or attached to sediments.

Primary Nutrients

Various hazards are posed by primary nutrients present in animal manure. Nitrogen and phosphorus have received the greatest attention in the scientific literature. Actual or anticipated levels of potassium in groundwater and surface water are unlikely to pose hazards to aquatic life or human health (Wetzel, 1983). Potassium does contribute to salinity, however, and applications of high-salinity manure are likely to reduce the fertility of the soil.

Ecology and Human Health

Eutrophication

Eutrophication occurs when phosphorus and nitrogen over-enrich a waterbody and disrupt the balance of life in that waterbody. Perhaps the best documented impact of nutrient pollution is the increase in surface water eutrophication (nutrient enrichment) and its effects on aquatic ecosystems (Vallentyne, 1974). Although nutrients are essential for the growth of phytoplankton (free-floating algae), periphyton (attached algae), and aquatic plants, which form the base of the aquatic food web, an overabundance of nutrients can lead to harmful algal blooms and other adverse effects, such as the following (Carpenter et al., 1998):

- Increased biomass of phytoplankton
- Shifts in phytoplankton to bloom-forming species that may be toxic or inedible
- Changes in macrophyte species composition and biomass
- Death of coral reefs and loss of coral reef communities
- Decreases in water transparency
- Taste, odor, and water treatment problems
- Oxygen depletion
- Increased incidence of fish kills
- Loss of desirable fish species
- Reductions in harvestable fish and shellfish
- Decrease in aesthetic value of the waterbody

The type of waterbody impacted may dictate which nutrient (nitrogen or phosphorus) will have the most impact. In estuaries and coastal marine waters, nitrogen is typically the limiting nutrient (i.e., in these waters, phosphorus levels are sufficiently high compared to nitrogen such that small changes in nitrogen concentrations have a greater effect on plant growth). In freshwaters, phosphorus is typically the limiting nutrient (Robinson and Sharpley, 1995; Wendt and Corey, 1980). There can be exceptions to this generalization, however, especially in water bodies with heavy pollutant loads. For example, estuarine systems may become phosphorus limited when nitrogen concentrations are high. In such cases, excess phosphorus will produce algal blooms. Thus, both nitrogen and phosphorus loads can contribute to eutrophication in either water type.

Algae and Other Toxic Microorganisms

Eutrophication causes the enhanced growth and subsequent decay of algae, which can lower the dissolved oxygen content of a waterbody to levels insufficient to support fish and invertebrates. In some cases, this situation can produce large areas devoid of life because of a lack of sufficient dissolved oxygen. One extreme example is the "Dead Zone," an area of hypoxic water larger than 10,000 km^2 that spreads off the Louisiana coast in the Gulf of Mexico each summer. The Dead Zone is believed to be caused by excess chemical fertilizer; however, nutrients from animal waste have also contributed to the problem. This condition has been attributed to excess nutrients delivered primarily by the Mississippi and Atchafalaya river systems (Atwood et al., 1994). The problem in the Gulf demonstrates that pollutant discharges can have far-reaching downstream impacts. In fact, the nutrient loadings to the Gulf originate from sources over a large land area covering approximately 41% of the conterminous United States (Goolsby et al., 1999).

Eutrophication can also affect phytoplankton and zooplankton population diversity, abundance, and biomass and increase the mortality rates of aquatic species. For example, floating algal mats can prevent sunlight from reaching submerged aquatic vegetation, which serves as habitat for fish spawning, juvenile fish, and fish prey (e.g., aquatic insects). The resulting reduction in submerged aquatic vegetation adversely affects both fish and shellfish populations (USEPA, 2000). Another effect of eutrophication is increased incidence of harmful algal blooms, which release toxins as they die and can severely impact wildlife as well as humans. In marine ecosystems, blooms known as red or brown tides have caused significant mortality in marine mammals (Carpenter et al., 1998). In freshwater, cyanobacterial toxins have caused many incidents of poisoning of wild and domestic animals that have consumed impacted waters (Health Canada Environmental Health Programs, 1998). Published reports of wildlife poisoning from these blooms include amphibians, fish, snakes, waterfowl, raptors, and deer (USFWS, 2000).

Eutrophication is also associated with blooms of other toxic organisms, such as the estuarine dinoflagellate *Pfiesteria piscicida*. *Pfiesteria* has been implicated as the primary causative agent of many major fish kills and fish disease events in North Carolina estuaries and coastal area (Burkholder, 1997), as well as in Maryland and Virginia tributaries to the Chesapeake Bay (USEPA, 1997). *Pfiesteria* (nicknamed "the cell from hell" because of its aggressive, flesh-eating nature) often lives as a nontoxic predatory animal, becoming toxic in response to human influences, including excessive nutrient enrichment (Burkholder, 1997). Although nutrient-enriched conditions are not required for toxic outbreaks to occur, excessive nutrient loadings are a concern because they help create an environment rich in microbial prey and organic matter that *Pfiesteria* uses as a food supply. By increasing the concentration of *Pfiesteria*, nutrient loads increase the likelihood of a toxic outbreak when adequate numbers of fish are present (Citizens *Pfiesteria* Action Commission, 1997). Researchers have documented stimulation of *Pfiesteria* growth by human sewage and swine effluent spills and have shown that the organism's growth can be highly stimulated by both inorganic and organic nitrogen and phosphorus enrichments (Burkholder, 1997). Increased algal growth can also raise the pH of water bodies, as algae consume dissolved carbon dioxide to support photosynthesis. Many biological processes, including reproduction, cannot function in water that is very acidic or alkaline (USEPA, 2000).

Nitrates and Nitrites

Nitrites can also pose a risk to aquatic life. If sediments are enriched with nutrients, the concentrations of nitrites in the overlying water may be raised enough to cause nitrite poisoning or "brown blood disease" in fish (USDA, 1992). In addition, excess nitrogen can contribute to water quality decline by increasing the acidity of surface waters.

DID YOU KNOW?

Brown blood disease is named for the color of the blood of dead or dying fish, indicating that the hemoglobin has been converted to methemoglobin. Brown blood disease occurs in fish when water contains high nitrite concentrations (Durborow and Crosby, 2003). Nitrite enters a fish culture system after feed is digested by fish and the excess nitrogen is converted into ammonia, which is then excreted as waste into the water. Total ammonia nitrogen is then converted to nitrite that, under normal conditions, is quickly converted to nontoxic nitrate by naturally occurring bacteria. Uneaten (wasted) feed and other organic material also break down into ammonia, nitrite, and nitrate in a similar manner.

The main hazard to human health from primary nutrients is elevated nitrate levels in drinking water. In particular, infants are at risk from nitrate poisoning (also referred to as methemoglobinemia or "blue baby syndrome"), which can be fatal. This poisoning results in oxygen starvation and is due to nitrite (a metabolite of nitrate), which is formed in the environment, foods, and the human digestive system. Compared to adults and older children, infants under 6 months experience elevated nitrite production because their digestive systems have a higher concentration of nitrate-reducing bacteria. Nitrite oxidizes iron in the hemoglobin of red blood cells to form methemoglobin, which cannot carry sufficient oxygen to the body's cells and tissues. Although methemoglobin is continually produced in humans, an enzyme in the human body reduces methemoglobin back to hemoglobin. In most individuals, this conversion occurs rapidly. Infants, however, have a low concentration of methemoglobin-reducing enzyme, as do individuals with an enzyme deficiency. In these people, methemoglobin is not converted to hemoglobin as readily (Skipton and Hay, 1998).

Because infants under six months have a higher concentration of digestive bacteria that reduce nitrates and a lower concentration of methemoglobin-reducing enzyme, they are at greater risk for methemoglobinemia (Skipton and Hay, 1998). To protect infant health, the USEPA set drinking water maximum contaminant levels (MCLs) of 10 mg/L for nitrate-nitrogen and 1 mg/L for nitrite-nitrogen. MCLs are the maximum permissible levels of pollutants allowed in water delivered to public drinking water systems. Once a water source is contaminated, the costs of protecting consumers from nitrate exposure can be significant. Nitrate is not removed by conventional drinking water treatment processes. Its removal requires additional, relatively expensive treatment units.

Although reported cases of methemoglobinemia are rare, the incidence of actual cases may be greater than the number reported. Studies in South Dakota and Nebraska have indicated that most cases of methemoglobinemia are not reported (Meyer, 1994, Michel et al., 1996). For example, in South Dakota between 1950 and 1980, only two cases were reported, while at least 80 were estimated to have occurred (Meyer, 1994). There are at least two reasons for this underreporting. First, methemoglobinemia can be difficult to detect in infants because its symptoms are similar to other conditions; also, doctors are not always required to report it (Michel et al., 1996).

In addition to blue baby syndrome, low blood oxygen due to methemoglobinemia has also been linked to birth defects, miscarriages, and general poor health in humans and animals. These effects are exacerbated by concurrent exposure to many species of bacteria in water (IRIS, 2000). Studies in Australia found an increased risk of congenital malformations with consumption of high-nitrate groundwater (Bruning-Fann and Kaneene, 1993). Multi-generation animal studies have found decreases in birth weight, postnatal growth, and organ weights among mammals prenatally exposed to nitrite (IRIS, 2000). Nitrate- and nitrite-containing compounds may also cause hypotension or circulatory collapse (Bruning-Fann and Kaneene, 1993).

High nitrate levels in drinking water have also been implicated in higher rates of stomach and esophageal cancer, although a National Research Council report concluded that exposures to nitrate and nitrite concentrations in drinking water are unlikely to contribute to human cancer risks (National Research Council, 1995). However, nitrate metabolites such as *N*-nitroso compounds (especially nitrosamines) have been linked to severe human health effects such as gastric cancer (Bruning-Fann and Kaneene, 1993). The formation of *N*-nitroso compounds occurs in the presence of catalytic bacteria (e.g., those found in the stomach) or thiocyanate.

Generally, people drawing water from domestic wells are at greater risk of nitrate poisoning than those drawing from public wells (Nolan and Ruddy, 1996), because domestic wells are typically shallower and not subject to wellhead protection or monitoring requirements. Reported cases of methemoglobinemia are most often associated with wells that were privately dug and that may have been badly positioned through the disposal of human and animal excreta (Addiscott et al., 1991). Furthermore, people served by public systems are better protected even if the water becomes contaminated, due to water quality monitoring and treatment requirements.

Phosphorus

Animal manure also contributes to increased phosphorus concentrations in water supplies. Previous evaluations of phosphorus have not identified significant adverse human health effects, but phosphate levels greater than 1.0 mg/L may interfere with coagulation in drinking water treatment plants and thereby increase treatment costs.

Eutrophication and Algal Blooms

To the extent that nitrogen and phosphorus contribute to algal blooms in surface water through accelerated eutrophication as previously described, these nutrients can reduce the aesthetic and recreational value of surface water resources. Algae can affect drinking water by clogging treatment plant intakes, producing objectionable tastes and odors. Algae can also increase the production of harmful chlorinated byproducts (e.g., trihalomethanes) by reacting with chlorine used to disinfect drinking water. These impacts result in increased costs of drinking water treatment, reduced drinking water quality, and increased health risks.

Eutrophication can also affect human health by enhancing growth of harmful algal blooms that release toxins as they die. In marine ecosystems, harmful algal blooms such as red tides can result in human health impacts via shellfish poisoning and recreation contact (Thomann and Mueller, 1987). In freshwater, blooms of cyanobacteria (blue–green algae) may pose a serious health hazard to humans via water consumption. When cyanobacterial blooms die or are ingested, they release water-soluble compounds that are toxic to the nervous system and liver (Carpenter et al., 1998).

In addition, eutrophication is associated with blooms of a variety of other organisms that are toxic to humans, such as the estuarine dinoflagellate *Pfiesteria piscicida*. Although *Pfiesteria* is primarily associated with fish kills and fish disease events, the organism has also been linked with human health impacts through dermal or inhalation exposure. Researchers working with dilute toxic cultures of *Pfiesteria* exhibited symptoms such as skin sores, severe headaches, blurred vision, nausea/vomiting, sustained difficulty breathing, kidney and liver dysfunction, acute short-term memory loss, and severe cognitive impairment (Burkholder, 1997). People with heavy environmental exposure have exhibited symptoms as well. In one study, such environmental exposure was definitively linked with cognitive impairment and less consistently linked with physical symptoms (Morris et al., 1998).

Ammonia

Ecology

Ammonia exerts a direct biochemical oxygen demand (BOD) on the receiving water. As ammonia is oxidized, dissolved oxygen is consumed. Moderate depressions of dissolved oxygen are associated with reduced species diversity, while more severe depressions can produce fish kills. In fact, ammonia is a leading cause of fish kills (USDA, 1992). Ammonia-induced fish kills are a potential consequence of the discharge of animal wastes directly to surface waters. For example, in a 1997 incident in Wabasha County, Minnesota, ammonia in a dairy cattle manure discharge killed 16,5000 minnows and white suckers (CWAA, 1998). Additionally, ammonia loadings can contribute to accelerated eutrophication of surface waters, which can significantly impact aquatic ecosystems in a number of ways, as noted above.

Human Health

Ammonia is a nutrient form of nitrogen that can have several impacts. First, volatized ammonia is of concern because of direct localized impacts on air quality. Ammonia produces an objectionable odor and can cause nasal and respiratory irritation. In addition, ammonia contributes to eutrophication of surface waters. This phenomenon, as stated previously, is primarily a hazard to aquatic life but is also associated with human health impacts. As previously mentioned, eutrophication reduces the aesthetic and recreational value of water bodies. Additionally, the associated algae blooms can affect drinking water by clogging treatment plant intakes, producing objectionable tastes and odors, and increasing production of harmful chlorinated byproducts. These impacts result in increased drinking water treatment costs, reduced drinking water quality, and increased health risks. Eutrophication can also impact human health by enhancing the growth of toxic algae and other toxic organisms.

Pathogens

Ecology

Animal wastes carry pathogens, bacteria, and viruses, many of which have the potential to be harmful to wildlife (Jackson et al., 1987; USDA, 1992). Some bacteria in livestock waste cause avian botulism and avian cholera, which have killed thousands of migratory waterfowl in the past (USEPA, 1993b). Avian botulism is a food poisoning caused by ingestion of a neurotoxin produced by the bacterium *Clostridium botulinum* type C and *Salmonella* spp., both of which naturally occur in the intestinal tract of warm-blooded animals (USFWS, 2000). Pathogens in surface water can adhere to the skin of fish or be taken up internally when present at high enough concentrations. In a controlled experiment, Fattal et al. (1992) detected significant bacterial concentrations in fish exposed to *Escherichia coli* and other microorganisms for up to 48 hours. The data suggest that harmful pathogens could be taken up by fish-eating carnivores feeding in contaminated surface waters. Shellfish are filter feeders that pass large volumes of water over their gills. As a result, they can concentrate a broad range of microorganisms in their tissues (Chai et al., 1994). This provides a pathway for pathogen transmission to higher trophic organisms. Little information is available, however, to assess the health effects of contaminated shellfish on wildlife receptors.

Human Health

Pathogens may be transmitted to humans through contaminated surface water or groundwater used for drinking, or by direct contact with contaminated surface water through recreational uses. About 20% of the human population (especially infants, the elderly, and those with compromised immune systems) can be classified as particularly vulnerable to the health effects of pathogens (Mulla, 1999). Over 150 pathogens in livestock manure are associated with risks to humans (CAST, 1992). Table 10.4 presents a list of several of these pathogens and their associated diseases, including salmonellosis, cryptosporidiosis, and giardiasis. Other pathogens that have been associated with livestock waste include those that cause cholera, typhoid fever, and polio (USEPA, 1993b). Many of these pathogens are transmitted to humans via the fecal–oral route. In the water environment, humans may be exposed to pathogens through consumption of contaminated drinking water (although adequate drinking water treatment of public supplies is assumed) or by incidental ingestion during activities in contaminated waters.

Although a wide range of organisms may cause disease in humans, relatively few microbial agents are responsible for the majority of human disease outbreaks from water-based exposure routes. This point is illustrated by Table 10.5, which presents reports of waterborne disease outbreaks and their causes (if known) in the United States from 1989 to 1996. Intestinal infections are the most common type of waterborne infection, and they affect the most people.

As presented in Table 10.5, most reported outbreaks were associated with protozoa and bacteria. As noted in Table 10.4, *Cryptosporidium parvum* can produce gastrointestinal illness, with symptoms such as severe diarrhea. Relatively low doses of both *C. parvum* as well as *Giardia* species are needed to cause infection (Stehman, 2000). Although healthy people typically recover relatively quickly (within 2 to 10 days) from this type of illness, these diseases can be fatal in people with weakened immune systems. These individuals typically include children, the elderly, people with human immunovirus (HIV) infection, chemotherapy patients, and those taking medications that suppress the immune system.

Table 10.5 shows that infections caused by *Giardia* species and *Cryptosporidium parvum* (considered the two most important waterborne protozoa) were the leading causes of infectious waterborne disease outbreaks in which an agent was identified, both for total cases and for number of outbreaks (Mulla et al., 1999; Stehman, 2000). In 1993 in Milwaukee, Wisconsin, *C. parvum* (the bug that made Milwaukee famous) contamination of a public water supply caused more than 100 deaths and an estimated 403,000 illnesses (Casman, 1996; Smith, 1994). The outbreak cost an estimated $37 million in lost wages and productivity (Smith, 1994). The source of the oocysts was not identified, but speculated sources include runoff from cow manure application sites, wastewater from a slaughterhouse and meat packing plant, and municipal wastewater treatment plant effluent (Casman, 1996). Four

TABLE 10.4
Some Diseases and Parasites Transmittable to Humans from Animal Manure[a]

Disease	Responsible Organism	Symptoms
	Bacteria	
Anthrax	*Bacillus anthracis*	Skin sores, fever, chills, lethargy, headaches, nausea, vomiting, shortness of breath, cough, nose/throat congestion, pneumonia, joint stiffness, joint pain
Brucellosis	*Brucella abortus*, *Brucella melitensis*, *Brucella suis*	Weakness, lethargy, fever, chills, sweating, headache
Colibaciliosis	*Escherichia coli* (some serotypes)	Diarrhea, abdominal gas
Coliform mastitismetritis	*Escherichia coli* (some serotypes)	Diarrhea, abdominal gas
Erysipelas	*Erysipelothrix rhusiopathiae*	Skin inflammation, rash, facial swelling, fever, chills, sweating, joint stiffness, muscle aches, headache, nausea, vomiting
Leptospirosis	*Leptospira pomona*	Abdominal pain, muscle pain, vomiting, fever
Listeriosis	*Listeria monocytogenes*	Fever, fatigue, nausea, vomiting, diarrhea
Salmonellosis	*Salmonella* species	Abdominal pain, diarrhea, nausea, chills, fever, headache
Tetanus	*Clostridium tetani*	Violent muscle spasms, "lockjaw" spasms of jaw muscles, difficulty breathing
Tuberculosis	*Mycobacterium tuberculosis*, *Mycobacterium avium*	Cough, fatigue, fever, pain in chest, back, and/or kidneys
	Rickettsia	
Q fever	*Coxiella burneti*	Fever, headache, muscle pains, joint pain, dry cough, chest pain, abdominal pain, jaundice
	Viruses	
Foot and mouth	Virus	Rash, sore throat
Swine cholera	Virus	—
New Castle	Virus	—
Psittacosis	Virus	Pneumonia
	Fungi	
Coccidioidomycosis	*Coccidioides immitus*	Cough, chest pain, fever, chills, sweating, headache, muscle stiffness, joint stiffness, rash, wheezing
Histoplasmosis	*Histoplasma capsulatum*	Fever, chills, muscle ache, muscle stiffness, cough, rash, joint pain, joint stiffness
Ringworm	Various *Microsporum* and *Trichophyton*	Itching, rash

(continued)

TABLE 10.4 (continued)
Some Diseases and Parasites Transmittable to Humans from Animal Manure[a]

Disease	Responsible Organism	Symptoms
	Protozoa	
Balantidiasis	*Balatidium coli*	—
Coccidiosis	*Eimeria* species	Diarrhea, abdominal gas
Cryptosporidiosis	*Cryptosporidium parvum*	Watery diarrhea, dehydration, weakness, abdominal cramping
Giardiasis	*Giardia lamblia*	Diarrhea, abdominal pain, abdominal gas, nausea, vomiting, headache, fever
Toxoplamosis	*Toxoplasma* species	Headache, lethargy, seizures, reduce cognitive function
	Parasites/Metazoa	
Ascariasis	*Ascaris lumbricoides*	Worms in stool or vomit, fever, cough, abdominal pain, bloody sputum, wheezing, skin rash, shortness of breath
Sarcocystiasis	*Sarcosystis* species	Fever, diarrhea, abdominal pain

Sources: Diseases and organisms were compiled from USDA/NRCS (1996) and Ohio EPA (1998). Symptom descriptions were obtained from various medical and public health service Internet sites.

[a] Pathogens in animal manure are a potential source of disease in humans and other animals. This list represents a sampling of diseases that may be transmittable to humans.

documented cases of cryptosporidiosis occurring since 1984 have been linked to nonpoint source agricultural pollution (Mulla et al., 1999). Two outbreaks of *C. parvum* were also traced to contamination of drinking water by cow manure in England (Stehman, 2000). The mandated treatment of public water supplies helps reduce the risk of infection via drinking water, but the first step in providing safe drinking water is source water protection, especially because *C. parvum* is resistant to conventional treatment.

Escherichia coli is an important cause of bacterial waterborne infection in untreated and recreational water (Stehman, 2000). Infection can be life threatening, especially in the young and in the elderly. It can cause bloody diarrhea and, if not treated promptly, can result in kidney failure and death (Shelton, 2000). In particular, *Escherichia coli* O157:H7 is emerging as the second most important cause of bacterial waterborne disease after *Shigella* species, which are associated with human feces. *Escherichia coli* O157:H7 was unknown until 1982, when it was associated with a multistate outbreak of hemorrhagic colitis (Shelton, 2000). In 1999, an *Escherichia coli* outbreak occurred at the Washington Country Fair in New York State. This outbreak was possibly the largest waterborne outbreak of *Escherichia coli* O157:H7 in U.S. history. It took the lives of two fair attendees and sent 71 others to the hospital. An investigation identified 781 persons with confirmed or suspected illness related to this outbreak. The outbreak is thought to have been caused by contamination of

TABLE 10.5
Etiology of Waterborne Disease Outbreaks Causing Gastroenteritis, 1989–1996

Type of Organism	Etiologic Agent	Total Number of Outbreaks	Outbreaks Associated with Drinking Water		Outbreaks Associated with Recreational Water	
			Surface	Ground	Natural	Pool/Park
Protozoa	*Giardia* spp.	27	12	6	4	5
	Cryptosporidium parvum	21	4	4	2	11
Bacteria with potential for infecting multiple species	*Escherichia coli* O157:H7	11	—	3	7	1
	Campylobacter jejuni	3	3	—	—	—
	Salmonella typhimurium	1	—	1	—	—
	Salmonella java	1	—	—	—	1
	Leptospira grippotyphosa	1	—	—	1	—
Bacterial infections associated with humans	*Shigella sonnei*	17	—	7	10	—
	Shigella flexneri	2	—	1	1	—
Human viruses	Hepatitis A	3	—	—	—	3
	Norwalk virus	1	—	1	—	—
	Norwalk-like virus	1	—	—	—	1
	Small round structured virus	1	1	—	—	—
Acute gastroenteritis	Unidentified etiology	60	8	44	7	1
Other	Cyanobacteria-like bodies	1	1	—	—	—

Source: Adapted from Stehman, S.M., in *Managing Nutrients and Pathogens from Animal Agriculture: Proceedings of a Conference for Nutrient Management Consultants, Extension Educators, and Produce Advisors*, Natural Resource, Agriculture, and Engineering Service, Ithaca, NY, 2000, pp. 93–107.

DID YOU KNOW?

Researchers at Cornell University concluded that problems with *Escherichia coli* O157:H7 could be eliminated by feeding cattle hay three days before slaughter; however, cattle on hay diets defecate much more frequently, and the manure is much less viscous (Price, 1999).

a well at the fair, either by a dormitory septic system or manure runoff from the nearby Youth Cattle barn (NYSDOH, 2000). In 2000, an outbreak of *Escherichia coli* O157:H7 in Walkerton, Ontario, resulted in at least seven deaths and 1000 cases of intestinal problems; public health officials theorize that one possible cause was floodwaters washing manure contaminated with *Escherichia coli* into the town's drinking water well (Brook, 2000). An outbreak of *Escherichia coli* O157:H7 was reported in Canada from well water potentially contaminated by manure runoff (Stehman, 2000).

Cow manure has specifically been implicated as a causative factor in the high bacteria levels and ensuing swimming restrictions on Tainter Lake, Wisconsin (Behm, 1989). Contact recreation can result in infections of the skin, eye, ear, nose, and throat (Juranek, 1995; Stehman, 2000). The USEPA's recommended ambient water quality standard for human health protection in contact-recreational freshwaters is either 120 *Escherichia coli* bacteria per 100 mL, or 33 *Enterococcus* bacteria per 100 mL. (This standard replaced the earlier standard of 200 fecal coliform bacteria per 100 mL.) About 8% of U.S. outbreaks of *Escherichia coli* O157:H7 between the years 1982 and 1996 occurred as a result of swimming (Griffin, 1998). Certain regions, in particular, may be adversely impacted. For example, pathogen impairment of surface waters is a great problem in most rural areas of southern Minnesota, which causes many rivers and lakes to be unsuitable for swimming (Mulla et al., 1999).

Most human infections caused by bacteria such as *Escherichia coli* O157:H7, *Salmonella* species, *Campylobacter jejuni*, and *Leptospira* species are spread by foodborne or direct contact (Stehman, 2000). Many pathogens might be transmitted through shellfish (Stelma and McCabe, 1992), which are filter feeders prone to accumulating bacteria and viruses. Others may be transmitted through inhalation. In particular, there is concern that pathogens may also be introduced to the air directly from animal feeding houses or during spray application of wastes. Flies and other vectors also serve as potential pathways for disease transmission.

A final concern is exposure to pathogens via consumption of raw foods improperly subjected to manure application. Cieslak et al. (1993) suggest that a 1993 *Escherichia coli* outbreak in Maine was the result of manure applications to a vegetable garden. Additionally, three *Escherichia coli* outbreaks (Montana in 1995, Illinois in 1996, and Connecticut in 1996) were traced to organic lettuce growers. It is suspected that the lettuces were contaminated by infected cattle manure (Nelson, 1997). In another incident in Maine, a few hundred children were sickened by *Cryptosporidium parvum*. The source was fresh-pressed apple cider made from apples gathered from a cattle pasture (Milliard et al., 1994).

Organic Matter

Ecology

Increased organic matter loading to surface waters supports increased microbial population and activity; as these organisms aerobically degrade the organic matter, dissolved oxygen is consumed, reducing the amount available for aquatic organisms. This impact is exacerbated in warm waters compared to colder waters, because the dissolved oxygen saturation level is lower and because the higher temperatures support increased microbial metabolism.

As a result of dissolved oxygen depletion, aquatic species may suffocate (USEPA, 1993a) or be driven out of areas that lack sufficient oxygen. This phenomenon can occur rapidly, particularly with loadings of high-strength waste such as those that may result from catastrophic lagoon breaches (Goldman and Horne, 1983). There are many examples nationwide of fish kills resulting from manure discharges from animal feeding operations. In Nebraska in 1995, 50% of all agriculture-related fish kills investigated were due to livestock waste. In 1996, that percentage rose to 75%. In 1997 and 1998, 100% of agriculture-related fish kills were traced to livestock waste (USFWS, 2000).

Oxygen-stressed aquatic systems may also experience decreases in species richness or community structure as sensitive species are driven out or die off. Organisms living in borderline hypoxic (low-oxygen) water are also likely to experience physiological stress, which may increase the potential for diseases, decrease feeding rates, or increase predation. Livestock has been widely reported to cause significant decreases in wildlife species and numbers (Mulla et al., 1999). For example, a reduction in biodiversity due to AFOs was documented in a study of three Indiana stream systems (Hoosier Environmental Council, 1997). That study showed that waters downstream of animal feedlots (mainly swine and dairy operations) contained fewer fish and a limited number of species of fish in comparison with reference sites. Excessive algal growth, altered oxygen content, and increased levels of ammonia, turbidity, pH, and total dissolved solids were also observed. High oxygen depletion rates due to microbial activity have been reported in manure-amended agricultural soils as well. In soils, elevated microbial populations can affect crop growth by competing with plant roots for soil oxygen and nutrients (USDA, 1992).

Human Health

The release of organic matter to surface waters is a human health concern insofar as it can impact drinking water sources and recreational waters. As aquatic bacteria and other microorganisms degrade organic matter in manure, they consume dissolved oxygen. This can lead to foul odors and ecological impacts, reducing the water's value as a source of drinking water and recreation. Additionally, increased organic matter in drinking water sources can lead to excessive production of harmful chlorinated byproducts, resulting in higher drinking water treatment costs and higher health risks. Pathogen growth is another concern, as large inflows of nutrient-rich organic matter, under the right environmental conditions, can cause rapid increases in microbial populations.

Salts and Trace Elements

Ecology

Salts in manure can impact the water and soil environment. In freshwaters, increasing salinity can disrupt the balance of the ecosystem. Drinking water high in salt content was shown to inhibit growth and cause slowed molting in mallard ducklings (Industrial Economics, Inc., 1993). On land, salts can accumulate and become toxic to plants and reduce crop yields. Salts can damage soil quality by increasing acidity, reducing permeability, and deteriorating soils structure. Salty soils cause plants to become nutrient deficient because they are not able to pick up enough of the nutrients they need, such as nitrogen and phosphorus (Spellman, 1998). Trace elements in manure can impact plants, aquatic organisms, and terrestrial organisms. Although many of the trace elements are essential nutrients at low concentrations, they can have significant ecotoxicological effects at elevated concentrations; for example, metals such as zinc (a feed additive) can accumulate in soil and become toxic to plants at high concentrations. Arsenic, copper, and selenium are other feed additives that can produce aquatic and terrestrial toxicity at elevated concentrations. Bottom-feeding birds can be quite susceptible to metal toxicity because they are attracted to shallow feedlot wastewater ponds and waters adjacent to feedlots. Metals can remain in aquatic ecosystems for long periods of time because of adsorption to suspended or bed sediments or uptake by aquatic biota.

Several of the trace elements in manure are regulated in treated municipal sewage biosolids (but not manure) by the Clean Water Act's Part 503 rule. Total concentrations of trace elements in animal manures have been reported as comparable to those in some municipal biosolids, with typical values well below the maximum concentrations allowed by Part 503 for land-applied sewage sludge (Sims, 1995). Based on this information, trace elements in agronomically applied manures should pose little risk to human health and the environment. However, repeated application of manures above agronomic rates could result in exceedance of the cumulative metal loading rates established in Part 503, thereby potentially impacting human health and the environment (USFWS, 1991).

In 1991, suspected impacts were reported from a large number of cattle feedlots on Tierra Blanca Creek, upstream of the Buffalo Lake National Wildlife Refuge in the Texas Panhandle (USFWS, 1991). Elevated concentrations of the feed additives copper and zinc were found in the creek sediment (as well as elevated aqueous concentrations of ammonia, chemical oxygen demand, chlorophyll *a*, coliform bacteria, chloride, conductivity, total Kjeldahl nitrogen, and volatile suspended solids). The relative contribution of these contaminants from various sources (e.g., runoff from facilities without containment lagoons, lagoon discharges, lagoon leachate) was not assessed.

In 1998, copper and zinc were found in wetlands fed by wastewater from a nearby swine production operation in Nebraska (USFWS, 2000). Concentrations of copper exceeded both a proposed aquatic life criterion of 43 μg/L and the current least-protective criterion of 121 μg/L. Zinc concentration exceeded the concentrations recommended for the protection of aquatic life.

Human Health

Salts from manure can impact surface and groundwater drinking water sources. Salt load into the Chino Basin in California from local dairies is over 1500 tons per year, and the cost to remove that salt by the drinking water treatment system ranges from $320 to $690 for every ton (USEPA, 1993b). At lower levels, salts can increase blood pressure in salt-sensitive individuals, increasing the risk of stroke and heart attack. Salts can also make drinking water unpalatable and unsuitable for human consumption. Some of the trace elements in manure are essential nutrients required for human physiology; however, they can induce toxicity at elevated concentrations. These include zinc, arsenic, copper, and selenium, which are feed additives (Sims,

1995). Although these elements are typically present in relatively low concentrations in manure, they are of concern because of their ability to persist in the environment and to bioconcentrate in plant and animal tissues. These elements could pose a hazard if manure is over-applied to land, due to insufficient acreage available to accommodate manure from increasingly concentrated AFOs. Over-applied manure increases the likelihood of pollutants reaching surface water and ultimately being ingested.

Trace elements are associated with a variety of illnesses. For example, arsenic is carcinogenic to humans, based on evidence from human studies; some of these studies have found increased skin cancer and mortality from multiple internal organ cancers in populations that consumed drinking water with high levels of inorganic arsenic. Arsenic is also linked with non-cancer effects, including hyperpigmentation and possible vascular complications. Selenium is associated with liver dysfunction and loss of hair and nails, and zinc can result in changes in copper and iron balances, particularly copper deficiency anemia (IRIS, 2000).

Solids

Excessive silting and sedimentation are prime agents responsible for the long-term degradation of rivers, streams, and lakes. Major sources of siltation include runoff from agricultural, urban, and forest lands and other nonpoint sources (USEPA, 1992b). Solids entering surface water can degrade aquatic ecosystems to the point of nonviability. Suspended particles can reduce the depth to which sunlight can reach, decreasing photosynthetic activity (and the resulting oxygen production) by plants and phytoplankton. The increased turbidity also limits the growth of desirable aquatic plants that serve as critical habitat for fish, crabs, and other aquatic organisms. In addition, suspended particles can clog fish gills, degrade feeding areas, and reduce visibility for sight feeders (USEPA, 1992c) and can disrupt migration by interfering with a fish's ability to detect chemical communication signals in water (Goldman and Horne, 1983). Sediment can smother eggs, interrupt the reproductive process, and alter or destroy habitat for fish and benthic organisms. Solids can also degrade drinking water sources, thereby increasing treatment costs. Furthermore, solids provide a medium for the accumulation, transport, and storage of other pollutants, including nutrients, pathogens, and trace elements. Sediment-bound pollutants often have a long history of interaction with the water column through cycles of deposition, resuspension, and redeposition.

Antibiotics and Antibiotic Resistance

Antibiotic-resistant strains of bacteria develop as a result of continual exposure to antibiotics. Use of antibiotics in raising animals, especially broad-spectrum antibiotics, is increasing. As a result, more strains of antibiotic-resistant pathogens are emerging, along with strains that are increasingly resistant (Mulla et al., 1999). Antibiotic-resistant forms of *Salmonella*, *Campylobacter*, *Escherichia coli*, and *Listeria* are known or suspected to exist. An antibiotic-resistant strain of the bacterium *Clostridium perfringens* was detected in the groundwater below plots of land treated with swine manure, but it was nearly absent beneath unmanured plots.

Antibiotic resistance poses a significant health threat. In 2000, the *New England Journal of Medicine* published an article that discussed the case of a 12-year-old boy infected with a strain of *Salmonella* that was resistant to no fewer than 13 antimicrobial agents (Fey et al., 2000). The cause of the child's illness was believed to be exposure to the cattle on his family's Nebraska ranch.

The Centers for Disease Control and Prevention, the Food and Drug Administration, and the National Institutes of Health issued a draft action plan in 2000 to address the increase in antibiotic-resistant diseases (CDC, 2000). The plan was intended to combat antimicrobial resistance through survey, prevention and control activities, research, and product development. One of the action items involves conducting pilot studies to assess the impact of environmental contamination by antimicrobial drug residues and by drug-resistant organisms that enter the soil or water from human and animal waste.

Case Study 10.1. Antibiotics and Poultry Production

The following, for illustrative purposes, presents a research project conducted under the auspices of the USGS (Fales et al., 2001) that was concerned with predicting the sorption, mobility, accumulation, and degradation potential of antibiotics in Iowa's soil/water environment.

Research Problem

Approximately 32.6 billion pounds of antibiotics are used in the production of poultry (10.6 million pounds), hogs (10.3 million pounds), and cattle (3.7 million pounds) in the United States each year (Mellon et al., 2001). Over three-fourths of these antibiotics (24.6 million pounds) are given to healthy animals in low doses to promote growth (Levy, 1997). Most of the antibiotics given to farm animals are not metabolized in the body; rather, they are excreted in the active form (Lee et al., 2000). The fate of antibiotics introduced into soil and aquatic environments with manure and other animal wastes is largely unknown. However, there is much concern that the presence and persistence of low levels of antibiotics in soil and aquatic environments could encourage the buildup of existing antibiotic-resistant bacterial populations and promote the development of new populations (Henry, 2000).

In Iowa, earthen waste storage structures (lagoons) are widely used for temporary storage of liquid animal wastes with the intent of protecting surface and groundwater from contamination and allowing farmers to use the wastes in a timely fashion. Liquid animal wastes are generally spread on agricultural soils both as a means of disposal of the wastes and as a nutrient source for crop production. The Iowa Department of Public Health found relatively high concentrations of chlorotetracycline (11 to 540 μg/L) and erythromycin (10 to 275 μg/L) in such liquid animal wastes. The report also indicated that many of the 18 *Escherichia coli* isolates, all three *Salmonella* species, and an isolate of *Enterococcus* demonstrated resistance to one or more of the antibiotics.

The antibiotics most commonly added to livestock feed as growth promoters (1 to 1000 mg per head per day) are chlorotetracycline (Aureomycin©), oxytetracycline (Terramycin©), and macrolide (erythromycin) (Herman et al., 1995; Muirhead, 1998; Sewell, 1993). The fate of these compounds in Iowa soils depends on sorption and desorption of the antibiotics on soils, leaching, and the rates of chemical, photochemical, and microbial decomposition of the antibiotics. The basic hypothesis of the study is that the fate (sorption/desorption, leaching, and decomposition) of antibiotics in soil environments is strongly influenced by the chemical reactions between the antibiotics and soils constituents.

Specific Objectives

1. Characterize three common Iowa soils and isolate and characterize reactive soil components (clay–humic complexes, clay minerals, and humic materials) from these soils.
2. Quantify sorption of tetracycline and chlorotetracycline on the soils and soil components.
3. Determine the effects of saturating cation (Ca vs. K) and ionic strength (I = 0.05 and I = 0.005) on sorption of tetracycline and chlorotetracycline on the soils and soil components.
4. Quantify the influence of sorption on tetracycline and chlorotetracycline degradation rates.
5. Quantify the mobility of tetracycline and chlorotetracycline in soil column.

Methodology

Soil samples, surface (0 to 15 cm) and subsurface (≥15 cm), were collected from three sites representing three different soil series and a range of soil physical and chemical properties. Both the studied soils and the general sampling locations had been previously characterized (McBride et al., 1987). Based on interviews with the landowners or operators, specific sampling sites that had never received manure applications were selected. The soils were characterized using standard analytical procedures to determine pH in $CaCl_2$, pH in KCl, pH in water, organic C, organic H, organic N, percent sand, percent coarse silt, percent fine silt, percent clay, and extractable cations (Ca, Mg, Na, and K).

Soil components were physically and chemically separated from the soils and prepared for the sorption and degradation studies. Clay–humic complexes were isolated from the soils by sedimentation (<2 μm equivalent spherical diameter [esd]). Portions of the clay–humic complexes were K or Ca saturated by washing in 1-*M* KCl or 0.5-*M* $CaCl_2$ and then dialyzed against distilled water and freeze dried. Other portions of the clay–humic complexes were treated with 30% H_2O_2 for removal of the humic materials before being K and Ca saturated, dialyzed against distilled water, and freeze-dried. Humic materials were separated from the three soils by hydrolyzing Na-saturated samples in 0.1-*M* NaOH under an N_2 purge. After the hydrolysis, the humic materials were separated by centrifugation, neutralized to pH 7, K or Ca saturated, dialyzed, and freeze-dried.

A batch equilibration technique was used to measure sorption of tetracycline and chlorotetracycline on the various soils and prepared soil components. High-performance liquid chromatography (HPLC) was used to quantify tetracycline and chlortetracycline in the supernatant solutions and sorption was determined by difference. Variables tested included soil components (clay–humic complexes, clay minerals, and humic substances), saturating cations (K vs. Ca), and ionic strength ($I = 0.05$ and $I = 0.005$). Previous research had demonstrated the importance of pH; hence, in this study, the pH was carefully controlled at 6.5. The data were used to prepare four point sorption isotherms with three replications for each point.

> ***Note:*** HPLC is used for separating, identifying, purifying, and quantifying various compounds.

After the sorption studies were complete, tetracycline was incubated under both sterile and nonsterile conditions in aqueous controls and with soil components exhibiting both high and low sorption. Tetracycline was also incubated under both sterile and nonsterile conditions with the soil exhibiting the highest sorption capacity. Degradation kinetics were quantified for these systems by extracting tetracycline from the samples at various times during the incubations and quantifying parent and degradation products by HPLC.

The final stage of the research was a column leaching study. Intact soil columns treated with tetracycline and chlorotetracycline were leached with high and low ionic strength solutions with different ratios of K and Ca. The ionic strength and the K:Ca ratios of the leaching solutions were selected to both encourage and discourage colloid mobility. Leachate was analyzed by HPLC.

Principal Findings and Significance

Initial analysis of the samples collected provided basic soil characterization data. After that, however, the project was reorganized and refocused. Specifically, the research group performed a literature review, refined the hypotheses being tested, developed new specific objectives, and designed three major sets of experiments focused on testing those hypotheses. From the literature review, it was apparent that the effects of pH on sorption and degradation of tetracyclines in soil environments have been carefully studied. However, little information was available distinguishing whether tetracyclines are dominantly sorbed on soil clays or soil humic materials and on the effects of saturating cation and ionic strength on sorption, degradation, and mobility of tetracyclines in soils. Therefore, the focus of the project became targeted on filling these knowledge gaps.

The soils sampled for the study are listed in Table 10.6. Clay content of the sampled soils ranged from 19.2% in the Nicollet surface sample to 34.6% in the Clarinda subsoil sample. Organic C content ranged from 0.44% for the Fayette subsoil sample to 1.65 for the Fayette surface soil sample. Total exchangeable cations ranged from

TABLE 10.6
Soils Sampled for the Study

Sample Site	Soil Series	Classification
Tama Co.	Fayette	Fine-silty mixed superactive mesic, Typic Hapludalfs
Boone Co.	Nicollet	Fine-loamy mixed superactive mesic, Aquic Hapludolls
Clarke Co.	Clarinda	Fine smectitic mesic, Vertic Agriaquolls

13.6 $cmol_c/kg$ for the Nicollet surface soil to 19.7 $cmol_c/kg$ for the Fayette subsoil. The pH values in KCl ranged from 4.5 in the Clarinda subsoil sample to 6.5 in the Fayette surface soil sample. In general, the properties of the sampled soils were sufficiently diverse to allow a reasonable assessment of the influence of soil properties and soil components on the fate of antibiotics.

Preliminary chemical characterizations of tetracycline, chlorotetracycline, and oxytetracycline were performed. Ultraviolet–visible light (UV-VIS) absorbance spectra of the antibiotics dissolved in water and various concentrations of KCl, $CaCl_2$, $MgCl_2$, and $AlCl_3$ were obtained using a UV-VIS spectrophotometer (Varian Instruments, Cary 50 Bio model, Walnut Creek, CA). Calibration curves for quantifying concentrations of the various antibiotics dissolved in water were developed for two wavelengths, near 270 nm (Wl) and 370 nm (W2). Solubility of the oxytetracycline in water was measured by determining the concentration where the absorbance–concentration relationship deviated from Beer's law.

Potentiometric titrations indicated two and possibly three pKa values for the tetracyclines. The solubility of oxytetracycline was found to be approximately 300 mg/L. UV-VIS spectroscopy revealed two prominent absorption maxima near 280 nm and 360 nm for oxytetracycline and tetracycline and two prominent sorption maxima near 280 nm and 370 nm for chlorotetracycline. Absorption spectra for all three tetracyclines were only slightly affected by background $CaCl_2$ (0 to 50 meq/L) and $MgCl_2$ (0 to 40 meq/L) concentrations. By contrast, the presence of as little as 2 meq/L $AlCl_3$ substantially altered the absorbance spectra for all three tetracyclines. The cause of change in the absorbance spectra in the presence of $AlCl_3$ is not clear but may indicate either a pH effect or the formation of aluminum–tetracycline complexes. More work is needed to resolve the cause of this effect. The results demonstrate that tetracycline, chlorotetracycline, and oxytetracycline concentrations in water and both $CaCl_2$ and $MgCl_2$ solutions can be quantified by UV-VIS spectroscopy with linear responses for the 0- to 20-mg/L concentration range. The presence of Al in aqueous solution, however, may cause problems with spectrometric analysis.

Major accomplishments during the study included the physical separation and chemical preparation of cation-saturated soil components (clay–humic complexes, clay minerals, and humic materials) from the studied soils and the development and testing of an HPLC method for quantification of tetracyclines. Considerable effort was expended to develop the HPLC technique. The tetracyclines are not well behaved in HPLC because they have three ionizable moieties (i.e., portions of a molecular structure having some property of interest) and are zwitterions (i.e., compounds that carry both positive and negative charges in solution) over a large pH range. Several published HPLC methods performed poorly on this HPLC system, and considerable refinement of one of those methods was necessary to obtain high quality analytical data.

Hormones and Endocrine Disruption

The presence of estrogen and estrogen-like compounds in surface water has caused much concern. Their ultimate fate in the environment is unknown, although early studies indicated that no common soil or fecal bacteria can metabolize estrogen (Shore et al., 1995). When present in high concentrations, hormones in the environment are linked to reduced fertility, mutations, and the death of fish, and there is evidence that fish in some streams are experiencing endocrine disruptions (Mulla et al., 1999; Shore et al., 1995). Estradiol, an estrogen hormone, was found in runoff from a field receiving poultry litter at concentrations up to 3.5 μg/L. Fish exposed to 0.25 μg/L of estradiol often have gender changes; exposures at levels above 10 μg/L can be fatal (Mulla et al., 1999). Estrogen levels of 10 μg/L have been shown to affect trout (Shore et al., 1995).

Endocrine disruptors have also been the subject of increasing concern because they alter hormone pathways that regulate reproductive processes in both human and animal populations. Estrogen hormones have been implicated in the drastic reduction in sperm counts among European and North American men (Sharpe and Skakkebaek, 1993) and widespread reproductive disorders in a variety of wildlife (Colburn et al., 1993). A number of agricultural chemicals have also been demonstrated to cause endocrine disruption as well, including pesticides (Shore et al., 1995). The effects of these chemicals on the environment and their impacts on human health through environmental exposures are not completely understood. They are currently being studied for neurobiological, developmental, reproductive, and carcinogenic effects (Tetra Tech, 2000a). The USEPA is not aware of any studies done on the human health impact of hormones from watersheds that have impairment from animal manure.

Other Pollutants of Concern

Gas Emissions

Odor sources include animal confinement buildings, waste lagoons, and land application sites. As animal waste decomposes, various gases are produced. The primary gases associated with aerobic decomposition include carbon dioxide and ammonia. Gases associated with anaerobic conditions, which dominate in typical, unaerated animal waste lagoons, include methane, carbon dioxide, ammonia, hydrogen sulfide, and over 150 other odorous compounds (Bouzaher et al., 1993; O'Neill and Phillips, 1992; USDA, 1992). These include volatile fatty acids, phenols, mercaptans, aromatics, sulfides, and various esters, carbonyls, and amines. The decomposition process is desirable because it reduces the biochemical oxygen demand and pathogen content of the waste. However, many of the end products can have negative impacts, including strong odors. Heavy odors are the most common complaint from neighbors of swine operations in particular (Agricultural Animal Waste Task Force, 1996).

Odor is itself a significant concern because of its documented effect on moods, such as increased tension, depression, and fatigue (Schiffman et al., 1995). Odor also has the potential for vector attraction and affects property values. Additionally, many of the odor-causing compounds can cause physical health impacts; for example, hydrogen sulfide is toxic, and ammonia gas is a nasal and respiratory irritant.

In 1996, the Minnesota Department of Health found levels of hydrogen sulfide gas at residences near AFOs that were high enough to cause symptoms such as headaches, nausea, vomiting, eye irritation, respiratory problems (including shallow breathing and coughing), achy joints, dizziness, fatigue, sore throats, swollen glands, tightness in the chest, irritability, insomnia, and blackouts (Addis et al., 1999). In an Iowa study, neighbors within two miles of a 4000-sow swine facility reported more physical and mental health symptoms than a control group (Thu, 1998). These symptoms included chronic bronchitis, hyperactive airways, mucus membrane irritation, headache, nausea, tension, anger, fatigue, and confusion.

Methane and carbon dioxide are greenhouse gases that contribute to global warming. Methane also contributes to the formation of tropospheric ozone (a component of photochemical smog). Based on various USEPA estimates (USEPA, 1989, 1992a), methane emissions from U.S. animal wastes are a very small contributor to the global warming effects.

Particulates

Sources of particulate emissions from AFOs may include dried manure, feed, skin, hair, and possibly bedding. The airborne particles make up an organic dust, which includes endotoxins (the toxic protoplasm liberated when a microorganism dies and disintegrates), adsorbed gases, and possibly steroids (Thu, 1995). At least 50% of dust emissions from swine production facilities are believed to be respirable. The main impact downwind appears to be respiratory irritation due to the inhalation of organic dusts. Studies indicate that the associated microbes generally are not infectious but may induce inflammation (Thu, 1995).

Pesticides

Pesticides may pose risks to the environment, such as chronic aquatic toxicity, and human health effects, such as systemic toxicity. In a few studies, common herbicides have been shown to cause endocrine disruption. There is some evidence that fish in some streams are experiencing endocrine disruption and that contaminants including pesticides may be the cause (Mulla et al., 1999).

HUMAN-GENERATED BIOLOGICAL CONTAMINANTS IN SURFACE WATER

Like animal-generated contaminants, human-generated contaminants also can end up in surface water sediments and can harm aquatic life, threaten human health, or result in the loss of recreational or aesthetic potential. Human-generated surface water contaminants come from industrial sources, nonpoint sources, municipal sources, and background sources, among others. The question is, what are the contaminants of concern for surface water systems? The eight chief pollutants are biochemical oxygen demand, nutrients, suspended solids, pH, oil and grease, pathogenic microorganisms, toxic pollutants, and nontoxic pollutants. Many of the human-generated contaminants are the same as animal-generated contaminants, but there are also different types and levels of contaminants. Much of the material presented in the following was discussed earlier in animal-generated contamination sources.

Biochemical Oxygen Demand

Organic matter (dead plants, animal debris, and wild animal and bird feces), human sewage, food-processing wastes, chemical plant wastes, slaughterhouse wastes, pulp- and papermaking operations wastes, and tannery wastes discharged to a water body are degraded by oxygen-requiring microorganisms. The amount of oxygen consumed during microbial utilization of organics is the biochemical oxygen demand (BOD). BOD is classified as a *conventional pollutant* because it is amenable to treatment by a municipal sewage treatment plant. Although some natural BOD is almost always present, BOD is often an indication of the presence of sewage and other organic waste. High levels of BOD can deplete the oxygen in water. Fish and other aquatic organisms present in such waters with low oxygen conditions may die.

Nutrients

Elements such as carbon, nitrogen, phosphorus, sulfur, calcium, iron, potassium, manganese, cobalt, and boron are called *nutrients* (or biostimulants); they are essential to the growth and reproduction of aquatic plants and animals that depend on the surrounding water to provide their nutrients. However, just as too much of any good thing can have serious side effects for all of us, so it is the case with too many nutrients in water. For example, when fertilizers composed of nutrients enter surface water systems, over-enrichment with nitrogen and phosphorus may result. A rich supply of such nutrients entering a lake may hasten *eutrophication*, a process during which a lake evolves into a bog or marsh and eventually disappears. Excess nutrients also can stimulate a very abundant dense growth of aquatic plants (bloom), especially algae. Again, the two nutrients that concern us in this text are nitrogen and phosphorus.

Nitrogen (N_2), an extremely stable gas, is the primary component of the Earth's atmosphere (78%). The nitrogen cycle is composed of four processes. Three of the processes—fixation, ammonification, and nitrification—convert gaseous nitrogen into usable chemical forms. The fourth process—denitrification—converts fixed nitrogen back to the unusable gaseous nitrogen state. Nitrogen occurs in many forms in the environment and takes part in many biochemical reactions. Major sources of nitrogen include runoff from animal feedlots, fertilizer runoff from agricultural fields, municipal wastewater discharges, and certain bacteria and blue–green algae that obtain nitrogen directly from the atmosphere. Certain forms of acid rain can also contribute nitrogen to surface waters.

Nitrogen in water is commonly found in the form of nitrate (NO_3), the presence of which indicates that the water may be contaminated with sewage. Nitrates can also enter the groundwater from chemical fertilizers used in agricultural areas. Excessive nitrate concentrations in drinking water pose an immediate health threat to infants, both human and animal, and can cause death. The bacteria commonly found in the

DID YOU KNOW?

Lakes and reservoirs usually have less than 2 mg/L of nitrate measured as nitrogen. Higher nitrate levels are found in groundwater ranging up to 20 mg/L, but much higher values are detected in shallow aquifers polluted by sewage or excessive use of fertilizers.

intestinal tract of infants can convert nitrate to highly toxic nitrites (NO_2). Nitrite can replace oxygen in the bloodstream which results in oxygen starvation and causes a bluish discoloration of the infant ("blue baby" syndrome).

Phosphorus (P) is an essential nutrient that contributes to the growth of algae and the eutrophication of lakes, although its presence in drinking water has little effect on health. In aquatic environments, phosphorus is found in the form of phosphate and is a limiting nutrient. If all of the phosphorus is used up, plant growth ceases, no matter the amount of nitrogen available. Many bodies of freshwater currently experience influxes of nitrogen and phosphorus from outside sources. The increasing concentration of available phosphorus allows plants to assimilate more nitrogen before the phosphorus is depleted. If sufficient phosphorus is available, high concentrations of nitrates will lead to phytoplankton (algae) and macrophyte (aquatic plant) production. Major sources of phosphorus include phosphates in detergents, fertilizer and feedlot runoff, and municipal wastewater discharges. The 1976 USEPA water quality standards recommended a phosphorus criterion of 0.10 µg/L (elemental) phosphorus for marine and estuarine waters but offered no freshwater criterion.

pH

pH refers to the acidity or alkalinity of water; when it exceeds regulatory limits it is considered to be a conventional pollutant. A low pH may mean a water body is too acidic to support life optimally. Some water bodies are naturally acidic, but others are made so by acidic deposition or acid runoff from mining operations. pH is a measure of the hydrogen ion (H^+) concentration. Solutions range from very acidic (having a high concentration of H^+ ions) to very basic (having a high concentration of OH^- ions). The pH scale ranges from 0 to 14, with 7 being the neutral value. The pH of water is important to the chemical reactions that take place within water, and pH values that are too high or low can inhibit the growth of microorganisms. High pH values are considered basic, and low pH values are considered acidic. Stated another way, low pH values indicate a high H^+ concentration, while high pH values indicate a low H^+ concentration. Because of this inverse logarithmic relationship, there is a tenfold difference in H^+ concentration. Natural water varies in pH depending on its source. Pure water has a neutral pH, with an equal number of H^+ and OH^-. Adding an acid to water causes additional positive ions to be released, so that the H^+ ion concentration goes up and the pH value goes down; for example,

$$HCl \Leftrightarrow H^+ + Cl^-$$

When preparing freshwater for potable water uses, waterworks operators test for the hydrogen ion concentration of the water to determine the water's pH. In coagulation tests, as more alum (acid) is added, the pH value lowers. If more lime (alkali) is added, the pH value rises. This relationship is important in water treatment—if a good floc is formed, the pH should then be determined and maintained at that pH value until the raw water changes. Pollution can change a water's pH, which in turn can harm animals and plants living in the water. For example, water coming out of

DID YOU KNOW?

Seawater is slightly more basic (the pH value is higher) than most natural freshwater. Neutral water (such as distilled water) has a pH of 7, which is in the middle of being acidic and alkaline. Seawater happens to be slightly alkaline (basic), with a pH of about 8. Most natural water has a pH range of 6 to 8, although acid rain can have a pH as low as 4.

an abandoned coal mine can have a pH of 2, which is very acidic and would definitely affect any fish crazy enough to try to live in it. On a logarithmic scale, this mine drainage water would be 100,000 times more acidic than neutral water—so stay out of abandoned mines.

Solids

Natural water can contain a number of solid substances (what we may call impurities) or constituents. The concentrations of various solid substances in water in dissolved, colloidal, or suspended form are typically low but can vary considerably. A hardness value of up to 400 ppm of calcium carbonate, for example, is sometimes tolerated in public supplies, whereas 1 ppm of dissolved iron would be unacceptable. When a particular solids constituent can affect the good health of the water user or the environment, it is referred to as a contaminant or pollutant. These solid contaminants or pollutants are considered to be conventional pollutants. Other than gases, all contaminants of water contribute to the solids content. Natural water carries many dissolved and undissolved solids; these are considered to be conventional pollutants. The undissolved solids are nonpolar substances and consist of relatively large particles of materials, such as silt, that will not dissolve. Classified by their size and state, their chemical characteristics, and their size distribution, solids can be dispersed in water in both suspended and dissolved forms.

Solids in water can be classified as suspended solids, settleable, colloidal, or dissolved. Total solids are those suspended and dissolved solids that remain behind when the water is removed by evaporation. Solids are also characterized as volatile or nonvolatile. The distribution of solids is determined by computing the percentage of filterable solids by size range. Solids typically include inorganic solids such as silt and clay from riverbanks and organic matter such as plant fibers and microorganisms from natural or manmade sources.

Suspended solids are physical pollutants and may consist of inorganic or organic particles or of immiscible liquids. Inorganic solids such as clay, silt, and other soil constituents are common in surface water. Organic materials—plant fibers and biological solids—are also common constituents of surface waters. These materials are often natural contaminants resulting from the erosive action of water flowing over surfaces. Fine particles from soil runoff can remain suspended in water and increase its turbidity or cloudiness. This can stunt the growth of aquatic plants by limiting the amount of sunlight reaching them. Effluents from wastewater treatment plants and industrial plants and runoff from forestry and agricultural operations are sources of suspended solids. Note that because of the filtering capacity of the soil, suspended solids are seldom a constituent of groundwater. *Colloidal solids* are extremely fine suspended solids (particles) that are less than 1 micron (µm) in diameter; they are so small (though they still make water cloudy) that they will not settle even if allowed to sit quietly for days or weeks.

Solids in water affect how clear it is. Water's clarity, its *turbidity*, is one of the first characteristics people notice. Turbidity in water is caused by the presence of suspended matter, which results in the scattering and absorption of light rays. The greater the amount of total suspended solids (TSS) in the water, the murkier it appears and the higher the measured turbidity. Thus, in plain English, turbidity is a measure of the light-transmitting properties of water. Natural water that is very clear (low turbidity) allows you to see images at considerable depths. High-turbidity water, on the other hand, appears cloudy. Keep in mind that water of low turbidity is not necessarily without dissolved solids. Dissolved solids do not cause light to be scattered or absorbed, so the water looks clear. High turbidity causes problems for waterworks operators—components that cause high turbidity can also cause taste and odor problems and will reduce the effectiveness of disinfection.

Color in water can be caused by a number of solids (contaminants) such as iron, which changes in the presence of oxygen to yellow or red sediments. The color of water can be deceiving. In the first place, color is considered an aesthetic quality of water with no direct health impact. Second, many of the colors associated with water are not true colors but the result of colloidal suspension (apparent color). This apparent color can often be attributed to iron and to dissolved tannins extracted from decaying plant material. True color is the result of dissolved chemicals (most often organics) that cannot be seen. True color is distinguished from apparent color by filtering the sample.

Fats, Oil, and Grease

Fats, oil, and grease (FOG) are household wastes (conventional pollutants) that are routinely disposed of improperly down kitchen drains or flushed down toilets. Putting these greasy materials down a sink can lead to sewer clogs, and the clogs can lead to wastewater backing up into a home or business, spilling out onto the streets, and even making its way into storm drains and to the beaches.

Oil spills in or near surface water bodies can have a devastating effect on fish, other aquatic organisms, birds, and mammals. Note that spills are not the only source of oil in water; for example, oil leaking from automobiles and other vehicles or released during accidents is washed off roads with rainwater and into water bodies. Improper disposal of used oil from vehicles is another source; motor and other recreational boats release unburned fuel into water bodies.

DID YOU KNOW?

It is advisable to mix fats, oils, and grease with absorbent waste such as paper towels, kitty litter, coffee grounds, or shredded newspaper before discarding them in the trash.

Pathogenic Organisms

From the perspective of human use and consumption, the biggest concern associated with microorganisms is infectious disease. Microorganisms are naturally found in water (and elsewhere in the environment) and can cause infections. However, organisms that are not native to aquatic systems are of greatest concern—native or not, they can be transported by natural water systems. These organisms usually require an animal host for growth and reproduction. Nonpoint sources of these microorganisms include runoff from livestock operations and stormwater runoff. Point sources include improperly operating sewage treatment plants. When the surface water body provides drinking water to a community, the threat of infectious microorganism contamination is very real and may be life threatening. People who live in industrial nations with generally safe water supplies think of pathogenic contamination as a Third World problem. However, several problems in industrial nations (e.g., contamination of the local water supply in Sydney, Australia, in 1998) have alerted us to the very real possibility of dangerous contamination in our own water supplies.

Other pathogenic contamination problems in water and humans have manifested themselves via certain waterborne protozoans that can cause disease. Of particular interest to water pollution practitioners are *Entamoeba histolytica* (amebic dysentery and amebic hepatitis), *Giardia lamblia* (giardiasis), *Cryptosporidium* (cryptosporidiosis), and the emerging *Cyclospora* (cyclosporosis). Sewage contamination transports eggs, cysts, and oocysts of parasitic protozoa and helminthes (tapeworms, hookworms, etc.) into raw water supplies, leaving water treatment (in particular, filtration) and disinfection as the means by which to diminish the danger of contaminated water for the consumer.

To prevent the occurrence of *Giardia* and *Cryptosporidium* spp. in surface water supplies and to address increasing problems with waterborne diseases, the USEPA implemented its Surface Water Treatment Rule (SWTR) in 1989. The rule requires both filtration and disinfection of all surface water supplies as a means of primarily controlling *Giardia* spp. and enteric viruses. Since implementation of its Surface Water Treatment Rule, the USEPA has also recognized that *Cryptosporidium* species are agents of waterborne disease. In its 1996 series of surface water regulations, the USEPA included *Cryptosporidium*.

To test the need for and the effectiveness of the Surface Water Treatment Rule, LeChevallier et al. (1991) conducted a study on the occurrence and distribution of *Giardia* and *Cryptosporidium* organisms in raw water supplies to 66 surface water filter plants. These plants were located in 14 states and a Canadian province. A combined immunofluorescence test indicated that cysts and oocysts were widely dispersed in the aquatic environment. *Giardia* was detected in more than 80% of the samples, and *Cryptosporidium* was found in 85% of the sample locations. Taking into account several variables, 97% of the raw water samples indicated the presence of *Giardia* or *Cryptosporidium*. After evaluating their data, the researchers concluded that the Surface Water Treatment Rule might have to be upgraded (subsequently, it has been) to require additional treatment.

Giardia

Giardia lamblia (also known as the hiker'/traveler's scourge or disease) is a microscopic parasite that can infect warm-blooded animals and humans. Although *Giardia* was discovered in the 19th century, not until 1981 did the World Health Organization (WHO) classify *Giardia* as a pathogen. An outer shell called a *cyst* allows *Giardia* to survive outside the body for long periods of time. If viable cysts are ingested, *Giardia* can cause the illness known as *giardiasis*, an intestinal illness that can cause nausea, anorexia, fever, and severe diarrhea.

In the United States, *Giardia* is the most commonly identified pathogen in waterborne disease outbreaks. Contamination of a water supply by *Giardia* can occur in two ways: (1) by the activity of animals in the watershed area of the water supply, or (2) by the introduction of sewage into the water supply. Wild and domestic animals are major contributors to the contamination of water supplies. Studies have also shown that, unlike many other pathogens, *Giardia* is not host specific. In short, *Giardia* cysts excreted by animals can infect and cause illness in humans. Additionally, in several major outbreaks of waterborne diseases, the *Giardia* cyst source was sewage-contaminated water supplies.

Treating the water supply, however, can effectively control waterborne *Giardia*. Chlorine and ozone are examples of two disinfectants known to effectively kill *Giardia* cysts. Filtration of the water can also effectively trap and remove the parasite from the water supply. The combination of disinfection and filtration is the most effective water treatment process available today for prevention of *Giardia* contamination.

In drinking water, *Giardia* is regulated under the Surface Water Treatment Rule (SWTR). Although the SWTR does not establish a maximum contaminant level (MCL) for *Giardia*, it does specify treatment requirements to achieve at least 99.9% (3-log) removal or inactivation of *Giardia*. This regulation requires all drinking water systems using surface water or groundwater under the direct influence of surface water to disinfect and filter the water. The Enhanced Surface Water Treatment Rule (ESWTR), which includes *Cryptosporidium* and further regulates *Giardia*, was established in 1996.

Giardiasis

Giardiasis is recognized as one of the most frequently occurring waterborne diseases in the United States. *Giardia lamblia* cysts have been discovered in places as far apart as Estes Park, Colorado (near the Continental Divide); Missoula, Montana; Wilkes-Barre, Scranton, and Hazleton, Pennsylvania; and Pittsfield and Lawrence, Massachusetts, just to name a few. Giardiasis is characterized by intestinal symptoms that usually last one week or more and may be accompanied by one or more of the following: diarrhea, abdominal cramps, bloating, flatulence, fatigue, and weight loss. Although vomiting and fever are commonly listed as relatively frequent symptoms, people involved in waterborne outbreaks in the United States have not commonly reported them. Most *Giardia* infections persist for only 1 or 2 months, but some people experience a more chronic phase that can follow the acute phase or may become manifest without an antecedent acute illness. Loose stools and increased

DID YOU KNOW?

Three prescription drugs are available in the United States to treat giardiasis: quinacrine, metronidazole, and furazolidone. Based on a review of drug trials in which the efficacies of these drugs were compared (Davidson, 1984), quinacrine produced a cure in 93% of patients, metronidazole cured 92%, and furazolidone cured about 84% of patients.

abdominal gassiness with cramping, flatulence, and burping characterize the chronic phase. Fever is not common, but malaise, fatigue, and depression may ensue (Weller, 1985). For a small number of people, the persistence of infection is associated with the development of marked malabsorption and weight loss. Similarly, lactose (milk) intolerance can be a problem for some people. This can develop coincidentally with the infection or be aggravated by it, causing an increase in intestinal symptoms after ingestion of milk products. Some people may have several of these symptoms without evidence of diarrhea or have only sporadic episodes of diarrhea every three or four days. Still others may not have any symptoms at all. The problem, then, may not be one of determining whether or not someone is infected with the parasite but how harmoniously the host and the parasite can live together. When such harmony does not exist or is lost, it then becomes a problem of how to get rid of the parasite, either spontaneously or by treatment.

Giardiasis occurs worldwide. In the United States, *Giardia* is the parasite most commonly identified in stool specimens submitted to state laboratories for parasitologic examination. During a 3-year period, approximately 4% of 1 million stool specimens submitted to state laboratories tested positive for *Giardia* (CDC, 1979). Other surveys have demonstrated *Giardia* prevalence rates ranging from 1 to 20%, depending on the location and ages of persons studied. Giardiasis ranks among the top 20 infectious diseases that cause the greatest morbidity in Africa, Asia, and Latin America; it has been estimated that about 2 million infections occur per year in these regions (Walsh and Warren, 1979). People who are at highest risk for acquiring *Giardia* infection in the United States may be placed into five major categories:

1. People in cities whose drinking water originates from streams or rivers and whose water treatment process does not include filtration, or where filtration is ineffective because of malfunctioning equipment
2. Hikers, campers, and those who enjoy the outdoors
3. International travelers
4. Children who attend daycare centers, daycare center staff, and parents and siblings of children infected in daycare centers
5. Homosexual men

People in categories 1, 2, and 3 have in common the same general source of infection; that is, they acquire *Giardia* from fecally contaminated drinking water. City residents usually become infected because the municipal water treatment process does not include the filter necessary to physically remove the parasite from the water. The number of people in the United States at risk (i.e., the number who receive municipal drinking water from unfiltered surface water) is estimated to be 20 million. International travelers may also acquire the parasite from improperly treated municipal waters in cities or villages in other parts of the world, particularly in developing countries. In Eurasia, only travelers to Leningrad appear to be at increased risk. In prospective studies, 88% of U.S. and 35% of Finnish travelers to Leningrad who had negative stool tests for *Giardia* on departure to the Soviet Union developed symptoms of giardiasis and had positive tests for *Giardia* after they returned home (Brodsky et al., 1974). With the exception of visitors to Leningrad, however, *Giardia* has not been implicated as a major cause of traveler's diarrhea, as it has been detected in fewer than 2% of travelers who develop diarrhea. However, hikers and campers risk infection every time they drink untreated raw water from a stream or river. Persons in categories 4 and 5 become exposed through more direct contact with feces or an infected person—exposure to the soiled diapers of an infected child in cases associated with daycare centers or through direct or indirect anal–oral sexual practices in the case of homosexual men.

Although community waterborne outbreaks of giardiasis have received the greatest publicity in the United States during the past decade, about half of the *Giardia* cases discussed with the staff of the Centers for Disease Control and Prevention over a 3-year period had a daycare exposure as the most likely source of infection. Numerous outbreaks of *Giardia* in daycare centers have been reported in recent years. Infection rates for children in daycare center outbreaks range from 21 to 44% in the United States and from 8 to 27% in Canada (Black et al., 1981). The highest infection rates are usually observed in children who wear diapers (1 to 3 years of age).

Local health officials and managers of water utility companies need to realize that sources of *Giardia* infection other than municipal drinking water exist. Armed with this knowledge, they are less likely to make a quick (and sometimes wrong) assumption that a cluster of recently diagnosed cases in a city is related to municipal drinking water. Of course, drinking water must not be ruled out as a source of infection when a larger than expected number of cases is recognized in a community, but the possibility that the cases are associated with a daycare center outbreak, drinking untreated stream water, or international travel should also be entertained.

To understand the finer aspects of *Giardia* transmission and strategies for control, drinking water practitioners must become familiar with several aspects of the parasite's biology. Two forms of the parasite exist: a *trophozoite* and a *cyst*, both of which are much larger than bacteria. Trophozoites live in the upper small intestine, where they attach to the intestinal wall by means of a disc-shaped suction pad on their ventral surface. Trophozoites actively feed and reproduce at this location. At some time during the trophozoite's life, it releases its hold on the bowel wall and floats in the fecal stream through the intestine. As it makes this journey, it undergoes a morphologic transformation into an egg-like structure called a cyst. The cyst (about 6 to 9 nm in diameter and 8 to 12 μm in length) has a thick exterior wall that protects the parasite against the harsh elements that it will encounter outside the body. This cyst form of parasite is infectious to other people or animals. Most people become infected either directly (by hand-to-mouth transfer of cysts from the feces of an infected individual) or indirectly (by drinking feces-contaminated

DID YOU KNOW?

Giardia has an incubation period of 1 to 8 weeks.

water). Less common modes of transmission include ingestion of fecally contaminated food and hand-to-mouth transfer of cysts after touching a fecally contaminated surface. After the cyst is swallowed, the trophozoite is liberated through the action of stomach acid and digestive enzymes and becomes established in the small intestine.

Although infection after ingestion of only one *Giardia* cyst is theoretically possible, the minimum number of cysts shown to infect a human under experimental conditions is 10 (Rendtorff, 1954). Trophozoites divide by binary fission about every 12 hours. What this means in practical terms is that if a person swallowed only a single cyst, reproduction at this rate would result in more than 1 million parasites 10 days later and 1 billion parasites by day 15.

The exact mechanism by which *Giardia* causes illness is not yet well understood, but it is not necessarily related to the number of organisms present. Nearly all of the symptoms, however, are related to dysfunction of the gastrointestinal tract. The parasite rarely invades other parts of the body, such as the gall bladder or pancreatic ducts. Intestinal infection does not result in permanent damage.

Data reported by the CDC indicate that *Giardia* is the most frequently identified cause of diarrheal outbreaks associated with drinking water in the United States. The remainder of this section is devoted specifically to waterborne transmissions of *Giardia*. *Giardia* cysts have been detected in 16% of potable water supplies (lakes, reservoirs, rivers, springs, groundwater) in the United States at an average concentration of 3 cysts per 100 L (Rose et al., 1991). Waterborne epidemics of giardiasis are a relatively frequent occurrence. In 1983, for example, *Giardia* was identified as the cause of diarrhea in 68% of waterborne outbreaks in which the causal agent was identified. From 1965 to 1982, more than 50 waterborne outbreaks were reported (CDC, 1984). In 1984, about 250,000 people in Pennsylvania were advised to boil their drinking water for 6 months because of *Giardia*-contaminated water.

Many of the municipal waterborne outbreaks of *Giardia* have been subjected to intense study to determine their cause. Several general conclusions can be made from data obtained in those studies. Waterborne transmission of *Giardia* in the United States usually occurs in mountainous regions where community drinking water obtained from clear running streams is chlorinated but not filtered before distribution. Although mountain streams appear to be clean, fecal contamination upstream by human residents or visitors, as well as by *Giardia*-infected animals such as beavers, has been well documented. Water obtained from deep wells is an unlikely source of *Giardia* because of the natural filtration of water as it percolates through the soil to reach underground cisterns. Wells that pose the greatest risk of fecal contamination are poorly constructed or improperly located ones. A few outbreaks have occurred in towns that included filtration in the water treatment process, where the filtration was not effective in removing *Giardia* cysts because of defects in filter construction, poor maintenance of the filter media, or inadequate pretreatment of the water before filtration. Occasional outbreaks have also occurred because of accidental cross-connections between water and sewage systems. Two major ingredients are necessary for a waterborne outbreak: (1) *Giardia* cysts must be present in untreated source water, and (2) the water purification process must fail to either kill or remove *Giardia* cysts from the water.

Although beavers are often blamed for contaminating water with *Giardia* cysts, the fact that they are responsible for introducing the parasite into new areas seems unlikely. Far more likely is that they are also victims: *Giardia* cysts may be carried in untreated human sewage discharged into the water by small-town sewage disposal plants or they may originate from cabin toilets that drain directly into streams and rivers. Backpackers, campers, and sports enthusiasts may also deposit *Giardia*-contaminated feces in the environment, which are subsequently washed into streams by rain. In support of this concept is a growing amount of data indicating a higher *Giardia* infection rate in beavers living downstream from U.S. national forest campgrounds when compared with beavers living in more remote areas that have a near zero rate of infection.

Although beavers may be unwitting victims of the *Giardia* story, they still play an important part in the contamination scheme, because they can (and probably do) serve as amplifying hosts. An *amplifying host* is one that is easy to infect, serves as a good habitat for the parasite to reproduce, and, in the case of *Giardia*, returns millions of cysts to the water for every one ingested. Beavers are especially important in this regard, because they tend to defecate in or very near the water, which ensures that most of the *Giardia* cysts excreted are returned to the water.

The microbial quality of water resources and the management of the microbially laden wastes generated by the burgeoning animal agriculture industry are critical local, regional, and national problems. Animal wastes from cattle, hogs, sheep, horses, poultry and other livestock, and commercial animals can contain high con centrations of microorganisms, such as *Giardia*, that are pathogenic to humans.

The contribution of other animals to waterborne outbreaks of *Giardia* is less clear. Muskrats (another semiaquatic animal) have been found in several parts of the United States to have high infection rates (30 to 40%) (Frost et al., 1980). Studies have shown that muskrats can be infected with *Giardia* cysts from humans and beavers. Occasional *Giardia* infections have been reported in coyotes, deer, elk, cattle, dogs, and cats (but not in horses and sheep) encountered in mountainous regions of the United States. Naturally occurring *Giardia* infections have not been found in most other wild animals (e.g., bear, nutria, rabbit, squirrel, badger, marmot, skunk, ferret, porcupine, mink, raccoon, river otter, bobcat, lynx, moose, bighorn sheep).

Scientific knowledge about what is required to kill or remove *Giardia* cysts from a contaminated water supply has increased considerably. We know, for example, that cysts can survive in cold water (4°C) for at least 2 months, and they are killed instantaneously by boiling water (100°C) (Bingham et al., 1979). We do not know how long the cysts will remain viable at other water temperatures (e.g., at 0°C or in a canteen at 15 to 20°C), nor do we know how long the parasite will survive on various environment surfaces, such as under a pine tree, in the sun, on a diaper-changing table, or in carpets in a daycare center.

The effect of chemical disinfection (chlorination, for example) on the viability of *Giardia* cysts is an even more complex issue. The number of waterborne outbreaks of *Giardia* that have occurred in communities where chlorination was employed as a disinfectant process demonstrates that the amount of chlorine used routinely for municipal water treatment is not effective against *Giardia* cysts. These observations have been confirmed in the laboratory under experimental conditions (Jarroll et al., 1979). This does not mean that chlorine does not work at all. It does work under certain favorable conditions. Without getting too technical, gaining some appreciation of the problem can be achieved by understanding a few of the variables that influence the efficacy of chlorine as a disinfectant:

- *Water pH*—At pH values above 7.5, the disinfectant capability of chlorine is greatly reduced.
- *Water temperature*—The warmer the water, the higher the efficacy. Chlorine does not work in ice-cold water from mountain streams.
- *Organic content of the water*—Mud, decayed vegetation, or other suspended organic debris in water chemically combines with chlorine, making it unavailable as a disinfectant.
- *Chlorine contact time*—The longer that *Giardia* cysts are exposed to chlorine, the more likely it is that the chemical will kill them.
- *Chlorine concentration*—The higher the chlorine concentration, the more likely it is that chlorine will kill *Giardia* cysts. Most water treatment facilities try to add enough chlorine to give a free (unbound) chlorine residual at the customer tap of 0.5 mg per liter of water.

These five variables are so closely interrelated that improving one can often compensate for another; for example, if chlorine efficacy is expected to be low because water is obtained from an icy stream, the chlorine contact time or chlorine concentration, or both, could be increased. In the case of *Giardia*-contaminated water, producing safe drinking water with a chlorine concentration of 1 mg per liter and contact time as short as 10 minutes might be possible if all the other variables are optimal—a pH of 7.0, water temperature of 25°C, and total organic content of the water close to zero. On the other hand, if all of these variables are unfavorable—pH of 7.9, water temperature of 5°C, and high organic content—chlorine concentrations in excess of 8 mg/L with several hours of contact time may not be consistently effective. Because water conditions and water treatment plant operations (especially those related to water retention time and, therefore, to chlorine contact time) vary considerably in different parts of the United States, neither the USEPA nor the CDC has been able to identify a chlorine concentration that would be safe yet effective against *Giardia* cysts under all water conditions. For this reason, the use of chlorine as a preventive measure against waterborne giardiasis generally has been used under outbreak conditions when the amount of chlorine and contact time have been tailored to fit specific water conditions and the existing operational design of the water utility.

In an outbreak, for example, the local health department and water utility may issue an advisory to boil water, may increase the chlorine residual at the consumer's tap from 0.5 mg/L to 1 or 2 mg/L, and, if the physical layout and operation of the water treatment facility permit, increase the chlorine contact time. These are emergency procedures intended to reduce the risk of transmission until a filtration device can be installed or repaired or until an alternative source of safe water (a well, for example) can be made operational.

The long-term solution to the problem of municipal waterborne outbreaks of giardiasis involves improvements in and more widespread use of filters in the municipal water treatment process. The sand filters most commonly used in municipal water treatment today cost millions of dollars to install, which makes them unattractive for many small communities. The pore sizes in these filters are not sufficiently small to remove *Giardia* (6 to 9 μm by 8 to 12 μm). For the sand filter to remove *Giardia* cysts from the water effectively, the water must receive some additional treatment before it reaches the filter. The flow of water through the filter bed must also be carefully regulated.

An ideal prefilter treatment for muddy water would include sedimentation (a holding pond where large suspended particles are allowed to settle out by the action of gravity) followed by flocculation or coagulation (the addition of chemicals such as alum or ammonium to cause microscopic particles to clump together). The sand filter easily removes the large particles resulting from the flocculation–coagulation process, including *Giardia* cysts bound to other microparticulates. Chlorine is then added to kill the bacteria and viruses that may escape the filtration process. If the water comes from a relatively clear source, chlorine may be added to the water before it reaches the filter.

The successful operation of a complete waterworks operation is a complex process that requires considerable training. Troubleshooting breakdowns or recognizing the potential problems in the system before they occur often requires the skills of an engineer. Unfortunately, most small water utilities with water treatment facilities that include filtration cannot afford the services of a full-time engineer. Filter operation or maintenance problems in such systems may not be detected until a *Giardia* outbreak is recognized in the community. The bottom line is that, although filtration is the best protection against waterborne giardiasis that water treatment technology has to offer for municipal water systems, it is not infallible. For municipal water filtration facilities to work properly, they must be properly constructed, operated, and maintained.

Whenever possible, persons outdoors should carry drinking water of known purity with them. When this is not practical and when water from streams, lakes, ponds, or other outdoor sources must be used, time should be taken to properly disinfect the water before drinking it.

Cryptosporidium

Ernest E. Tyzzer first described the protozoan parasite *Cryptosporidium* in 1907. Tyzzer frequently found a parasite in the gastric glands of laboratory mice. Tyzzer identified the parasite as a sporozoan but of uncertain taxonomic status, and he named it *Cryptosporidium muris*. Later, in 1910, after more detailed study, he proposed *Cryptosporidium* as a new genus and *muris* as the type of species. Amazingly, except for developmental stages, Tyzzer's original description of the life cycle was later confirmed by electron microscopy. In 1912, Tyzzer described another new species, *Cryptosporidium parvum* (Tyzzer, 1912).

For almost 50 years, Tyzzer's discovery of the genus *Cryptosporidium* remained (like himself) relatively obscure because it appeared to be of no medical or economic importance. Slight rumblings of the importance of the genus began to be felt in the medical community when Slavin (1955) wrote about a new species, *Cryptosporidium melagridis*, which was associated with illness and death in turkeys. Interest remained slight even when *Cryptosporidium* was found to be associated with bovine diarrhea (Panciera et al., 1971).

Not until 1982 did worldwide interest focus on the study of organisms in the genus *Cryptosporidium*. At that time, the medical community and other interested parties were beginning a full-scale, frantic effort to find out as much as possible about acquired immune deficiency syndrome (AIDS), and the CDC reported that 21 AIDS-infected males from six large cities in the United States had severe protracted diarrhea caused by *Cryptosporidium*. It was in 1993, though, that *Cryptosporidium*—the "pernicious parasite"—made itself and Milwaukee famous (Mayo Foundation, 1996).

Note: The *Cryptosporidium* outbreak in Milwaukee caused the deaths of 100 people—the largest episode of waterborne disease in the United States in the 70 years since health officials began tracking such outbreaks.

The massive waterborne outbreak in Milwaukee (more than 400,000 persons developed acute and often prolonged diarrhea or other gastrointestinal symptoms) increased interest in *Cryptosporidium* at an exponential level. The Milwaukee incident spurred both public interest and the interest of public health agencies, agricultural and environmental agencies and groups, and suppliers of drinking water. This increase in interest level and concern spurred on new studies of *Cryptosporidium*, with an emphasis on developing methods for recovery, detection, prevention, and treatment (Fayer et al., 1997). The USEPA is particularly interested in this pathogen. In its reexamination of regulations on water treatment and disinfection, the USEPA issued a maximum contaminant level goal (MCLG) and contaminant candidate list (CCL) for *Cryptosporidium*. Its similarity to *Giardia lamblia* and the need for an efficient conventional water treatment capable of eliminating viruses forced the USEPA to regulate surface water supplies in particular. The Enhanced Surface Water Treatment Rule (ESWTR) includes regulations ranging from watershed protection to specialized operation of treatment plants (certification of operators and state overview) and effective chlorination. Protection against *Cryptosporidium* includes control of waterborne pathogens such as *Giardia* and viruses (De Zuane, 1997).

Cryptosporidium Basics

Cryptosporidium is one of several single-celled protozoan genera in the phylum Apircomplexa (all referred to as coccidian). *Cryptosporidium* along with other genera in the phylum Apircomplexa develop in the gastrointestinal tract of vertebrates through all of their life cycle; in short, they live in the intestines of animals and people. This microscopic pathogen causes a disease called *cryptosporidiosis*. The dormant (inactive) form of *Cryptosporidium* is called an *oocyst* and is excreted in the feces (stool) of infected humans and animals. The tough-walled oocysts survive under a wide range of environmental conditions.

TABLE 10.7
Valid Named Species of *Cryptosporidium*

Species	Host	Species	Host
C. baileyi	Chicken	*C. nasorium*	Fish
C. felis	Domestic cat	*C. parvum*	House mouse
C. meleagridis	Turkey	*C. serpentis*	Corn snake
C. murishouse	House mouse	*C. wrairi*	Guinea pig

Source: Fayer, R. et al., in *Cryptosporidium and Cryptosporidiosis*, Fayer, R., Ed., CRC Press, Boca Raton, FL, 1997, pp. 1–41.

Several species of *Cryptosporidium* were incorrectly named after the host in which they were found, and subsequent studies have invalidated many species. Now, eight valid species of *Cryptosporidium* (see Table 10.7) have been named. Upton (1997) reported that *C. muris* infects the gastric glands of laboratory rodents and several other mammalian species but is not known to infect humans (even though several texts state otherwise). *C. parvum*, however, infects the small intestine of an unusually wide range of mammals, including humans, and is the zoonotic species responsible for human cryptosporidiosis. In most mammals, *C. parvum* is predominately a parasite of neonate (newborn) animals. Upton (1997) pointed out that, even though exceptions occur, older animals generally develop poor infections, even when unexposed previously to the parasite. Humans are the one host that can be seriously infected at any time in their lives, and only previous exposure to the parasite results in either full or partial immunity to challenge infections.

Oocysts are present in most surface bodies of water across the United States, many of which supply public drinking water. Oocysts are more prevalent in surface waters when heavy rains increase runoff of wild and domestic animal wastes from the land or when sewage treatment plants are overloaded or break down. Only laboratories with specialized capabilities can detect the presence of *Cryptosporidium* oocysts in water. Unfortunately, current sampling and detection methods are unreliable. Recovering oocysts trapped on the material used to filter water samples is difficult. When a sample has been obtained, however, determining whether the oocyst is alive and if it is *C. parvum* and thus can infect humans can be easily accomplished by looking at the sample under a microscope.

The number of oocysts detected in raw (untreated) water varies with location, sampling time, and laboratory methods. Water treatment plants remove most, but not always all, oocysts. Low numbers of oocysts are sufficient to cause cryptosporidiosis, but the low numbers of oocysts sometimes present in drinking water are not considered cause for alarm in the public. Protecting water supplies from *Cryptosporidium* demands multiple barriers. Why? Because *Cryptosporidium* oocysts have tough walls that can withstand many environmental stresses and are resistant to chemical disinfectants such as chlorine that are traditionally used in municipal drinking water systems.

Physical removal of particles, including oocysts, from water by filtration is an important step in the water treatment process. Typically, water pumped from rivers or lakes into a treatment plant is mixed with coagulants, which help settle out particles suspended in the water. If sand filtration is used, even more particles are removed. Finally, the clarified water is disinfected and piped to customers. Filtration is the only conventional method now in use in the United States for controlling *Cryptosporidium*.

Ozone is a strong disinfectant that kills protozoa if sufficient doses and contact times are used, but ozone leaves no residual for killing microorganisms in the distribution system, as does chlorine. The high costs of new filtration or ozone treatment plants must be weighed against the benefits of additional treatment. Even well-operated water treatment plants cannot ensure that drinking water will be completely free of *Cryptosporidium* oocysts. Water treatment methods alone cannot solve the problem; watershed protection and monitoring of water quality are critical. Land use controls such as septic system regulations and best management practices to control runoff can help keep human and animal wastes out of water.

Under the Surface Water Treatment Rule of 1989, public water systems must filter surface water sources unless water quality and disinfection requirements are met and a watershed control program is maintained. This rule, however, did not address *Cryptosporidium*. The USEPA has now set standards for turbidity (cloudiness) and coliform bacteria (which indicate that pathogens are probably present) in drinking water. Frequent monitoring must occur to provide officials with early warning of potential problems to enable them to take steps to protect public health. Unfortunately, no water quality indicators can reliably predict the occurrence of cryptosporidiosis. More accurate and rapid assays of oocysts will make it possible to notify residents promptly if their water supply is contaminated with *Cryptosporidium* and thus avert outbreaks.

Bottom line: The collaborative efforts of water utilities, government agencies, healthcare providers, and individuals are needed to prevent outbreaks of cryptosporidiosis.

Cryptosporidiosis

> *Cryptosporidium parvum* is an important emerging pathogen in the U.S. and a cause of severe, life-threatening disease in patients with AIDS. No safe and effective form of specific treatment for cryptosporidiosis has been identified to date. The parasite is transmitted by ingestion of oocysts excreted in the feces of infected humans or animals. The infection can therefore be transmitted from person-to-person, through ingestion of contaminated water (drinking water and water used for recreational purposes) or food, from animal to person, or by contact with fecally contaminated environmental surfaces. Outbreaks associated with all of these modes of transmission have been documented. Patients with human immunodeficiency virus infection should be made more aware of the many ways that *Cryptosporidium* species are transmitted, and they should be given guidance on how to reduce their risk of exposure.
>
> **Juranek (1995)**

Since the Milwaukee outbreak, concern about the safety of drinking water in the United States has increased, and new attention has been focused on determining and reducing the risk of acquiring cryptosporidiosis from community and municipal water supplies. Cryptosporidiosis is spread by putting something in the mouth that has been contaminated with the stool of an infected person or animal. In this way, people swallow the *Cryptosporidium* parasite. As mentioned earlier, a person can become infected by drinking contaminated water or eating raw or undercooked food contaminated with *Cryptosporidium* oocysts, by direct contact with the droppings of infected animals or stools of infected humans, or by hand-to-mouth transfer of oocysts from surfaces that may have become contaminated with microscopic amounts of stool from an infected person or animal.

Symptoms may appear 2 to 10 days after infection by the parasite. Although some persons may not have symptoms, others have watery diarrhea, headache, abdominal cramps, nausea, vomiting, and low-grade fever. These symptoms may lead to weight loss and dehydration. In otherwise healthy persons, these symptoms usually last 1 to 2 weeks, at which time the immune system is able to defeat the infection. In persons with suppressed immune systems, such as persons who have AIDS or who recently have had an organ or bone marrow transplant, the infection may continue and become life threatening.

Currently, no safe and effective cure for cryptosporidiosis exists. People with normal immune systems improve without taking antibiotic or antiparasitic medications. The treatment recommended for this diarrheal illness is to drink plenty of fluids and to get extra rest. Physicians may prescribe medication to slow the diarrhea during recovery. The best way to prevent cryptosporidiosis is to

- Avoid water or food that may be contaminated.
- Wash hands after using the toilet and before handling food.
- Be sure, if you work in a daycare center, to wash your hands thoroughly with plenty of soap and warm water after every diaper change, even if you wear gloves when changing diapers.

During community-wide outbreaks caused by contaminated drinking water, drinking water practitioners should inform the public to boil drinking water for 1 minute to kill the *Cryptosporidium* parasite.

Cyclospora

Cyclospora organisms, which until recently were considered blue–green algae, were discovered at the turn of the 19th century. The first human cases of *Cyclospora* infection were reported in the 1970s. In the early 1980s, *Cyclospora* was recognized as a pathogen in patients with AIDS. We now know that *Cyclospora* is endemic in many parts of the world and appears to be an important cause of traveler's diarrhea. *Cyclospora* are two to three times larger than *Cryptosporidium* but otherwise have similar features. *Cyclospora* diarrheal illness in patients with healthy immune systems can be cured by a week of therapy with timethoprim–sulfamethoxazole (TMP–SMX).

So, what exactly is *Cyclospora*? In 1998, the CDC described *Cyclospora cayetanensis* as a unicellular parasite previously known as a cyanobacterium-like (blue–green algae-like) or coccidian-like body. The disease is known as *cyclosporiasis*. *Cyclospora* infects the small intestine and causes an illness characterized by diarrhea with frequent stools. Other symptoms can include loss of appetite, bloating, gas, stomach cramps, nausea, vomiting, fatigue, muscle ache, and fever. Some individuals infected with

Cyclospora may not show symptoms. Since the first known cases of illness caused by *Cyclospora* infection were reported in the medical journals in the 1970s, cases have been reported with increasing frequency from around the world (in part because of the availability of better techniques for detecting the parasite in stool specimens).

Huang et al. detailed what they believe is the first known outbreak of diarrheal illness associated with *Cyclospora* in the United States. The outbreak, which occurred in 1990, consisted of 21 cases of illness among physicians and others working at a Chicago hospital. Contaminated tap water from a physicians' dormitory at the hospital was the probable source of the organisms. The tap water probably picked up the organism while in a storage tank at the top of the dormitory after the failure of a water pump (Huang et al, 1995).

The transmission of *Cyclospora* is not a straightforward process. When infected persons excrete the oocyst state of *Cyclospora* in their feces, the oocysts are not infectious and may require from days to weeks to become so (i.e., to sporulate). Thus, transmission of *Cyclospora* directly from an infected person to someone else is unlikely; however, indirect transmission can occur if an infected person contaminates the environment and oocysts have sufficient time, under appropriate conditions, to become infectious. For example, *Cyclospora* may be transmitted by ingestion of water or food contaminated with oocysts. Outbreaks linked to contaminated water, as well as outbreaks linked to various types of fresh produce, have been reported (Herwaldt et al., 1997). The various modes of transmission and sources of infection are not yet fully understood nor is it known whether animals can be infected and serve as sources of infection for humans.

Persons of all ages are at risk for infection. Persons living or traveling in developing countries may be at increased risk, but infection can be acquired worldwide, including in the United States. In some countries of the world, infection appears to be seasonal. Based on currently available information, avoiding water or food that may be contaminated with stool is the best way to prevent infection. Reinfection can occur.

> ***Note:*** Pathogenic parasites are not easily removed or eliminated by conventional treatment and disinfection unit processes (De Zuane, 1997). This is particularly true for *Giardia lamblia, Cryptosporidium,* and *Cyclospora*. Filtration facilities can be adjusted with regard to depth, prechlorination, filtration rate, and backwashing to become more effective in the removal of cysts. The pretreatment of protected watershed raw water is a major factor in the elimination of pathogenic protozoa.

Toxic Pollutants

There are hundreds of potentially toxic water pollutants. Of these, the USEPA, under the Clean Water Act, regulates more than 100 pollutants of special concern. These include arsenic and the metals mercury, lead, cadmium, nickel, copper, and zinc. Organic toxic pollutants include benzene, toluene, and many pesticides, herbicides, and insecticides. Many of the most prevalent and persistent chemical contaminants, their fate in freshwater systems, and their sediment interface are discussed in the next chapter.

REFERENCES AND RECOMMENDED READING

Addis, P.G., Blaha, T., Crooker, B., Diez, F., Feirtag, J. et al. (1999). *Generic Environmental Impact Statement on Animal Agriculture: A Summary of the Literature Related to the Effects of Animal Agriculture on Human Health (K)*. St. Paul: Minnesota Environmental Quality Board.

Addiscott, T.M., Whitmore, A.P., and Powlson, D.S. (1991). *Farming, Fertilizers, and the Nitrate Problem*. Wallingford, Oxfordshire, U.K.: CAB International.

Agricultural Animal Waste Task Force. (1996). *Policy Recommendations for Management of Agricultural Animal Waste in North Carolina*. Durham, NC: Nicholas School of the Environment, Duke University.

Altekruse, S.F. (1998). *Campylobacter jejuni* in foods. *Journal of American Veterinary Medical Association*, 213(12): 1734–1735.

Aneja, V., Roelle, P.A., Murray, G.C., Southerland, J., Erisman, J.W., Fowler, D., Asman, W.A., and Patni, N. (2001). Atmospheric nitrogen compounds. II. Emissions, transport, transformation, deposition, and assessment. *Atmospheric Environment*, 35(11): 1903–1911.

Anon. (1999). State settles with farmers in 1997 manure spill. *Macomb Journal*, August 8.

ASAE. (1999). *Manure Production and Characteristics*, ASAE D384.1. St. Joseph, MI: American Society for Agricultural Engineers.

ASAE. (2005). *Manure Production and Characteristics*, ASAE D384.2. St. Joseph, MI: American Society for Agricultural Engineers.

Atwood, D.K., Bratkovich, A., Gallagher, M., and Hitchcock, G., Eds. (1994). Introduction to the dedicated issue. *Estuaries*, 17(4): 729–911.

Barrette, M. (1996). Hog-tied by feedlots. *Zoning News*, October, pp. 1–4.

Behm, D. (1989). Ill waters: the fouling of Wisconsin's lakes and streams. *The Milwaukee Journal*, November 5–10, p. 2.

Berner, R.A. (1980). *Early Diagensis: A Theoretical Approach*. Princeton, N.J., Princeton University Press.

Bingham, A.K., Jarroll, E.L, Meyer, E.A., and Radulescu, S. (1979). Introduction to *Giardia* excystation and the effect of temperature on cyst viability compared by eoxin exclusion and *in vitro* excystation. In: *Waterborne Transmission of Giardiasis* (Jakubowski, J. and Hoff, H.C., Eds.), EPA-600/9-79-001. Washington, DC: U.S. Environmental Protection Agency.

Black, R.E., Dykes, A.C., Anderson, K.E., Wells, J.G., Sinclair, S.P., Gary, G.W., Hatch, M.H., and Ginagaros, E.J. (1981). Handwashing to prevent diarrhea in day-care centers. *American Journal of Epidemiology*, 113: 445–451.

Bodek, I., Lyman, W.J., Reehl, W.F., and Rosenblatt, D., Eds. (1988). *Environmental Inorganic Chemistry: Properties, Processes, and Estimation Methods*. New York: Pergamon Press.

Bouzaher, A., Lakshminarayan, P.G., Johnson, S.R., Jones, T., and Jones, R. (1993). *The Economic and Environmental Indicators for Evaluating the National Pilot Project on Livestock and the Environment*, Livestock Series Report 1, Staff Report 93-SR 64. Ames: Iowa State University, Center for Agricultural and Rural Development; Stephenville, TX: Tarleton State University, Texas Institute for Applied Environmental Research.

Brady, N. (1990). *The Nature and Properties of Soils*, 10th ed. New York: Macmillan.

Brodsky, R.E., Spencer, H.C., and Schultz, M.G. (1974). Giardiasis in American travelers to the Soviet Union. *Journal of Infectious Diseases*, 130: 319–323.

Brook, J. (2000). Few left untouched after deadly *E. coli* flows through an Ontario town's water. *The New York Times*, July 10.

Bruning-Fann, C.S. and Kaneene, J.B. (1993). The effects of nitrate, nitrite, and *N*-nitroso compounds on human health: a review. *Veterinary and Human Toxicology*, 35(6): 521–538.

Burkholder, J.M. (1997). NCSU Aquatic Botany Laboratory *Pfiesteria piscicida* homepage, www2.ncsu.edu/unity/lockers/project/aquatic_botany/pfiest.

Cantrell, P., Perry, R., and Sturtz, P. (1999). Economic disaster: boom or bust? In: *Hog Wars: The Corporate Grab for Control of the Hog How Citizens Are Fighting Back*. Columbia: Missouri Rural Crisis Center (http://www.inmotionmagazine.com/hwdisas.html).

Cantrell, P., Perry, R., and Sturtz, P. (2004). *Hog Wars*. Part 4. *The Environment (... and Factory Farms)*. Columbia: Missouri Rural Crisis Center (http://inmotionmagazine.com/hwenv.html).

Carpenter, S., Caraco, N.F., Correll, D.L., Howarth, R.W., Sharpley, A.N., and Smith, V.H. (1998). Nonpoint pollution of surface waters with phosphorus and nitrogen. *Ecological Applications*, 8(3): 559–568.

Casman, E.A. (1996). *Chemical and Microbiological Consequences of Anaerobic Digestion of Livestock Manure: A Literature Review*, ICPRB Report #96-6. Rockville, MD: Interstate Commission on the Potomac River Basin.

CAST. (1992). *Water Quality: Agriculture's Role*, Report No. 120. Ames, IA: Council for Agricultural Science and Technology.

CDC. (1979). *Intestinal Parasite Surveillance, Annual Summary 1978*. Atlanta, GA: Centers for Disease Control and Prevention.

CDC. (1983). *Water-Related Disease Outbreaks Surveillance, Annual Summary 1983*. Atlanta, GA: Centers for Disease Control and Prevention.

CDC. (1995). *Giardiasis*. Atlanta, GA: Centers for Disease Control and Prevention.

CDC. (1998). *Report to the State of Iowa Department of Public Health on the Investigation of the Chemical and Microbial Constituents of Ground and Surface Water Proximal to Large-Scale Swine Operations*. Atlanta, GA: Centers for Disease Control and Prevention.

CDC. (2000). *A Public Health Action Plan to Combat Antimicrobial Resistance, Draft*. Atlanta, GA: Centers for Disease Control and Prevention (http://www.cdc.gov/drugresistance/actionplan/aractionplan.pdf).

Chai, T.J., Han, T., and Cockey, R.R. (1994). Microbiological quality of shellfish-growing waters in Chesapeake Bay. *Journal of Food Protection*, 57: 229–234.

Cieslak, P.R., Barret, T.J., Griffin, P.M., Gensheimer, K.F., Beckett, G., Buffington, J., and Smith, M.G. (1993). *Escherichia coli* O157:H7 infection from a manured garden. *Lancet*, 342(8867): 367.

Citizens *Pfiesteria* Action Commission. (1997). Final Report of the Citizens *Pfiesteria* Action Commission, Governor Harry R. Hughes, Commission Chairman.

Colburn, T., vom Saal, F.S., and Soto, A.M. (1993). Developmental effects of endocrine disrupting chemicals in wildlife and humans. *Environmental Health Perspectives*, 101: 378–384.

Collins E.R. et al. (1975). Effect of anaerobic swine lagoons on groundwater quality in high water table soils. In: *Managing Livestock Wastes: Proceedings of the 3rd International Symposium on Livestock Wastes*, pp. 303–305. St. Joseph, MI: ASAE.

Coyne, M.S. and Blevins, R.L. (1995). Fecal bacteria in surface runoff from poultry manured fields. In: *Animal Waste and the Land–Water Interface* (Steele, K., Ed.), pp. 77–87. Boca Raton, FL: Lewis Publishers.

Crane, S.R., Westerman P.W., and Overcash, M.R. (1980). Die-off of fecal indicator organisms following land application of poultry manure. *Journal of Environmental Quality*, 9: 531–537.

CWAA. (1998). *Minnesota Manure Spills and Runoff*. Minneapolis: Clean Water Action Alliance.

Daniel, T.C., Edwards, D.R., and Nichols, D.J. (1995). Edge-of-field losses of surface-applied animal manure. In: *Animal Waste and the Land–Water Interface* (Steele, K., Ed.), pp. 89–98. Boca Raton, FL: Lewis Publishers.

Davidson, R.A. (1984). Issues in clinical parasitology: the treatment of giardiasis. *American Journal of Gastroenterology*, 79(4): 256–261.

Dazzo, F., Smith, P., and Hubbell, D. (1973). The influence of manure slurry irrigation on the survival of fecal organisms in Scranton fine sand. *Journal of Environmental Quality*, 2: 470–473.

De Zuane, J. (1997). *Handbook of Drinking Water Quality*. New York: John Wiley & Sons.

Derbyshire, J.B. and Brown, E.G. (1978). Isolation of animal viruses from farm livestock waste, soil, and water. *Journal of Hygiene*, 81: 295–302.

Derbyshire, J.B., Clark, M.C., and Jessett, D.M. (1966). Observations on the fecal excretion of adenoviruses and enteroviruses in conventional and "minimal disease" pigs. *Veterinary Record*, 79: 595–599.

Dimura, J. (2000). New York State Department of Environmental Conservation, personal communication with Patricia Harrigan, USEPA, Washington, DC, on direct access of CAFO animals to surface water, via e-mail. August 10.

Dunne, T. and Leopold, L.B. (1978). *Water in Environmental Planning*. San Francisco, CA: W.H. Freeman.

Durborow, R.M. and Crosby, M.D. (2003). *Brown Blood Disease*, Information Sheet 1390. Mississippi State University: Mississippi Cooperative Extension Service.

Edwards, D.R. and Daniel, T.C. (1992). Potential runoff quality effects of poultry manure slurry applied to fescue plots. *Transactions of the ASAE*, 35(6): 1827–1832.

Edwards, D.R. and Daniel, T.C. (1993a). Effects of poultry litter application rate and rainfall intensity on quality of runoff from fescue grass plots. *Journal of Environmental Quality*, 22(2): 361–365.

Edwards, D.R. and Daniel, T.C. (1993b). Runoff quality impacts of swine manure applied to fescue plots. *Transactions of the ASAE*, 36(1): 81–86.

Edzwald, J.K., Upchurch, J.B., and O'Meilia, C.R. (1974). Coagulation in estuaries. *Environmental Science & Technology*, 8(1): 58–63.

Elder, D.J. and Fowler, S.W. (1977). Polychlorinated biphenyls: penetration into the deep ocean by zooplankton fecal pellet transport. *Science*, 197(4302): 459–461.

Elder, R.O., Keen, J.E., Siragusa, G.R., Barkocy-Gallagher, G.A., Koohmaraie, M., and Laegreid, W.W. (2000). Correlation of enterohemorrhagic *Escherichia coli* O157 prevalence in feces, hides, and carcasses of beef cattle during processing. *PNAS*, 97: 2999–3003.

Fales, S. et al. (2001). Predicting sorption, mobility, accumulation, and degradation potential of antibiotics in Iowa's soil/water environment, Project No. 2001IA1261B. In: *Iowa State Water Resources Research Institute Annual Technical Report FY 2001*, pp. 41–45. Ames: Iowa State University, Iowa State Water Resources Institute (http://water.usgs.gov/wrri/AnnualReports/2001/FY2001_IA_Annual_Report.pdf).

Fattal, B., Dotan, A., and Tchorsh, Y. (1992). Rates of experimental microbiological contamination of fish exposed to polluted water. *Water Resources Bulletin*, 26(12): 1621–1627.

Fayer, R., Speer, C.A., and Dubey, J.P. (1997). The general biology of *Cryptosporidium*. In: *Cryptosporidium and Cryptosporidiosis* (Fayer, R., Ed.), pp. 1–41. Boca Raton, FL: CRC Press.

Fey, P.D., Safranek, T.J., Rupp, M.E., Dunn, E.F., Ribot, E., Iwen, P.C., Bradford, P.A., Angulo, F.J., and Hinrichs, S.H. (2000). Ceftriaxone-resistant *Salmonella* infection acquired by a child from cattle. *The New England Journal of Medicine*, 342(17): 1242–1249.

Follet, R.R. (1995). *Fate and Transport of Nutrients: Nitrogen*, Working Paper No. 7. Fort Collins, CO: U.S. Department of Agriculture, Agricultural Research Service.

Frost, F., Plan, B., and Liechty, B. (1980). *Giardia* prevalence in commercially trapped mammals. *Journal of Environmental Health*, 42: 245–249.

Gerritse, R.F. and Zugec, I. (1977). The phosphorus cycle in pig slurry measured from $^{32}PO_4$ distribution rates. *Journal of Agricultural Science*, 88(1): 101–109.

Giddens, J., and Barnett, A.P. (1980). Soil loss and microbiological quality of runoff from land treated with poultry litter. *Journal of Environmental Quality*, 9: 518–520.

Goldman, C. and Horne, A. (1983). *Limnology*. New York: McGraw-Hill.

Goolsby, D.A., Battaglin, W.A., Lawrence, G.B., Artz, R.S., Aulenbach, B.T., Hooper, R.P., Keeney, D.R., and Stensland, G.J. (1999). Flux and sources of nutrients in the Mississippi–Atchafalaya river basin, Series No. 17, Topic 3. In: *Hypoxia in the Gulf of Mexico: Progress Towards the Completion of an Integrated Assessment*. Silver Spring, MD: National Oceanic and Atmospheric Administration (www.nos.noaa.gov/products/pubs_hypox.html).

Gresham, C.W. Janke, R.R., and Moyer, J. (1990). *Composting of Poultry Litter, Leaves, and Newspaper*. Maxatawny, PA: Rodale Research Center.

Griffin, P.M. (1998). Epidemiology of *Shiga* toxin-producing *Escherichia coli* infections in humans in the United States. In: *Escherichia coli O157:H7 and Other Shiga Toxin-Producing Escherichia coli Strains* (Kasper, J.B. and O'Brien A.D., Eds.), pp. 15–22. Washington, DC: ASM Press.

Health Canada Environmental Health Programs. (1998). *Blue–Green Algae (Cyanobacteria) and Their Toxins*. Ottawa: Health Canada.

Hegg, R.O., King, T.G., and Ianzen, I.I. (1981). *Four-Year Study of the Effect on Groundwater from a Dairy Lagoon in the Piedmont*. St. Joseph, MI: American Society of Agricultural and Biological Engineers.

Hegg, R.O., King, T.G., and Wilson, T.V. (1978). *The Effect on Groundwater from Seepage of Livestock Manure Lagoons*, WRRI Tech. Report 78. Clemson, SC: Clemson University.

Henry, C.M. (2000). Antibiotic resistance. *Chemical & Engineering News*, 78(10): 41–58.

Herman, T., Baker, S., and Stokka, G.I. (1995). *Medicated Feed Additives for Beef Cattle and Calves*, Publ. No. MF-2043. Manhattan: Kansas State University.

Herwaldt, F.L. et al. (1997). An outbreak in 1996 of cyclosporiasis associated with imported raspberries. *New England Journal of Medicine*, 336: 1548–1556.

Himathongkham, S., Bahari, S. Riemann, H., and Cliver, D. (1999). Survival of *Escherichia coli* O157:H7 and *Salmonella typhimurium* in cow manure and cow manure slurry. *FEMS Microbiology Letters*, 178: 251–257.

Hoosier Environmental Council. (1997). www.envirolink.org/orgs/hecweb/monitorspring97/confined.htm.

Hrubant, G.R. (1973). Characterization of the dominant aerobic microorganism in cattle feedlot waste. *Journal of Applied Microbiology*, 26: 512–516.

Humenick, F.J., Overcash, M.R., Baker, J.C., and Western, P.W. (1980). Lagoons: state of the art. In: *Proceedings of the 4th International Symposium on Livestock Wastes*, Amarillo, TX, April 15–17.

IDNR. (1998). *Prohibited Discharges at Iowa Livestock Operations Resulting in Monetary Penalties and/or Restitution of Fish Kill Being Proposed, Collected or Pending—1992–Present*. Des Moines: Iowa Department of Natural Resources.

Ikerd, J. (1998). *Large Scale, Corporate Hog Operations: Why Rural Communities Are Concerned and What They Should Do*. Columbia: University of Missouri (http://web.missouri.edu/~ikerdj/papers/top-10h.htm).

Industrial Economics, Inc. (1993). *Irrigation Return Flow Fee Feasibility Study*. Washington, DC: U.S. Environmental Protection Agency.

IRIS. (2000). *Chemical Files and Background Documents and Papers*. Washington, DC: Integrated Risk Information System.

ISA. (1997). *The Adverse Impacts of CAFOs Continue in Illinois without Regulation*. Springfield: Illinois Stewardship Alliance.

Jackson, G., Keeney, D., Curwen, D., and Webendorfer, B. (1987). *Agricultural Management Practices to Minimize Groundwater Contamination*. Madison: University of Wisconsin, Environmental Resources Center.

Jarroll, Jr., E.L., Gingham, A.K, and Meyer, E.A. (1979). *Giardia* cyst destruction: effectiveness of six small-quantity water disinfection methods. *American Journal of Tropical Medicine and Hygiene*, 29: 8–11.

Juranek, D.D. (1995). Cryptosporidiosis: sources of infection and guidelines for prevention. *Clinical Infectious Diseases*, 21(Suppl. 1): S37–S61.

Keller, D. and Miller, D. (1997). Neighbor against neighbor. *Progressive Farmer*, 112(11): 16–18.

Kudva. I., Blanch, K., and Hovde, C. (1998). Analysis of *Escherichia coli* O157:H7 survival in ovine or bovine manure and manure slurry. *Applied and Environmental Microbiology*, 64(9): 3166–3174.

LeChevallier, M.W., Norton, W.D., and Less, R.G. (1991). Occurrences of *Giardia* and *Cryptosporidium* spp. in surface water supplies. *Applied and Environmental Microbiology*, 57: 2610–2616.

Lee, W., Li, Z., Vakulenko, S., and Mobashery, S. (2000). A light-activated antibiotic. *Journal of Medicinal Chemistry*, 43: 128–132.

Levy, S.B. (1997). Antibiotic resistance: an ecological imbalance. In: *Antibiotic Resistance: Origins, Evolution, Selection and Spread* (Chadwick, D.J. and Goode, F., Eds.), pp. 1–14. Chichester: John Wiley & Sons.

Loehr, R.C. (1972). Animal waste management: problems and guidelines for solutions. *Journal of Environmental Quality*, (1(1): 71–78.

Lush, D.L. and Hynes, H.B.N. (1973). The formation of particles in freshwater leachates of dead leaves. *Limnology and Oceanography*, 18(5): 968–977.

Maule, A. (2000). Survival of verocytotoxigenic *Escherichia coli* O157 in soil, water and on surfaces. *Journal of Applied Microbiology*, 88: 71S–78S.

Mayo Foundation. (1996). *The "Bug" That Made Milwaukee Famous*. Rochester, MN: Mayo Foundation.

McBride, J.F., Horton, R., and Thompson, M.L. (1987). Evaluation of three Iowa soil materials as liners for hazardous-waste landfills. *Proceedings of the Iowa Academy of Science*, 94: 73–77.

McBride, W.D. (1997). *Change in U.S. Livestock Production, 1969–92*, AER-754. Washington, DC: U.S. Department of Agriculture.

McFall, W. (2000). USEPA Region X, personal communication with Patricia Harrigan, USEPA, Washington, DC, on direct access of CAFO animals to surface water via e-mail, August 2.

Meadows, R. (1995). Livestock legacy. *Environmental Health Perspectives*, 103(12): 1096–1100.

Mellon, M., Benbrook, C., and Benbrook, K. (2001). *Hogging It! Estimates of Antimicrobial Abuse in Livestock*. Cambridge, MA: Union of Concerned Scientists.

Meyer, M. (1994). How Common Is Methemoglobinemia from Nitrate-Contaminated Wells? A South Dakota Perspective, paper presented at the 39th Annual Midwest Groundwater Conference, Bismarck, ND, October 16–18.

Meyers, P.A., Leenheer, M.J., Eadie, B.J., and Maule, S.J. (1984). Organic geochemistry of suspended and settling particulate matter in Lake Michigan. *Geochimica et Cosmochimica Acta*, 48: 443–452.

Michel, K., Bacon, J.R., Gempesaw II, C.M., and Martin, Jr., J.H. (1996). *Nutrient Management by Delmarva Poultry Growers: A Survey of Attitudes and Practices*. Newark: University of Delaware, College of Agricultural Sciences.

Miller, H.H., Robinson, J.B., and Gillam, W. (1985). Self-sealing of earthen liquid manure storage ponds. I. A case study. *Journal of Environmental Quality*, 14: 533–538.

Milliard, P.S., Gensheimer, K.F., Addis, D.G., Sosin, D.M., Beckett, G.A., Houch-Jankoski, A., and Hudson, A. (1994). An outbreak of cryptosporidiosis from fresh-pressed apple cider. *JAMA*, 272(20): 1592–1596.

Morris, Jr., J.G., Matuszak, D.L., Taylor, J.L., Dickson, C., Benjamin, G.C., and Grattan, L.M. (1998). *Pfiesteria*!? *Maryland Medical Journal*, 47(3): 104–105.

Muirhead, S., Ed. (1998). *Feed Additive Compendium*. Minnetonka, MN: Miller Publishing.

Mulla, D., Sekely, A., Birr, J., Perry, B., Vondracek, E., Bean, E., Macbeth, S. Goyal, Wheeler, B., Alexander, C., Randall, G., Sands, G., and Linn, J. (1999). *Generic Environmental Impact Statement on Animal Agriculture: A Summary of the Literature Related to the Effects of Animal Agriculture on Water Resources (G)*. Minneapolis: University of Minnesota, College of Agriculture, Food, and Environmental Sciences.

National Research Council. (1993). *Soil and Water Quality: An Agenda for Agriculture*. Washington, DC: National Academy Press.

National Research Council. (1995). *Nitrate and Nitrite in Drinking Water*. Washington, DC: National Academy Press.

NCAES. (1982). *Best Management Practices for Agricultural Nonpoint Source Control: Animal Waste*. Raleigh: North Carolina State University, Biological and Agricultural Engineering Department.

Nelson, H. (1997). *The Contamination of Organic Produce by Human Pathogens in Animal Manures*. Montreal: McGill University.

Nolan, B.T. and Ruddy, B.C. (1996). *Nitrate in Ground Waters of the United States—Assessing the Risk*, USGS Fact Sheet FS-092-96. Reston, VA: U.S. Geological Survey.

Nowlin, M., Harris, D., and Rudek, J. (1997). Industrialization of animal production in North Carolina: impacts on the environment and public health, memorandum, August 19.

NRC. (1993). *Soil and Water Quality: An Agenda for Agriculture*. Washington, DC: National Research Council.

NRC. (1995). *Nitrate and Nitrite in Drinking Water*. Washington, DC: National Research Council.

NRC. (2000). *Clean Coastal Waters: Understanding and Reducing the Effects of Nutrient Pollution*. Washington, DC: National Research Council.

NYSDOH. (2000). Health Commissioner Releases *E. coli* Outbreak Report [press release]. Albany: New York State Department of Health.

ODNR. (1997). *Division of Wildlife Pollution Investigation Report—Manure Related Spills and Fish and Wildlife Kills*. Columbus: Ohio Department of Natural Resources.

Ohio EPA. (2008). *Concentrated Animal Feeding Operations*. Columbus: Ohio Environmental Protection Agency (http://www.epa.ohio.gov/dsw/cafo/index.aspx).

O'Neill, D.H. and Phillips. V.R. (1992). A review of the control of odor nuisance from livestock buildings. Part 3. Properties of the odorous substances which have been identified in livestock wastes or in the air around them. *Journal of Agricultural Engineering Research*, 53: 23–50.

Panciera, R.J., Thomassen, R.W., and Garner, R.M. (1971). Cryptosporidial infection in a calf. *Veterinary Pathology*, 8: 479.

Phillips, J.M., Scott, H.D., and Wolf, D.C. (1992). Environmental implications of animal waste application to pastures. In: *Proceedings from the South Pasture Forage Crop Improvement Conference*, pp. 30–38. Washington, DC: U.S. Department of Agriculture.

Poultry Water Quality Consortium. (1998). *Poultry Water Quality Handbook*, 2nd ed. Chattanooga, TN: Poultry Water Quality Consortium.

Price, D. (1999). The latest on *E. coli* O157:H7. *Beef Magazine*, November.

Proulx, A. (2002). *That Old Ace in the Hole*. New York: Scribner.

Rendtorff, R.C. (1954). The experimental transmission of human intestinal protozoan parasites. II. *Giardia lamblia* cysts given in capsules. *American Journal of Hygiene*, 59: 209–220.

Ritter, W.F. and Chirnside, A.E.M. (1990). Impact of animal waste lagoons on groundwater quality. *Biologic Wastes*, 34: 39–54.

Ritter, W.R., Humenik, F.J., and Skaggs, R.W. (1989). Irrigated agriculture and water quality in east. *Journal of Irrigation and Drainage Engineering*, 115(5): 807–821.

Ritter, W.R., Walpole, E.W., and Eastburn, R.P. (1980). An anaerobic lagoon for swine manure and its effect of the groundwater quality in sandy-loam soils. In: *Proceedings of the 4th International Symposium on Livestock Wastes*, Amarillo, TX, April 15–17.

Ritter, W.R., Walpole, E.W., and Eastburn, R.P. (1984). Effect of an anaerobic swine lagoon on groundwater quality in Sussex Country Delaware. *Agricultural Wastes*, 10: 267–284.

Robinson, J.S. and Sharpley, A.N. (1995). Release of nitrogen and phosphorus from poultry litter. *Journal of Environmental Quality*, 24(1): 62–67.

Rodecap, J. (2002). Nitrate in surface water—trends and concerns. *Water Watch*.

Rose, J.B., Gerb, C.P., and Jakubowski, W. (1983). Survey of potable water supplies for *Cryptosporidium* and *Giardia. Environmental Science & Technology*, 25: 1393–1399.

Schiffman, S.S., Sattely Miller, E.A., Suggs, M.S., and Graham, B.G. (1995). The effect of environmental odors emanating from commercial swine operations on the mood of nearby residents. *Brain Research Bulletin*, 37: 369–375.

Schwab, J. (1998). *Planning and Zoning for Concentrated Animal Feeding Operations*. Chicago, IL: APA Planning Advisory Service.

Sewell, H.B. (1993). *Feed Additives for Beef Cattle*, Agric. Publ. G02075. Columbia: University of Missouri, Department of Animal Sciences.

Sewell, J.I., Mulling, J.A., and Vaigneur, H.O. (1975). Dairy lagoon system and groundwater quality. In: *Managing Livestock Wastes: Proceedings of the 3rd International Symposium on Livestock Wastes*, pp. 286–288. St. Joseph, MI: ASAE.

Sharpe, R.M. and Skakkebaek, N.E. (1993). Are oestrogens involved in falling sperm count and disorders of the male reproductive tract? *Lancet*, 341: 1392–1395.

Sheldon, R.W., Sutcliffe, W.H., and Prakash, A. (1973). The production of particles in the surface waters of the ocean with particular reference to the Sargasso Sea. *Limnology and Oceanography*, 18(5): 719–733.

Shelton, D.R. (2000). Sources of pathogens in a watershed: humans, wildlife, farm animals? In: *Managing Nutrients and Pathogens from Animal Agriculture: Proceedings of a Conference for Nutrient Management Consultants, Extension Educators, and Producer Advisors*, pp. 108–115. Ithaca, NY: Natural Resource, Agriculture, and Engineering Service.

Sherer, B.M., Miner, J.R., Moore, J.A., and Buckhouse, J.C. (1988). Resuspending organisms from a rangeland stream bottom. *Transactions of the ASAE*, 31: 1217–1222.

Sherer, B.M., Miner, J.R., Moore, J.A., and Buckhouse, J.C. (1992). Indicator bacterial survival in stream sediments. *Journal of Environmental Quality*, 21: 591–595.

Shore, L.S., Correll, D.L., and Chakraborty, P.K. (1995). Relationship of fertilization with chicken manure and concentrations of estrogens in small steams. In: *Animal Waste and the Land–Water Interface* (Steele, K., Ed.), pp. 155–163. Boca Raton, FL: Lewis Publishers.

Sierra Club. (2004). *Clean Water and Factory Farms: Reports and Factsheets: Air Pollution from Factory Farms*. Oakland, CA: Sierra Club (http://www.sierraclub.org/factoryfarms/factsheets/air.asp).

Sigleo, A.C., Helz, G.R., and Zoller, W.H. (1980). Organic-rich colloidal material in estuaries and its alteration by chlorination. *Environmental Science & Technology*, 14(6): 673–679.

Sigleo, A.C., Hoering, T.C., and Helz, G.R. (1982). Composition of estuarine colloidal material: organic components. *Geochimica et Cosmochimia Acta*, 46: 1619–1626.

Simpkins, W.W., Burkart, M.R., Helmke, M.F. et al. (1999). *Hydrogeologic Settings of Selected Earthen Waste Storage Structures Associated with Confined Animal Feeding Operations in Iowa in Earthen Waste Storage Structures in Iowa: A Study for the Iowa Legislature*, EDC-186. Ames: Iowa State University.

Sims, J.T. (1995). Characteristics of animal wastes and waste-amended soils: an overview of the agricultural and environmental issues. In: *Animal Waste and the Land–Water Interface* (Steele, K., Ed.), pp. 1–14. Boca Raton, FL: Lewis Publishers.

Skipton, S. and Hay, D. (1998). *Drinking Water: Nitrate and Methemoglobinemia ("Blue Baby" Syndrome)*, G98-1369. Lincoln: University of Nebraska–Lincoln Extension (http://digitalcommons.unl.edu/cgi/viewcontent.cgi?article=2429&context=extensionhist).

Slavin, D. (1955). *Cryptosporidium melagridis. Journal of Comparative Pathology*, 65: 262.

Smith, V. (1994). Disaster in Milwaukee: complacency was the root cause. *EPA Journal*, 20: 16–18.

Spellman, F.R. (1996). *Stream Ecology and Self-Purification: An Introduction for Wastewater and Water Specialists*. Boca Raton, FL: CRC Press.

Spellman, F.R. (1998). *The Science of Environmental Pollution*. Boca Raton, FL: CRC Press.

Stanley, K., Cunningham, R., and Jones, K. (1998). Isolation of *Campylobacter jejuni* from groundwater, *Journal of Applied Microbiology*, 85: 187–191.

Steen, W.C. and Karickhoff, S.W. (1981). Biosorption of hydrophobic organic pollutants by mixed microbial populations. *Chemosphere*, 10(1): 27–32.

Stehman, S.M. (2000). Ag-related waterborne pathogens. In: *Managing Nutrients and Pathogens from Animal Agriculture: Proceedings of a Conference for Nutrient Management Consultants, Extension Educators, and Produce Advisors*, pp. 93–107. Ithaca, NY: Natural Resource, Agriculture, and Engineering Service.

Stelma, Jr., G.N. and McCabe, L.J. (1992). Nonpoint pollution from animal sources and shellfish sanitation. *Journal of Food Protection*, 55(8): 649–656.

Sutton, A. and Humenik, F. (2003). *Technology Options to Comply with Land Application Rules*, Fact Sheet No. 24. Ames: Iowa State University, MidWest Plan Service.

Tetra Tech. (2000a). *Literature Review and Assessment of Pathogens, Heavy Metals, and Antibiotic Content of Waste and Wastewater Generated by CAFOs*, EPA Contract 68-C-99-263. Washington, DC: U.S. Environmental Protection Agency.

Tetra Tech. (2000b). *Development of Pollutant Loading Reductions from the Implementation of Nutrient Management and Best Management Practices*. Washington, DC: U.S. Environmental Protection Agency.

Thomann, R.V. and Mueller, J.A. (1987). *Principles of Surface Water Quality Modeling and Control*. New York: Harper Collins.

Thu, K., Ed. (1995). *Understanding the Impacts of Large-Scale Swine Production: Proceedings from an Interdisciplinary Scientific Workshop*, Des Moines, IA, June 29–30.

Thu, K. (1998). Presentation at the Centers for Disease Control Workshop on Public Health Issues Related to Concentrated Animal Feeding Operations, Washington, DC, June 23–24.

Tyzzer, E.E. (1912). *Cryptosporidium parvum* sp.: a coccidium found in the small intestine of the common mouse. *Archiv fur Protistenkunde*, 26: 394.

Upton, S.J. (1997). *Basic Biology of Cryptosporidium*. Manhattan: Kansas State University.

U.S. Senate. (1997). *Animal Waste Pollution in America: An Emerging National Problem*, report compiled by the Minority Staff of the U.S. Senate Committee on Agriculture, Nutrition and Forestry for Senator Tom Harkin.

USDA. (1992). *Agricultural Waste Management Field Handbook*. Washington, DC: U.S. Department of Agriculture.

USDA/ARS (1998). *Agricultural Uses of Municipal Animal, and Industrial Byproducts*, Report No. 44. Washington, DC: U.S. Department of Agriculture, Agricultural Research Service.

USDA/ARS (1999). *Agricultural Phosphorus and Eutrophication*, ARS-149. Washington, DC: U.S. Department of Agriculture, Agricultural Research Service.

USDA/NRCS (1996). *Agricultural Waste Management Field Handbook*, 210-AWMFH. U.S. Department of Agriculture, National Agricultural Statistics Service.

USDA/USEPA. (1999). Comprehensive nutrient management plan components. In: *Unified National Strategy for Animal Feeding Operations*, Chapter 3.3. Washington, DC: U.S. Department of Agriculture and U.S. Environmental Protection Agency.

USEPA. (1989). *Policy Options for Stabilizing Global Climate*. Washington, DC: U.S. Environmental Protection Agency.
USEPA. (1992a). *Global Methane Emissions from Livestock and Poultry Manure*, EPA 400/1-91-048. Washington, DC: U.S. Environmental Protection Agency.
USEPA. (1992b). *Managing Nonpoint Source Pollution: Final Report to Congress*. Washington, DC: U.S. Environmental Protection Agency.
USEPA. (1992c). *Human Health Risk Assessment for the Use and Disposal of Sewage Sludge: Benefits of Regulation*, EPA 822-B-92-003. Washington, DC: U.S. Environmental Protection Agency.
USEPA. (1993a). *Guidance Specifying Management Measures for Sources of Nonpoint Pollution in Coastal Waters*, EPA 840-B-92-002. Washington, DC: U.S. Environmental Protection Agency.
USEPA. (1993b). *The Reports of the EPA/State Feedlot Workgroup*. Washington, DC: U.S. Environmental Protection Agency.
USEPA. (1993c). *Standards for the Use of Disposal of Sewage Sludge*, 40 CFR 503. Washington, DC: U.S. Environmental Protection Agency.
USEPA. (1997). *Section 319 Success Stories*. Vol. II. *Highlights of State and Tribal Nonpoint Source Programs*, EPA-R-97-001. Washington, DC: U.S. Environmental Protection Agency.
USEPA. (2000). *National Water Quality Inventory: 1998 Report to Congress*, EPA 841-R-00-001. Washington, DC: U.S. Environmental Protection Agency.
USEPA. (2001). *Environmental Assessment of Proposed Revisions to the National Pollutant Discharge Elimination System Regulations and the Effluent Guidelines for Concentrated Animal Feeding Operations*, EPA-821-B-01-001. Washington, DC: U.S. Environmental Protection Agency.
USEPA. (2003). *What's the Problem?* Washington, DC. U.S. Environmental Protection Agency (www.epa.gov/region.09/crossprlanimalwaste/problem.html).
USFWS. (1991). *Contaminants in Buffalo Lake National Wildlife Refuge, Texas*. Arlington, TX: U.S. Fish and Wildlife Service.
USFWS. (2000). *Environmental Contaminants Associated with Confined Animal Feeding Operations and Their Impacts to a Service Waterfowl Production Area*. Lakewood, CO: U.S. Fish and Wildlife Service.
USGAO. (1997). *Drinking Water: Information on the Quality of Water Found at Community Water Systems and Private Wells*, GAO/RCED-97-123. Washington, DC: U.S. Government Accountability Office.
USGS. (1997). *Nutrients in the Nation's Waters—Too Much of a Good Thing?* Washington, DC: U.S. Geological Survey, National Water Quality Assessment Program.
Vallentyne, J. (1974). *The Algal Bowl: Lakes and Man*, Misc. Special Publ. No. 22. Quebec: Canadian Government Publishing Center Supply and Services.
Van Donsel, D., Geldreich, E., and Clark, H. (1967). Seasonal variations in the survival of indicator bacteria in soil and their contribution to storm water pollution. *Applied Microbiology*, 15: 1362–1370.
Vanderholm, D.H. (1975). Nutrient losses from livestock waste during storage, treatment and handling. In: *Managing Nutrients and Pathogens from Animal Agriculture: Proceedings of a Conference for Nutrient Management Consultants, Extension Educators, and Produce Advisors*, pp. 282–285. Ithaca, NY: Natural Resource, Agriculture, and Engineering Service.
Warrick, J. (1995). Hog farm is fined $110,000 for spill. *The News and Observer*, August 23.
Weller, P.F. (1985). Intestinal protozoa: giardiasis. *Scientific American Medicine*, 12(4): 554–558.
Walsh, J.D. and Warren, K.S. (1979). Selective primary health care: an interim strategy for disease control in developing countries. *New England Journal of Medicine*, 301: 974–976.

Wendt, R.C. and Corey, R.B. (1980). Phosphorus variations in surface runoff from agricultural lands as a function of land use. *Journal of Environmental Quality*, 9(1): 130–136.

Wetzel, R.G. (1983). *Limnology*, 2nd ed. London: Saunders College.

Zirbser, K. (1998). Memo to administrative record, feedlots point source category study, re: bacterial count information from Dorothea Paul, February 5. (Mr. Zirbser reported results of independent monitoring of a stream flowing through Ms. Paul's property, downstream of a hog farm.)

11 Chemically Contaminated Sediments

During a recent Mr. Rabbit and Ms. Grasshopper meeting, Mr. Rabbit observed that Ms. Grasshopper was about to quench her thirst or cool off or whatever at a dripping outdoor water faucet. Mr. Rabbit shouted to his friend Ms. Grasshopper: *"Don't drink that!"* Ms. Grasshopper, startled, her left tarsus scratching the spiracles of her abdomen, shot a glance at her friend Mr. Rabbit. "My dearest friend," he said, "I do not want you poisoned by the lead and other contaminants in that water. I know you and I have enjoyed water from that faucet before, but that was before a city administrator decided to change where the water came from. Now it comes from a local river instead of one of the Great Lakes." Mr. Rabbit had Ms. Grasshopper's full attention. "Unfortunately …," Mr. Rabbit paused a moment to thump one of his feet. He then continued, "You, see my dear, beloved friend, that human in charge did not test the river water before allowing it to flow in the pipes and taps. Corrosive water has leached lead from the old city pipes and contaminated the humans and our animal friends who drank the water and ingested dangerously high levels of harmful lead."

There was a moment of silence.

Finally, Ms. Grasshopper spoke: "Mr. Rabbit, you have always been my Knight in Furry White Armor and I want to thank you for warning me. I will simply fly and leap on over to the lake to take care of business."

There was another moment of silence.

Then Mr. Rabbit spoke up and said, "Well, Ms. Grasshopper, if I could fly, I would go with you right now. I am a bit thirsty, you know."

Ms. Grasshopper replied, "Well, my prince, if you were a bit smaller and lighter I would load you on my back and fly you out of here with me. But, I can't. I will think about you, though, when I get to that clean water and will gulp some for you. And, as far as understanding that human who switched the water supply from good to bad water, that is beyond my small brain to comprehend. I just know that doesn't make any sense, but I am not sure why."

Mr. Rabbit said, "Right on, Ms. Grasshopper."

Chuang Tzu and Hui Tzu were standing on the bridge across the Hao River. Chuang Tzu said, "Look how the minnows are shooting to and fro! How joyful they are!" "You are not a fish," said Hui Tzu. "How can you know that the fishes are joyful?" "You are not I," answered Chuang Tzu. "How can you know I do not know about the joy of fishes? I know it from my own joy of the water."

Ancient Chinese story

Contaminated sediment can cause lethal and sublethal effects in benthic (sediment-dwelling) and other sediment-associated organisms. In addition, natural and human disturbances can release pollutants to the overlying water, where pelagic (water column) organisms can be exposed. Sediment pollutants can reduce or eliminate species of recreational, commercial, or ecological importance, either through direct effects or by affecting the food supply that sustainable populations require. Furthermore, some sediment pollutants can bioaccumulate through the food chain and pose health risks to wildlife and human consumers even when sediment-dwelling organisms are not themselves impacted (USEPA, 2001).

INTRODUCTION

When persistent and highly sorptive chemical contaminants in air, water discharges, and soil find their way into lakes, streams, and rivers, sediments provide excellent repositories. Chemically contaminated sediments have been a legacy of rapid industrial and agricultural development in the United States since the mid-19th century. Many relatively nonbiodegradable chemicals originating from these developments have spread throughout the environment into water, soil, and sediment. Because chemicals may persist for indefinite periods, these chemically contaminated sediments can have substantial effects on the ecosystem, ranging from direct effects on benthic communities to substantial contributions to contaminant loads and effects on upper trophic levels* through food chain chemical contamination of various species. The major contaminant loads, organic and inorganic, are listed below; some of these contaminants have been discussed earlier. Organic water pollutants include the following:

- Detergents
- Disinfection byproducts found in chemically disinfected drinking water, such as chloroform
- Perchlorate
- Pharmaceutical drugs and their metabolites
- Volatile organic compounds, such as industrial solvents
- Chemical compounds found in personal hygiene and cosmetic products
- Insecticides and herbicides
- Petroleum hydrocarbons (fuels and lubricants)
- Chlorinated solvents (nonaqueous dense-phase liquids) including polychlorinated biphenyls (PCBs) and trichloroethylene
- Food processing waste, which can include oxygen-demanding substances, fats, and grease

Inorganic water pollutants include the following:

- Industrial byproduct chemical waste
- Sediment (silt) from runoff from construction sites, logging, slash and burn practices, or land clearing sites
- Ammonia from food processing waste
- Heavy metals from motor vehicles and acid mine drainage (yellow boy)
- Nutrients from fertilizers
- Industrial discharges

* Trophic levels are the feeding positions in a food web—that is, groups of organisms that are united by the fact that they all obtain their energy from the same part of a food web of a biological community. One example of such a group would be peregrine falcons, bald eagles, and other raptors (Orians, 1995).

Sediments are important sinks for such contaminants because of their enormous quantities and their abilities to pick up, or sorb, large amounts of a wide variety of contaminants. Because this chapter details sediments that eventually become deposited in surface water bodies it is essential to understand the mechanism by which the chemical contaminant is sorbed to sediment. Understanding the interactions between the sediment and water-column compartments helps to reveal how specific contaminants are transformed and transported within the surface water body.

CONTAMINANT SORPTION BY SEDIMENTS

Sorption to sediments is probably the most influential factor on the transport and fate of organic contaminants in the environment. The extent of the sorption to sediment affects not only the contaminant level in an ecosystem but also the movement and fate of the contaminant. For example, in a hydrogeologic system, the increased sorption of a contaminant to sediment reduces its level in the adjacent water column and thus decreases its exposure and transport to other parts of the ecosystem, such as fish and plankton (see Figure 11.1).

Sediment Mineral Matter vs. Sediment Organic Matter

The way in which contaminants are sorbed into sediment varies with the nature of the contaminant and the makeup of the sediment (Chiou et al., 1979; Karickhoff et al., 1979). The composition of sediment includes both mineral matter and organic matter as the primary constituents (see Figure 11.1). Under relatively dry conditions, the sediment mineral matter acts as an *adsorbent*, where the sorbed organic compounds are held on the surface of the mineral grains. The sediment organic matter

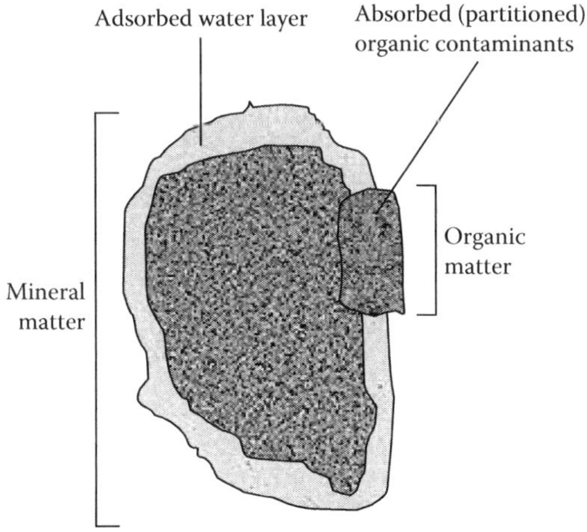

FIGURE 11.1 In the presence of water with many contaminants, water is adsorbed on the surface of mineral matter, whereas contaminants are absorbed into the organic matter by a partition process. (From USGS, *Contaminant Sorption by Soil and Bed Sediment*, Fact Sheet 087-00, U.S. Geological Survey, Reston, VA, 2000.)

(SOM) acts as a *absorbent*, or a partition medium, when the sorbed organic compounds dissolve (partition) into the matrix of the entire SOM (see Figure 11.1). The sediment, then, is characterized as a dual-function sorbent, in which the mineral matter sorbs the contaminant by adsorption while the SOM sorbs that contaminant by a partition process (Chiou and Shoup, 1985; Chiou et al., 1983, 1985).

Adsorption vs. Partition Process

Consider a natural water system with many organic contaminants present. Adsorption in sediment mineral matter occurs as a consequence of the competition among all species, including water. In the presence of water, the sediment mineral matter prefers to *adsorb* water because of their similar molecular polarities, while the soil organic matter prefers to *absorb* the contaminants (organic solutes) in water. This means that the (nonionic) organic contaminants are not significantly adsorbed to minerals and that the partition of a contaminant is not affected by water or by other contaminants. So two processes are at work: (1) the organic contaminants are competitively prevented by water from adhering to the surface of the mineral matter, and (2) the organic contaminants are able to partition independently into the SOM. Because so many environmental contaminants are transported by groundwater and surface water, it is important to understand the unique function of the organic matter within these aquatic system and how the partition processes affect the fate of common environmental contaminants.

Calculating the Distribution of Contaminants

In aquatic systems, a linear relationship exists between the concentration of a contaminant solute in sediment and the concentration of the contaminant in water at equilibrium. In other words, when the concentration of a contaminant in water is increased, the concentration of that contaminant in the sediment will also increase by a constant factor. This linear relationship is expressed mathematically via the solute distribution coefficient, which is the ratio of the solute concentration in sediment (C_s) to the solute concentration in water (C_w):

$$K_d = C_s/C_w \tag{11.1}$$

where K_d is the solute distribution coefficient, C_s is the solute concentration in sediment, and C_w is the solute concentration in water. Knowing the concentration of a contaminant in one compartment, either water or sediment, thus allows predicting the concentration of a contaminant in the other compartment. Because the contaminant sorption occurs predominantly by partition into the SOM, it is more useful to express the distribution coefficient in terms of the SOM content:

$$K_{oc} = K_d/f_{oc} \tag{11.2}$$

where K_{oc} is the partition coefficient normalized to the organic carbon of the sediment, K_d is the solute distribution coefficient, and f_{oc} is the organic carbon fraction of the sediment. By normalizing the partition coefficient to the sediment organic matter

TABLE 11.1
Classification of Human-Made Organic Compounds by Chemical Source

Groups	Common Examples or Subgroups
Hydrocarbons (coal- and petroleum-derived)	Crude oil (alkanes, cycloalkanes, aromatics)
	Refined petroleum products (petrol, kerosene, diesel, heating oil, asphalts)
	Combustion and conversion products (PAHs, carbon dioxide, synfuels, byproducts)
Synthetic organics	Halogenated hydrocarbons (PCBs, DDTs, kepone, mirex, dieldrin, chlordane, industrial solvents, other pesticides)
	Phthalate esters (plasticizers such as polyvinyl chloride [PVC])
	Others (surface-active agents, organophosphate insecticides, carbamate insecticides, many herbicides)
Municipal wastes	Sewage sludge (biosolids), secondary effluents, and trihalomethanes (THMs)
	Nutrients (phosphates, nitrogen, carbon, silica)

Source: Adapted from Olsen, C.R. et al., *Marine Chemistry*, 11(3), 501–533, 1982.

(organic carbon) content, scientists can then compare the relative partition properties of the organic matter from different geographic sources. Any variation between sediments from different geographic sources can then be attributed to the variation in SOM properties.

SEDIMENT INTERFACE/INTERACTION WITH PERSISTENT HUMAN-MADE ORGANIC COMPOUNDS*

To gain appreciable knowledge and understanding of freshwater sediment contamination it is important to gain some insight into the important physicochemical properties that are characteristic of different groups of anthropogenic (human-made) organic compounds. Moreover, an understanding of the transformation processes that influence the fate of human-made organic compounds is important. Combined, the physicochemical properties and the transformation processes determine the environmental persistence of certain human-made organic compounds. This section focuses on these combined system characteristics to examine the interaction of persistent organic compounds with sediment parts. The number of existing human-made compounds is large, and their environmental behavior is more readily reviewed from the standpoint of combined or compound groups rather than of individual chemicals. Table 11.1 categorizes organic compounds by source, and Table 11.2 categorizes them by a combination of chemical families and substance usage groups. These tables are intended to be representative of the more common human-made organic compounds and do not include all those that might have an adverse impact on water quality (Witkowski et al., 1987).

* Adapted from Wikowski, P.J. et al., *A Review of Surface-Water Sediment Fractions and Their Interactions with Persistent Manmade Organic Compounds*, USGS Circular 993, U.S. Geological Survey, Reston, VA, 1987.

DID YOU KNOW?

With regard to the sediment to groundwater interface, it is interesting to note that the amount of organic material stored on a aquifer's sediments can slow down natural processes that are often relied upon to restore groundwater quality in a wastewater-contaminated aquifer.

TABLE 11.2
Classification of Human-Made Organic Compounds by Chemical Structure and Chemical Usage

Groups	Common Examples or Subgroups
Halogenated aliphatic hydrocarbons	Chlorinated/brominated alkanes and chlorinated/brominated alkenes
Halogenated ethers	Chloroalkyl ethers and chlorophenylethers
Monocyclic aromatics	Chlorinated benzenes, alkyl benzenes, nitrotoluenes, and aminobenzenes
Phenols and cresols	Chlorinated phenols, nitrophenols, alkylphenols, and substituted cresols
Phthalate esters	Dialkyl phthalates
Polynuclear aromatic hydrocarbon	Naphthalene, anthracene, chrysenes, and benzo(*a*)pyrene
Nitrogen substituted compounds	Alkylmines, nitrosamines, and halogenated benzidines
Polychlorinated biphenyls	Isomers and Aroclors
Organochlorine insecticides	Aldrin, DDTs, endosulfan, heptachlor, chloradae, methoxychlor, and toxaphene
Organophosphorus insecticides	Parathion, malathion, and diazonon
Carbamate insecticides and herbicides	Carbaryl (Sevin), carbofuran, and aldicarb
Acid herbicides (phenoxy and benzoic)	2,4-D, silvex, 2,4,5-T, benzoic acid, and dicamba
Triazine herbicides	Atrazine, propazine, simazine, and prometryne
Substituted urea herbicides	Diuron, flometuron, and linouron
Other herbicides	Bipyridylium (diquat, paraquat)
	Dinitroanilines (trifuralin, profluralin)
	Others (alachlor, propachlor, picloram)
Other pesticides	Nematicides (ethylene dibromide, methyl bromide)
	Miticides (kelthane, chlorobenzilate, dichlore)
	Fungicides (captan, quinones, dithiocarbamic acid derivatives)

Physical and Chemical Properties

A number of common physicochemical properties of human-made organic substances contribute to environmental persistence and toxicity. Table 11.3 lists typical ranges of molecular weights, vapor pressures, water solubilities, and octanol–water partition coefficients for the compound groups and some subgroups from Table 11.2. The ranges are meant to be representative within each group (outliers certainly exist) and also to serve as benchmarks for comparison among different groups (Witkowski, et al., 1987). Water solubility is probably the most important property affecting the movement of compounds among the various environmental compartments, including air, the water column, aquatic biota, and sediment (Oschwald, 1972; Pionke and Chesters, 1973). Several of the organic compounds are highly water insoluble with solubilities in the range of low milligrams per liter to micrograms per liter. With specific chemical groups, a decreasing trend in water solubility is discernable as the degree of halogen substitution increases (for example, insecticides, aliphatic hydrocarbons, monocyclic aromatics, and phenols). In the same way, within specific chemical groups, water solubility decreases as the number of aromatic and cyclic ring substitutions increases (for example, low-solubility polycyclic aromatic hydrocarbons [PAHs], ethers, and nitrogen-substituted compounds).

In general, the low solubility of an organic contaminant in water is commonly reflected by its affinity for relatively nonpolar organic phases (solid and liquid). This lipophilic tendency is found in compounds possessing high octanol–water partition coefficients, where concentrations in the organic phase are 10^3 to 10^6 times greater than those in the water phase (e.g., PAHs, PCBs, phthalate esters). The octanol–water partition coefficient is a dimensionless concentration ratio whose magnitude expresses the distribution of a compound between equal volumes of two partially miscible solvents, *n*-octanol and water (Leo et al., 1971).

The partition behavior of a compound in a simple organic solvent–water system provides a useful estimate of the probable behavior of the compound in an aqueous environment (Haque and Freed, 1974; USEPA, 1979a). The use of the partition coefficient to estimate the environmental distribution of a compound is based on the recognition that much of the molecular driving force controlling distribution among water, sediment, and biota is analogous to that found in the octanol–water model system (Chiou, 1981).

The vapor pressure of an organic compound indicates the extent to which it will volatize from solution. Although vapor pressure exerts an important influence on the volatility of a compound as a pure substance, volatilization from an environmental system also depends on other physical properties, such as water solubility, and on system variables, such as temperature, wind speed, and water turbulence. Low-molecular-weight compounds (<200 amu), such as aliphatic hydrocarbons, monocyclic aromatics, and some nitrogen-substituted compounds, possess relatively high vapor pressures and limited water solubilities that in conjunction with certain environmental conditions make volatilization an important environmental pathway (USEPA, 1978, 1981).

DID YOU KNOW?

The magnitude of the partition coefficient is a useful index of the tendency of an organic compound to partition from water into sediment organic matter and into aquatic organisms (bioconcentration). The greater the magnitude of the partition coefficient, the greater the relative solubility of the compound in the organic phase compared to that in the aqueous phase.

TABLE 11.3
Physical and Chemical Properties of Selected Groups of Human-Made Organic Compounds

Groups	Subgroups	No. of Compounds	Molecular Weight Range	Log Vapor Pressure (torr)	Log Water Solubility (mg/L)	Log Octanol–Water Partition Coefficient	Refs.
Halogenated aliphatic hydrocarbons	1–4 Cl/Br	28	50–200	2 to 3	2 to 4	1 to 3	1,2,3,4,5,6
	>5 Cl/Br	3	200–300	–2 to -1	0 to 1	3 to 4	1,2,3,4,5,6
Halogenated ethers	Aliphatic	5	100–200	–1 to 1	3 to 5	1 to 3	1,2,3,4,5,6
Monocyclic aromatics	—	19	100–200	–1 to 2	2 to 3	2 to 4	1,2,3,4,5,6
Phenols	—	12	100–200	–1 to 0	3 to 5	1 to 4	1,2,3,4,5,6
Monocyclic aromatics and phenols	>5 Cl/Br	2	250–300	–5 to –4	–3 to 1	5 to 6	1,2,3,4,5,6,12
Phthalate esters	—	6	200–400	–3 to –1	0 to 4	2 to 9	1,2,3,4,5,6,16
Polynuclear aromatic hydrocarbons	2 benzene rings	4	100–150	–3 to –2	0 to 2	3 to 5	1,2,3,4,5,6,11
	3 benzene rings	3	175–200	–6 to –4	–2 to 0	4 to 6	1,2,3,4,5,6,11
	4 benzene rings	5	200–250	–11 to –6	–3 to –1	5 to 7	1,2,3,4,5,6,11
	>4 benzene rings	4	250–275	–11 to –9	–4 to –3	6 to 8	1,2,3,4,5,6,11
Nitrogen-substituted compounds	Aliphatic	10	30–125	1 to 3	>4	–1 to 1	1,2,3,4,5,6
	Aromatic	4	175–250	—	0 to 3	2 to 3	1,2,3,4,5,6

(continued)

Polychlorinated biphenyls	—	9	200–375	−5 to −3	−3 to 1	3 to 7	1,2,3,4,6,10,12
Organochlorine insecticides	—	13	300–400	−7 to −4	−3 to 0	3 to 6	1,2,3,4,5,6,9,12
Organophosphorus insecticides	—	10	275–375	−7 to −4	1 to 2	—	8,9
Carbamate insecticides and herbicides	—	8	175–275	−5 to −1	1 to 5	—	7,8,9,14,15
Acid herbicides	—	11	175–250	−7 to −3	2 to 4	—	7,8,9,12,15
Triazine herbicides	—	11	200–250	—	1 to 3	—	7,8,9,15
Substituted urea herbicides	—	7	150–275	−7 to −4	2 to 3	—	7,9,15
Bipyridylium herbicides	—	2	250–350	—	Miscible	—	7,9,10
Dinitroaniline herbicides	—	13	200–450	−8 to −4	−3 to 1	—	7,9,13

Sources: (1) USEPA, 1979a; (2) USEPA, 1981; (3) USEPA, 1982; (4) Verschueren, 1983; (5) USEPA, 1978; (6) USEPA, 1979b; (7) Bailey and White, 1965; (8) Paris and Lewis, 1973; (9) Wauchope, 1978; (10) Garten and Tralbalka, 1983; (11) Hassett et al., 1983; (12) Mackay et al., 1980; (13) Helling, 1976; (14) Schlagbauer and Schlagbauer, 1972; (15) Bailey et al., 1968; (16) Mathur, 1974

Note: 760 torr = 1 atmosphere. Base 10 raised to the −1 power = 0.1; base 10 raised to the 0 power = 1.0; base 10 raised to the +1 power = 10.

Although water solubility and octanol–water partitioning are of primary importance in controlling the environmental distribution of human-made organic compounds, other physicochemical properties also exert a significant influence on their overall distribution. These other properties include polarity, ionization constant, molecular charge, molecular size, and molecular configuration (USEPA, 1979a; Verschueren, 1983). Because their effects are exerted at the molecular level it is difficult to determine the exact contribution of each of these properties. However, some properties, such as the tendency to ionize in solution, determine which compounds respond to changes in pH (e.g., acid and bipyridylium herbicides) and which are not sensitive to its variation (e.g., chlorinated insecticides). The polarity of the organic compound influences its degree of chemical compatibility with polar solvents such as water and other less polar solvents such as sediment organic matter. Finally, fundamental properties such as molecular size and charge exert their influence on the behavior of a compound at all levels, from molecular interactions to the macroscopic scale of field processes.

Chlorinated insecticides, PCBs, PAHs, phthalate esters, highly halogenated aliphatics, and aromatics (the more environmentally persistent organic compounds) are generally characterized by low water solubility and high octanol–water partition coefficients (USEPA, 1982). Most of these chemical groups represent nonionic compounds that were synthesized for industrial or agricultural applications that require physical stability and specific properties (USEPA, 1979b; Verschueren, 1983). On the basis of low solubility and high partition coefficients, the distribution and fate of these persistent compounds in the environment largely depend on sorptive interactions with natural particles. Nonetheless, other physicochemical properties may become important under certain specific conditions; therefore, environmental behavior must be studied in terms of as many compound properties and environmental variables as possible.

Transformation Processes

Although physicochemical properties are indicative of the potential behavior of a compound in a water system, environmental variables are also important. Transformation processes play a role in determining the environmental fate of an organic compound. Additionally, how the physicochemical properties couple with the environmental variables to determine the primary transformation processes contributing to the persistence of certain human-made organic compounds is also important. During their movement through the water column and the surficial sediment layer, organic compounds participate in many interactions that determine their ultimate environmental fate. Pertinent physical, chemical, and biological transformation processes are defined and controlling factors that affect human-made organic compounds are briefly explained in the following:

- *Volatilization*—Evaporative loss of a chemical; depends on vapor pressure, water solubility of the chemical, and environmental variables such as wind, water turbulence, and temperature; potentially important for compounds with high vapor pressures, low solubilities, and high activity coefficients

- *Sorption*—General term encompassing surface attraction (absorption) and partition (solubilization); depends on the hydrophilic (water-loving) and lipophilic properties of the chemical and the composition of the sorbent; indicative parameters are solubility, octanol–water partition coefficient, and sorbent organic carbon content
- *Photolysis*—Nonmetallic degradation requiring light energy such that a chemical undergoes either a direct transformation reaction from the absorbed energy or an indirect change from a reaction (e.g., oxidation, with an excited chemical species or free radical); depends on the chemical's absorption spectrum coefficient in the ultraviolet to visible range as well as the sunlight intensity distribution for a given time of day, season, latitude, depth in water, and ozone thickness; also depends on the chemical's reaction quantum yield
- *Chemical oxidation*—Breaking down of the chemical bonds in organic compounds through a chemical reaction with photochemically derived oxidants (singlet oxygen or free radicals); depends on the number and types of possible reactive sites and on the presence of oxidants; rate constants are either measured directly or estimated from structure–activity relations
- *Hydrolysis*—Reaction of a chemical with water, hydrogen, or hydroxide ion, commonly resulting in the introduction of a hydroxyl group in exchange for the removal of another functional group; depends on the presence and number of hydrolyzable functional groups at neutral pH plus the catalytic effect of the addition of acids and bases at other pH values
- *Bioaccumulation*—Uptake and retention of chemicals in the water column by aquatic organisms through intake from water or diet; depends on the nature of the chemical (lipophilic) and the organism's fat content, metabolic, and depuration rates
- *Biotransformation and biodegradation*—Enzyme-catalyzed transformation of chemicals as a source of energy, carbon, and nutrients; depends on the refractory and toxic nature of the chemical, the presence of an acclimated microbial population, and a host of environmental factors including pH, temperature, dissolved oxygen, available nutrients, and cometabolites

Table 11.4 presents some of the measurable characteristics frequently used to assess the degree to which the behavior of a given compound is determined by a particular transformation process. The relative importance of the competing processes to each compound depends on the physicochemical properties of the organic chemical as well as the prevailing environmental conditions in the water column and in the surficial sediment layer.

Volatilization, as a transformation process, illustrates some of the information presented in Tables 11.3 and 11.4. It can be an important exchange process between a compound dissolved in solution and one vaporized in the atmosphere. Organic compounds with high vapor pressures, low water solubilities, and high activity coefficients are most susceptible to volatilization (for example, aliphatic hydrocarbons, monocyclic aromatics, and some nitrogen-substituted compounds). The magnitude of Henry's law constant is used as an indicator of the volatility of a compound from

TABLE 11.4
Physical and Chemical Characteristics Used to Assess the Distribution and Fate of Human-Made Organic Compounds

Characteristics	Definitions
Water solubility	Indicative of a chemical's hydrophobic/hydrophilic nature, limiting load in water, and potential behavior in solution. Units: mass per volume.
Vapor pressure	The partial pressure of vapor that is in equilibrium with a substance in pure state at a specified temperature. Relative or qualitative measure of the volatility of the chemical in its pure state. Units: torr.
Henry's law constant	Relative equilibrium concentration of a compound in air and water at standard temperature and pressure. Ratio of the vapor pressure divide by the chemical's water solubility; indicative of the compound's volatility. Units: torr per molar.
Reaeration rate ratio	Ratio of the first-order rate constant for loss of a chemical from aqueous solution divided by the rate constant for oxygen uptake by the same solution. Unitless. Estimate of the volatility of compounds with Henry's constants >3500 torr per molar for which the volatilization rate constant is limited by diffusion through the liquid-phase boundary layer.
Partition coefficient	Equilibrium distribution of a compound solubilized between two immiscible solvents. Unitless.
Octanol–water partition coefficient (K_{ow})	Equilibrium distribution of a compound between water and *n*-octanol. Indicative characteristic of the bioconcentration potential of compounds. Used to predict and model the migration of dissolved hydrophobic (water-fearing) organic compounds in soil and groundwater. Unitless.
Soil/sediment partition coefficient	Equilibrium distribution of a compound between water and a sediment/soil substrate. Mass of chemical per mass of sediment divided by the dissolved mass of chemical in solution per mass of solution volume. Unitless.
Soil/sediment partition constant	Soil/sediment partition coefficient divided by the sediment/soil percent organic carbon content. Normalizes the partition coefficient to a sediment/soil substrate that is 100% organic carbon, permitting comparison of partition coefficients between substrates of differing carbon content.
Absorption spectrum coefficient	Quantity of light absorbed by the chemical at a particular wavelength in the ultraviolet-visible range of the electromagnetic spectrum.
Reaction quantum yield	Efficiency of light utilization by a chemical. Ratio of number of moles of chemical transformed to quantity (flux) of light absorbed. Unitless.
Photolysis rate constant	First-order rate constant for direct photolysis. Units: reciprocal time.
Oxidation rate constant	Sum of the rate constants from each individual type of reactive site in the compound. Units: reciprocal time.
Hydrolysis rate constant	First-order rate constant at pH 7 and the second-order rate constants for acid- and base-promoted hydrolysis. Units: reciprocal time.
Bioconcentration factor	Concentration of a chemical in tissue on a dry-weight basis divided by the concentration in water. Also, the ratio of the uptake to depuration rates for a compound in a given organism. Unitless.
Biotransformation rate constant	Second-order rate constant dependent on the chemical concentration and the microbial biomass. Units: reciprocal time.

water because it takes into account the physical properties of water solubility and vapor pressure. Yet, alone it cannot be used as a quantitative measure of the volatility of a compound, because volatilization also depends on environmental variables such as wind velocity, water turbulence, and temperature (USEPA, 1978, 1979a). The balance of these environmental variables and physicochemical properties determines which groups of organic compounds are susceptible to volatilization and to what degree the exchange will alter the water column concentrations.

Photolysis is another example of a transformation process that depends on both the properties of a compound and the environmental conditions. The degree of photolysis depends on the amount of light a particular chemical can absorb (absorption spectrum coefficient) and the efficiency with which it uses that energy (reaction quantum yield). Given these properties of the organic compound, the actual degree of photolysis depends on climatic and meteorological conditions such as sunlight intensity, cloud cover, time of day, season, latitude, and even ozone thickness (USEPA, 1979a, 1981). In addition, the water turbidity determines the amount of incident sunlight that actually penetrates the water column to the depth of the organic compound. As a result of the large number of physicochemical properties and environmental variables that determine photolysis, the rate of photolysis for a given compound is commonly expressed by an overall parameter such as the photolysis rate constant. This parameter lumps most of the important factors into one measure, and it is useful for relative comparisons of photolysis rates between different organic compounds measured under similar conditions.

A closer examination of the important transformation processes for anthropogenic organic compounds indicates that different processes are spatially distributed throughout the water and sediment layers. The distribution of different transformation processes reflects preferences dictated by the driving forces for each process (Haque and Freed, 1974; Pionke and Chesters, 1973). At the air–water interface, energy supplied by light, wind, or surface tension effects is important. Volatilization, photolysis, and the surface microlayer concentration of organic compounds are the most influential processes near the air–water interface. In the bulk solution, factors such as pH, ionic strength, dissolved oxygen, and the polarity of the chemical relative to water control the behavior of compounds. Other processes, such as the biological and chemical transformations and sorptive interactions, depend more on the interaction of the bulk solution with natural particles and consequently can occur throughout the water column and surficial sediment layers (Olsen et al., 1982). At the sediment–water interface (Figure 11.2), the increase in organic matter, microbial populations, and other sediment components forces diagenetic reactions to take precedence (Oschwald, 1972).

Atmospheric influence is most pronounced at the air–water interface, where volatilization provides an exchange mechanism for compounds in solution that have relatively high air–water partition coefficients (Henry's law constants). Input of light energy into the water column is also highest near the air–water interface. Consequently, the importance of photolysis as a degradative process for certain organic compounds decreases with depth as light radiation decreases. Some compounds may be concentrated within the surface microlayer (top 1 mm of the water column) to levels much greater than those found in the bulk solution (MacIntyre,

FIGURE 11.2 Sediment–water interface without organic contamination. Sample was taken by author in Glacier National Park in Two Medicine River a few yards below Running Eagle Falls.

1974; Rice et al., 1982). This surface-excess effect is manifested at the air–water interface and results from the low affinity of highly nonpolar organic compounds for polar water molecules. Despite available exchange pathways and low water solubilities, many organic compounds still enter into aqueous solution at environmentally significant concentrations.

Once in the water column, organic compounds are subject to chemical and biological transformations or interactions with other sorbents. Reactions such as solvation (dissolution), hydrolysis, oxidation, and ionization may affect the chemical speciation of certain organic molecules in a manner consist with the oxidation–reduction potential (Eh), pH, and ionic strength of the bulk solution. Microbiologically mediated reactions mineralize parent organic molecules into more fundamental functional groups useful as nutrient and energy sources. Most of these reactions either transform the organic contaminant, sometimes to a more toxic form (e.g., aldrin to dieldrin), or remove it from the water column to another environmental compartment such as the atmosphere, sediment, or biota.

Research conducted on the geochemistry of interstitial waters has demonstrated that many of the processes that occur in the water column continue across the sediment–water interface and into the interstitial fluid of the bed sediment (Presley and Trefry, 1980). Broadly classified in diagenetic interactions, these physical, chemical, and biological processes include deposition, accumulation, compaction, mineralization, and decomposition of settled particulate matter along with its associated contaminant content (Baker and Feely, 1978; Krom and Sholkovitz, 1977). In the surficial sediment layer, chemical and microbial processes continue the degradation reactions that began in the water column. Biopolymers (proteins, lipids, and carbohydrates) are converted to monomers (amino acids, fatty acids, and sugars) and eventually to simple molecular components for recycling into solution by diffusion. Commonly, these reactions are incomplete because new sediment is deposited or resuspension results from turbulent eddies or biological activity (bioturbation). Human-made organic compounds, because of their sorptive interactions with natural organic materials, are incorporated into this biogeochemical cycling of elements; consequently, diagenetic interactions provide another mechanism for exchange of human-made organic compounds between the surficial sediment and the water column

Sorption and Bioconcentration

Sorption and bioconcentration are two other transformation processes that are of particular importance to the fate of human-made organic compounds. Sorption interactions within the water column increase the carrying capacity of the bulk solution beyond the solubility limit of the contaminant. Sorptive interactions within the surficial sediment layer provide the sediment with a mechanism for accumulating and releasing organic contaminants. Together, sorptive interactions in the water column and in the surficial sediment layer can significantly affect the transport and cycling of contaminants (Eisma, 1981; Means and Wijayaratne, 1982; Olsen et al., 1982). In a similar manner, the bioconcentration of organic compounds by biological organisms, in the water column and sediment, can affect the distribution and fate of contaminants. Biological organisms can progressively concentrate organic contaminants, and this can directly impact humans because of consumption within the food chain. In addition, because biological organisms do not necessarily move with the average flow of surface waters (many species are independently motile), contaminants stored within the organisms may differ in their transport characteristics from

DID YOU KNOW?

Based on years of experience in teaching environmental health and environmental science to upper undergraduate and graduate students, I have found that many students are confused about the exact meaning of the term *bioconcentration.* This is especially the case whenever students confuse the meanings of *bioconcentration*, *bioaccumulation*, and *biomagnification*, so let's define these three terms:

- *Bioconcentration* is a process leading to a higher concentration of a substance in an organism than in the environmental media to which it is exposed (International Union of Pure and Applied Chemistry, 1993).
- *Bioaccumulation* is the biological sequestering of a substance at a higher concentration than at which it occurs in the surrounding environment or medium (USGS, 2007).
- *Biomagnification* is the result of the process of bioaccumulation and biotransfer by which tissue concentrations of chemicals in organisms at one tropic level exceed tissue concentrations in organisms at the next lower trophic level in a food chain (USEPA, 2010).

those in the bulk solution. Finally, because biological organisms play an integral role in the exchange of matter between the water column and sediment (e.g., bioturbation), they can alter the distribution of contaminants by exchanging organic compounds with the water column or the sediment (Kenaga, 1975a,b).

For sorptive processes, the exact nature of the particle–organic-compound interaction is a subject of practical importance in understanding the environmental behavior of organic contaminants. Current attention has focused on whether the sorption of organic contaminants results from a surface adsorption process or from a partitioning process between water and an organic sorbent (Chiou et al., 1979, 1983, 1984; Mingelgrin and Gerstl, 1983; Pavlou and Dexter, 1979). The advocates of a partition model believe that the sediment organic matter functions as an immobilized amorphous organic phase capable of partitioning nonionic organic compounds from the bulk aqueous solution in a manner similar to the solubilization taking place with a conventional polymer. The surface adsorption concept treats the same results as adsorption wherein the mineral and organic fractions of the sediment have different adsorptive capacities. For metal ions and ionic organic solutes it is logical that surface adsorption by different colloidal particles can take place as a result of electrostatic attraction, covalent bonding, and hydrogen bonding. For nonionic solutes, available data indicate that sorption by sediment particles is more effectively accounted for by the partition mechanism of the solutes with the sediment organic phase.

This text deals primarily with nonionic organics (e.g., chlorinated insecticides, PCBs, PAHS), so it is pertinent to present evidence supporting the partition model. In addition to the recognized dependence of sorption on the sediment organic content, Chiou et al. (1979, 1983) showed that the sediment concentration of nonionic organic solutes increases linearly with aqueous solute concentration up to the solubility limit of the solute. A nonlinear relation would be expected in a surface adsorptive process because the available adsorption sites eventually become saturated as solute concentrations approach solubility. In experiments using binary solutes, Chiou et al. (1983) found that the sorption relation showed no competitive effects between different solutes and that the magnitudes of the sorptive capacities were similar to those for the single-solute cases. In a surface adsorption model, solutes compete for a fixed number of adsorption sites, and the overall sorptive capacities per contaminant are generally lower than those in single-solute cases. Thermodynamically, the heats of sorption were found by Chiou et al. (1983) to be small and less exothermic than the heats of solute condensation from water; this is in contrast to the large exothermic heats of adsorption characteristics of surface adsorptive processes. Chiou et al. (1985) suggested that the inorganic fraction of the sediment is relatively inert to nonionic organic compounds in the presence of water, presumably because water is preferentially adsorbed over the nonionic organic solutes.

Just as nonionic organic compounds partition into sediment organic matter, bioconcentration can be viewed as a transformation process in which organic compounds partition into the lipid fraction of aquatic organisms. Bioconcentration is the accumulated concentration of a given compound over that in the surrounding medium (water) by a particular aquatic organism resulting from its diet or from intake of contaminated water during respiration (Kenaga, 1975a,b). This process generates a reservoir of contaminants within the organisms that may enter the food chain. Biomagnification is commonly associated with bioconcentration; it is the progressive increase in the concentration of a compound moving from one trophic level to the next higher trophic level of the food chain. This discussion is limited to bioconcentration because biomagnification studies lack a consistent reference point for comparison among different trophic levels, especially aquatic as opposed to terrestrial organisms. Biomagnification has yet to be rigorously defined in a manner that accounts for the physiologic differences between species and that provides an unequivocal reference concentration for each organic contaminant. Once a reference standard is established, it should compensate for the different compositions and amounts of organic solvents present in the bodily fluids of each species of aquatic organism.

Chiou et al. (1985) demonstrated that the bioconcentration of nonionic organics by aquatic organisms results from the partitioning of compounds into the lipid fraction of the organism. Those compounds that are strongly bioconcentrated generally have a low water solubility and a relatively high affinity for organic phases, as is indicated by a high octanol–water partition coefficient. These properties are characteristic of many pesticides whose original function was to penetrate the cellular structure of the organism and disrupt the life-sustaining processes of the plants and insects. Once within the organism, these compounds may exhibit improved stability against decomposition and transformation.

From a mass balance standpoint, bioconcentration and biomagnification cycle contaminants, as well as vital elements such as carbon. From a toxicological standpoint, these processes are important because of the physiologic effects promoted by elevated concentrations of persistent contaminants. With regard to persistent substances, materials that are the products of modern scientific and engineering wizardry persist; they are not normally changed or degraded to harmless substances. These persistent substances include the pesticides DDT and chlordane, metals (e.g., mercury), and organic chemicals such as PCBs. These substances do not break down easily and tend to magnify throughout the food chain. Consequently, organisms at higher trophic levels suffer the most serious effects of these nonbiodegradable, persistent substances. In general, by the time most pesticides run off into surface waters, concentration levels are sufficiently low that the acute toxicity of the compound is low and frequently unnoticed (Neely et al., 1974). Bioconcentration raises the potential for chronic effects because it increases the residence time of contaminants within the biological reservoir. For these reasons, estimates of the bioconcentration potential of synthetic compounds are important in order to assess their possible impact prior to their widespread use and environmental release.

For persistent human-made organic compounds, several physicochemical properties and transformation processes play important roles in their environmental distribution. Low water solubilities and high octanol–water partition coefficients are common characteristics. These properties lead to sorptive interactions and bioconcentration as the primary processes controlling environmental distribution. They are not solely responsible for the distribution of human-made organics, but, because their relative contribution is greater than that of other transformation processes, sorptive processes are the focus of the next discussion of interactions between organic compounds and sediment fractions. First, however, the following text considers the contribution of environmental variables in determine the dominance of certain transformation processes under a given set of environmental conditions.

Environmental Behavior

The specific behavior of organic compounds in the environment is primarily a manifestation of the interaction between the physical and chemical properties of the compound and the sorbent. The effect of these inherent properties in the water column and sediment may be subject to changes by environmental variables such as pH, Eh (or pE, reduction potential), dissolved oxygen, dissolved solids, temperature, water hydrodynamics, and atmospheric forcings. Table 11.5 presents some important reaction mechanisms controlling the transformation and removal processes for the different organic compound groups. For some groups, such as halogenated aliphatic hydrocarbons, a single property dominates the transformation mechanism. For others, apparent physicochemical properties can be masked because they are more dependent on environmental conditions. For example, the extent of hydrolysis of the carbamates and phenoxy herbicides is largely a function of pH.

Note that the physicochemical properties of a compound alone may not adequately indicate its probable field behavior without also considering the effect of environmental variables. Physical and chemical characteristics are measured under controlled conditions, whereas in nature the behavior of organic compounds may be influenced by other physicochemical factors not identified in simple laboratory systems. This is especially true for periods before a steady-state condition is reached. For example, insecticides and herbicides may be applied in different modes such as waster solutions, emulsions, gases, wettable powders, and granular pellets. The form of introduction of the compound certainly affects its residence time and transit behavior in the different environmental compartments. For example, compounds that are sufficiently volatile in water, such as hexachlorobutadiene, are less likely to volatilize if they are initially sorbed to some soil or sediment particle. Nevertheless, the physicochemical properties of the compound and the sorbent provide an important criterion for predicting and evaluating the initial distribution tendency of a compound between different sorbent phases. For this reason, the physicochemical properties provide a useful initial assessment of the ultimate fate of organic contaminants.

The fate of many organic compound groups is directly affected by competing processes mediated by the properties of the compounds and the prevailing environmental conditions. For example, monocyclic (one ring, one cycle) aromatics are susceptible to either volatilization followed by atmospheric photolysis or sorption followed by bioaccumulation (USEPA, 1979a,b). The pair of transformation processes that control the behavior of the compound may vary with environmental factors such as wind speed, wave turbulences, temperature, and the availability of particulates for sorption in the water column. The reactivity of other organic groups, such as the substituted phenols, likewise depends on the location of the compounds in the water column. Near the air–water interface photolysis is the primary reaction mechanism, but closer to the sediment–water interface photolysis is greatly reduced and biodegradation reactions predominate. For other compounds that readily undergo enzymatically catalyzed microbial degradation (e.g., phenols, phenoxy herbicides, triazine herbicides), environmental fate strongly depends on the presence of a sufficiently large and acclimated microbial population.

Organic compound groups such as the organophosphorus insecticides (e.g., parathion, malathion) undergo rapid transformation to different chemical forms (paraoxon and malaoxon), which are more persistent and sometimes more toxic than the parent compounds. The dissipation of the parent compounds may appear to be deceptively rapid from an environmental standpoint, but their effect is manifested primarily through other reaction intermediates. Some other compounds (e.g., halogenated ethers) may persist in an aqueous medium despite dissipation by hydrolysis primarily because of their very high solubilities in aqueous solution. In other words, consideration of the physical and chemical properties of a compound should indicate its tendency for transport and partitioning, whereas the environmental fate of the chemical may depend on other system conditions.

DID YOU KNOW?

It is recognized that certain components of sediment (e.g., humic substances) do not exist as particles *per se*; consequently, use of the term "particle" is not intended to imply a specific physical or geometric form but rather to differentiate fractions, some of which behave as discrete entities (such as sand and colloids) when transported.

TABLE 11.5
Environmental Fate Processes for Selected Groups of Human-Made Organic Compounds

Chemical Group	Primary Mechanism	Secondary Mechanism	Refs.
Halogenated aliphatic hydrocarbons	Volatilization	—	1,2,4
Halogenated ethers			
Aliphatic	Hydrolysis	Volatilization and photooxidation in atmosphere	1,24
Aromatic	Sorption	Bioaccumulation and biodegradation	1,2,4
Monocyclic aromatics	Volatilization	Atmospheric oxidation/photolysis	1,2,4
	Sorption and bioaccumulation	Slow biodegradation (especially nitrogen compounds)	
Phenols	Photolysis	Near air–water surface	1,2,4
	Biodegradation	Near water sediment if sufficient microbes	
Monocyclic aromatics and phenols with 5 or more chlorines	Sorption Bioaccumulation	—	1,2,4
Phthalate esters	Sorption	Bioaccumulation Biotransformation Biodegradation	1,2,4
Polycyclic aromatic hydrocarbons	Sorption Bioaccumulation Biodegradation Volatilization	Photolysis (for compounds with low ring numbers)	1,2,3,4
Nitrogen compounds			1,2,4
Aliphatic	Photolysis	—	
Aromatic	Sorption	Biodegradation	

(continued)

Organochlorine insecticides and PCBs	Sorption	Volatilization (if chemical is not immediately sorbed)	1,2,3,4,5
	Bioaccumulation	Photooxidation	
	Biotransformation		
	Acid pH stable		
Organophosphorus insecticides	Alkaline pH—rapid chemical–microbial hydrolysis	Biological oxidation; some sorption slows down hydrolysis	3,5,7,8,9,10
Carbamate insecticides and herbicides	Chemical/microbial		5,6,11
	Hydrolysis	pH sensitive	
	Photooxidation	Some sorption to sediments	
Phenoxy acid herbicides	Photolytic decomposition under ultraviolet light	Slow microbial degradation	5,6,7
	Hydrolysis (pH sensitive)	Sorption more pronounced for esters than for acids	
Triazine herbicides	Hydrolysis (acid and base catalyzed)	Microbial degradation	5
Substituted urea herbicides	Sorption	Microbial, chemical, photochemical—slow (months)	12
	Degradation		
Bipyridylium herbicides	Sorption (clays)	Essentially irreversibly sorbed under field conditions	13,14
	Sorption (humic substances)	Strongly sorbed but unlike clays can undergo some microbial degradation	
Dinitroaniline herbicides	Highly sorptive	—	15

Sources: (1) USEPA, 1979a; (2) USEPA, 1981; (3) USEPA, 1978; (4) USEPA, 1982; (5) Paris and Lewis, 1973; (6) Norris, 1981; (7) Faust and Suffet, 1966; (8) Van Middelem, 1963; (9) Mulla and Mian, 1981; (10) Mulla et al., 1981; (11) Schlagbauer and Schlagbauer, 1972; (12) Maier-Bode and Hartel, 1981; (13) Akhavein and Linscott, 1968; (14) Hayes et al., 1975; (15) Helling, 1976.

TABLE 11.6
Estimates of Environmental Persistence of Selected Groups of Human-Made Organic Compounds

Groups	Estimated Half-Lives
Lead, arsenic, copper, and mercury pesticides	10 to 30 years
Organochlorine insecticides and PCBs	2 to >4 years
Bipyridylium herbicides	>2 years
Triazine herbicides	6 months to 2 years
Substituted urea herbicides and picloram	4 months to 1 year
Benzoic acid, dinitroaniline, and amide herbicides	3 months to 1 year
Phenoxy and toluidine herbicides	1 month to 6 months
Carbamates and aliphatic acid herbicides	2 weeks to 3 months
Organophosphorus insecticides	1 week to 2 months

Sources: Metcalf and Pitts (1969); Wauchope (1978); Stojanovic et al. (1972); Maier-Bode and Hartel (1981); Metcalf (1972); Paris and Lewis (1973).

The combination of the properties of a compound and available field data give a good indication of its potential persistence in the water column and sediment system. Table 11.6 presents estimates of different pesticide half-life ranges based on data collected from different ecosystems (rivers, lakes, soils, and estuaries). Information on the industrial organic compounds is less readily available; consequently, these compounds are not included in the list.

SUSPENDED AND SURFICIAL SEDIMENT FRACTIONS

This section discusses the properties of inorganic particles and humic substances. Recall that biological particles were discussed in Chapter 10. Although these three categories do not include all sediment components, they include those components currently thought to be most important with regard to influencing the behavior of human-made organic compounds.

Inorganic Particles

The most commonly identified sediment components are the mineral particles of clay, silt, sand, and gravel. These particles comprise the bulk of the transportable sediment on a mass basis. The clay, silt, sand, and gravel fractions are separated on the basis of size by sieving, settling, or centrifugation methods. Size is used as the distinguishing criterion because of its relation to the transport and settling properties of the various particle types. Unfortunately, the names assigned to the particle-size classes (sand, silt, and clay) also imply certain compositional properties. Although the implications are partially accurate, many particles falling within the sand-sized class may be composed of organic matter (shells, seeds, and detritus) rather than, say, quartz. Other particles falling within the clay-sized class may be metal precipitates (e.g., iron and manganese oxides and hydroxides) rather than particles of clay mineral composition.

Mixed within the various size classes of sand, silt, and clay are other less readily identifiable mineral fragments. They occur in various stages of mineralization and dissolution and include the opaline and calcareous exoskeletons of dead organisms. Although the sorptive capacity of these mineral particles for nonionic organic compounds in water is small relative to the sorptive capacity of organic matter (Chiou et al., 1985), it is not negligible. The mineral particles are commonly coated with organic sheaths that provide a microscale organic medium for sorption (partition) of nonionic organic solutes. Commonly, this partitioning capacity is mistakenly attributed to the mineral grains.

Organic matter and metal oxides are two groups of surface-active substances capable of altering the electrochemical properties of suspended mineral particles in surface waters. Acting individually or together, these groups are abundant enough to affect the interfacial properties of transported sediment matter (Hunter and Liss, 1982). Surface-active substances are sufficiently strong to remove the intrinsic differences in the laboratory-measured stability of clay minerals. Hunter and Liss (1982) found that the charge distribution on suspended particle samples collected from four estuaries was highly uniform despite the mixed composition of the suspended matter particles. They attributed the homogeneous electrical properties to both the sorbed organic matter and the metal oxides.

Although the percentage by weight of dissolved organic matter sorbed on suspended particles in lakes and rivers may be small, the extent of surface coverage by absorbed organic matter can be large. Davis (1982) found that within the typical pH range of natural waters, almost complete surface coverage by dissolved organic matter can be expected for hydrous aluminum oxides, hydrous iron oxides, and edge sites of aluminosilicates. The natural organic matter is characterized by weak, acidic functional groups that form surface complexes with the relatively basic surface hydroxyls of the mineral and metal oxides (Davis, 1982). Other mineral types (for example, silica) have relatively acidic surface hydroxyls that form weaker complexes that under natural conditions may not be sufficiently energetic to cause complete surface coverage. The extent of coverage of the oxide–water interface also depends on the pH, the availability of surface area sites and sorbable organics, the nature of the solid surface, and the inorganic electrolyte strength and composition of the bulk solution. Organic coatings may play a major role in the surface chemistry of some particulate-sized matter, especially for nonionic organic solutes; they are geochemically important in the interfacial processes of coagulation, sedimentation, and sorption.

Many inorganic particles fall within the colloidal range of the particle-size spectrum. They include clay mineral colloids as well as oxides and hydroxides of iron and manganese. Colloidal size, organic content, and composition give the clay particles cation-exchange capabilities for ionic species (e.g., metal ions) and a microorganic environment for sorption of nonionic organic solutes. Some expansive clays, such as montmorillonite, allow interlamellar penetration and entrapment of cationic organic species, such as paraquat (Adams, 1973). Because of the presence of organic matter in the form of organomineral colloids, the sorptive capacities measured for clay-sized samples are commonly much larger than those expected solely on the basis of a pure mineral sorbent.

Depending on water pH and redox potential, trace metals such as iron and manganese are present in suspension in precipitated forms. The oxides and hydroxides of iron and manganese occur as "dissolved" colloids, organometal colloids, or aggregated particles (Sigleo and Helz, 1981). In conjunction with humic materials, the metals coat the mineral particles that serve as nucleating surfaces for their precipitation reactions (Sholkovitz, 1976). The sorptive and scavenging ability of the metals is attributable to surface properties and is more pronounced for freshly precipitated flocs than for aged aggregates (Duinker, 1980). Sorption of ionic pesticides by iron and aluminum oxides has been difficult to verify in field situations because of the problem of sampling particle aggregates without altering or destroying them (Olsen et al., 1982).

Humic Substances

The river is 34.3 miles long and courses its way through the Allegheny Mountains of eastern West Virginia. It is a principal tributary of the Cheat River and, along with the Monongahela and Ohio rivers, is part of the watershed of the Mississippi River; it drains an area of more than 140 square miles. It is called the Blackwater River, and the name is fitting because of its color. The river's riparian corridor is lined with spruce and hemlock trees and their input of humic substances, mainly tannins contained in dead leaves and decaying tree debris, imparts an amber or tea color to its water.

Humic substances are important to this discussion. The importance of humic substances is reflected by the large quantity of research focused on the transport and fate of aquatic and terrestrial humic materials (e.g., Aiken et al., 1985; Chrisman and Gjessing, 1983a,b). As compounds from diverse origins and of complex chemical composition, humic substances can function in sediment and water chemistry as buffers, ion exchangers, surfactants, sorbents, and chelating agents (Jackson, 1975; Kononova, 1966). Research also indicates that humic substances interact with metal ions and their hydrous oxides, clay minerals, and organic pesticides, and they also serve as precursors to halogenated hydrocarbon formation following water-supply chlorination (Josephsen, 1982). A significant effect of the sorptive capacity of humic substances is the ability to concentrate toxic substances in solution to levels above their normal solubilities (Carter and Suffet, 1982; Chiou et al., 1986; Poirrier et al., 1972; Schnitzer and Khan, 1972; Wershaw et al., 1969).

The humic substances have been broadly defined as amorphous, brown or black, polydisperse substances with polar and acidic functional groups whose molecular weight ranges from several hundred to several million (Schnitzer, 1978). Keep in mind that this general definition does not describe the diverse physical forms humic substances take on in response to variations in pH and aqueous ionic strength. In solution, humic substances may be in the form of dissolved species, dispersed colloids, or precipitated flocs or may occur in association with mineral and hydrous metal aggregates.

Research has shown that the characteristics of a humic substance are strongly dependent on its origin. Differences have been distinguished between humic substances derived from terrigenous plants and planktonic algal sources (Josephsen, 1982). Functional group content analyses indicated that terrigenous humic substances have a higher percentage of aromatics (e.g., phenols) that are characteristic of lignin breakdown products derived from vascular plants (Plechanov et al., 1983). Planktonic humic substances are more aliphatic and are characterized by protein and carbohydrate residues (Ertel and Hedges, 1983). Terrigenous humic substances have higher carbon-to-nitrogen ratios than do planktonic ones because of the higher content of nitrogen-poor, lignin- and cellulose-derived compounds (Meyers et al., 1984). Also, chemical fingerprinting has demonstrated that aquatic humic substances can be formed *in situ*, entirely from degradation of plankton and without any contribution from terrigenous sources (Khan, 1980).

The categorization and chemical characterization of humic substances are difficult because they represent organic matter from diverse origins that react degradatively to form an even wider spectrum of breakdown compounds. Consequently, most of the analytical research conducted has been aimed at elucidating the gross structural features and major functional moieties active in the different groups of degradative reactions (Christman and Gjessing, 1983a,b). The most commonly used classification of humic substances is operationally defined and founded on the basis of solubility in different extraction solvents. The three classes identified on the basis of solubility/insolubility in acid and dilute alkali are fulvic acids, humic acids, and humins (see Figure 11.3). Further fractionation of the humic acids into hymatomelanic acid (alcohol soluble), gray humic acid (precipitated in electrolytic solution), and brown humic acid (soluble in electrolytic solution) can be, but is infrequently, performed.

Humic substances originate from the chemical and biological degradation of plant and animal residues. Attempts to identify the general breakdown mechanisms have not been entirely successful, but it is agreed that the more complex, higher-molecular-weight humic substances (>1000, amu) are degraded first (oxidized) to lower-molecular-weight materials such as the humic and fulvic acids (Khan, 1980). Consequently, the structure of the three groups is similar, but they differ in ultimate analysis, molecular weight, and functional group content. In general, the lower-molecular-weight humic substances are more prevalent in the sediment of rivers, lakes, and seawater, whereas the heavier humic substances predominate in soils (Josephsen, 1982). This difference in distribution reflects the importance of water column degradative processes that are continually active during the transport of the humic materials.

The fulvic acids are described as water-soluble compounds that impart the yellowish to brownish color to certain natural waters. In comparison with the humins and humic acids, the fulvic acids have progressed further through the chain of chemical and biological degradative processes. Consequently, they have lower molecular weights and more hydrophilic functional groups than do the heavier humic acids and humins (Khan, 1980).

In aqueous solution, the humic acids are less soluble than the fulvic acids and can be precipitated at pHs below neutral. Their molecular weight is intermediate between that of the fulvic acids and humins and changes with variations in pH and ionic

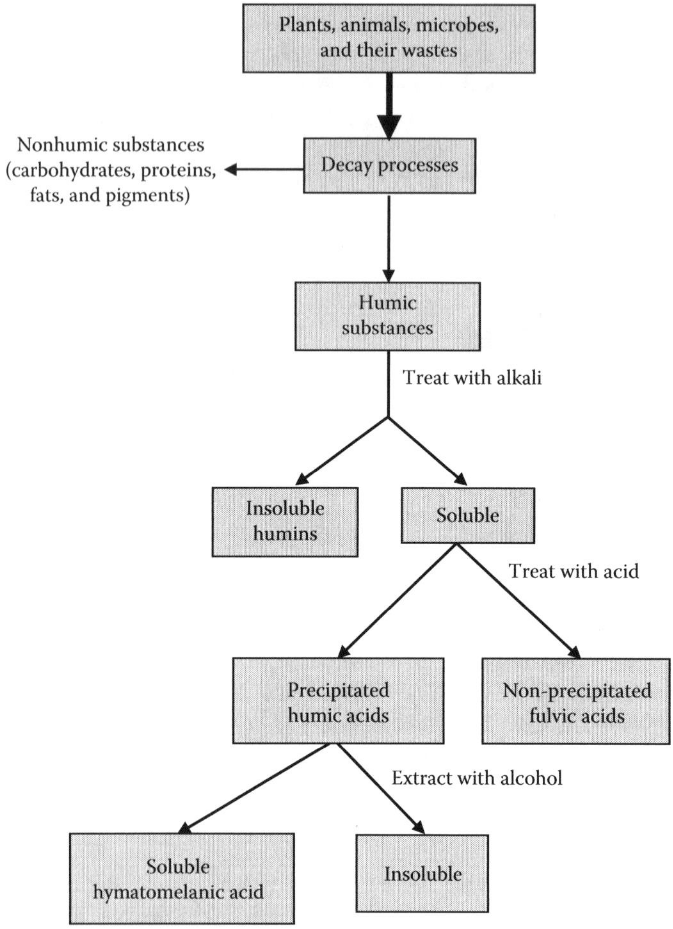

Humic acid—A long-chain polymer of high molecular weight that is dark brown to black and is soluble in alkaline solutions.

Ulmic acid—Also known as hymatomelanic acid, an alkaline-extractable, alcohol-soluble minor fraction of humic acids.

Fulvic acid—A short-chain polymer of low molecular weight that is yellow in color and soluble in both acid and alkali.

FIGURE 11.3 Fractionation of humic substances. (Adapted from Saar, R.A. and Weber, J.H., *Environmental Science & Technology*, 16(9), 510a–517a, 1982.)

strength (Plechanov et al., 1983). As a result of their sensitivity to pH, the molecular structure of the humic acid complex is more aggregated and coiled in acidic conditions and more dispersed and extended in an alkaline environment. The consequences of this reversible change in molecular configuration can potentially affect the ability of the humic acid molecule to interact with or sorb other chemical species in water. Note that we do not know what we do not know about the behavior of humic substances interacting with organic contaminants; the jury is still out on the subject.

The last and least studied of the humic substance fractions are the humins. In the gradation of bonding strengths found when progressing from the fulvic acids to the humic acids to the humins, the humins possess the most stable configuration. The relative insolubility of humins, even under a wide range of pH conditions, seems to be due to their association with inorganic constituents of soil and sediment (Schnitzer and Khan, 1972). Their molecular structure and ability to form stable metal and organomineral complexes contribute to their observed resistance to microbial and chemical degradation.

TRANSPORT OF CONTAMINATED SEDIMENTS

Whether discussing the transport of sediment or contaminated sediments, the mechanisms at work in the process are the same. One thing is certain: Uncontaminated or contaminated sediment particles cannot move in or settle in a stream or lake without the many physical processes that act in conjunction with the downstream flow of water (hydrodynamics) to move their difference sizes in the water column and in surficial locations. O'Melia (1980) and Hunt (1982) identified and investigated three major physical processes affecting particle transport and removal in the water column. First, gravity causes the sedimentation of particles on the basis of their size, shape, and density. The influence of gravity relative to other physical processes is greater in controlling the transport of silt and larger-sized particles. Second, and in conjunction with gravity, the fluid shearing forces generated by velocity differences between adjacent layers of moving water cause transport of silt- and clay-sized particles. Although these silt- and clay-sized particles settle vertically by gravity, they also are transported laterally and longitudinally by fluid shearing forces that arise from the mixing of water currents of different densities and velocities. The third physical process identified by O'Melia (1980) and Hunt (1982) is molecular diffusion (also called *Brownian diffusion*) and is only of importance for particles of submicron dimensions. Diffusion is the apparently random motion of molecules caused by minute thermal gradients in the fluid medium. Transport by diffusion is independent of both the fluid motion and gravitational forces. Its influence is only noticeable for very small colloids where gravity and fluid shearing forces exert a smaller relative effect on particle transport.

The description of sediment transport has historically been approached—in contrast to the studies by O'Melia (1980) and Hunt (1982)—by differentiating between the movement of coarse and fine particles. The movement of coarse mineral particles results from the combination of sedimentation, resuspension, and suspended and bedload transport. The description of the vertical settling of gravel, sand, and silt is approximated by Stokes' relation for ideal spherical bodies falling under quiescent conditions. The settling velocity depends on the balance between the upward buoyant forces and the downward gravitational forces for a particular particle geometry, particle density, and water density (Guy, 1969, 1970). For larger particles (sand and gravel) and for less spherical particle shapes, the velocities deviate progressively more from those predicted on the basis of an ideal Stokes' settling behavior. These deviations increase in magnitude as the inertial effects become more significant as a result of surface and form drag forces.

As the coarse particles settle, they are distributed longitudinally and laterally as suspended particulates caught in the advection of flow. When they have settled, the particles may resuspend when currents are sufficiently strong to cause turbulent eddies capable of scouring the surficial sediment layer. Transport of settled particles occurs chiefly by rolling or sliding within the bedload layer—the uppermost layer of the sediment, commonly only a few grain diameters in thickness. Overall, the transport of coarse sediment depends primarily on the flow characteristics, the particle properties and dimensions, and the streambed geometry.

Because the supply of coarse particles commonly exceeds the carrying capacity of a stream (Guy, 1969, 1970), the concentration of coarse sediment in suspension is a function of the flow and sediment characteristics. For fine sediment (clay- and colloid-sized particles), the suspended concentration depends less on flow characteristics and more on the available supply of particles from erosion and biological processes. For fine sediment, unlike coarse sediment, the carrying capacity of the stream is typically much greater than the supply of particles from upstream and adjacent hill slopes (Colby, 1963).

The transport of small particles (micron and submicron sized) of fine sediment is less dependent on the influence of gravity than are larger particles of silt and sand dimensions. Clay- and colloid-sized particles are more buoyant and more responsive than are larger-sized particles to fluid shear and Brownian diffusive forces within the water column. Mineral colloids settle orders of magnitude more slowly than sand grains, and organic colloids are even slower as their densities approach that of water. These properties give fine particles longer residence times in the water column and make them more susceptible to the shearing forces at the benthic boundary layer that resuspend sediment components. The combined effect of their transport characteristics and their natural abundance makes colloid-sized particles significant in terms of their potential impact on contaminant transport (Bilby and Likens, 1979).

Of all the particles found in the water column and sediment, the transport of organic particles, both dead and alive, and of organomineral assemblages has been the least accurately described in terms of particle dynamics. Discrete particles such as dead organisms vary in size, shape, and density, depending on their degree of fragmentation, decay, and dissolution (Lorenzen et al., 1983). The sizes range from that of the parent organisms to those below the 0.45-μm division between particulate and dissolved matter. Below this demarcation, the motion of particles is considered indistinguishable from the motion of water molecules (Sheldon et al., 1972).

The movement of living particles is further complicated by the various degrees of motility with which different species respond to environmental variables. Phytoplankton respond to changes in season, nutrient concentration, population age, time of day, and relative brightness of the sun. Burns and Rosa (1980) investigated the settling velocities of 10 different species of phytoplankton (size range of 10 to 64 μm) and found that values were as high as 0.8 m/day for flagellates. These velocities are for ascending and descending movement in the water column because light-dependent organisms such as phytoplankton follow diurnal cycles of respiration and photosynthesis. This type of behavior is important within the euphotic zone of the water column because it is the primary source of natural particles. These particles also are the main food source for higher trophic organisms and form the basis of detrital matter.

DID YOU KNOW?

Stokes' law deals with the drag force exerted on spherical objects in a viscous fluid. The law makes the following assumptions for the behavior of a particle in a fluid:

- Laminar flow
- Spherical, homogeneous particles
- Smooth surfaces
- Particles avoid each other

In an analysis of the settling behavior of organomineral aggregates in the 5- to 500-μm range, Chase (1979) found that measured sedimentation velocities were different from values predicted by Stokes' law. For aggregates from lake and marine waters, the settling velocities were as much as an order of magnitude faster than the Stokes' law predictions. Edzwald et al. (1974) found similar order of magnitude departures for the difference between discrete and flocculated minerals. The derivations from ideal behavior are attributable in part to approximating the actual shapes of aggregates by spheroids, but Chase (1979) showed that three other factors are important; natural surface coatings, the presence of solution electrolytes, and dissolved organic substances combining to increase the settling velocity of aggregates by reducing the skin friction of their outer layer. Although a more exact description of the transport of aggregates is hampered by their amorphous structure and composition, more evidence is accumulating on their integral role in the cycling of organic and mineral compounds, natural and human-made, in the water column and sediment.

Because the hydrodynamic behavior of colloid size particles depends largely on the balance within the aqueous solution (Stumm, 1977), surface coatings can be expected to affect the physical and chemical properties of colloid particles (e.g., their sorptive capacities). By masking differences among mineral colloids, surface coatings affect the flocculation behavior of particles and may help to explain the lack of field evidence verifying laboratory studies that suggest differences in the flocculation behavior of different minerals. From a hydrodynamic standpoint, surface coatings reduce the frictional drag on settling particles and increase their downward velocities (Chase, 1979). All of these results indicate the effects surface coatings can have on particle properties and also raise questions about the accuracy of laboratory studies that purposefully wash or purge particle coatings from suspended and surficial sediment (Davis, 1982; Schwarzenbach and Westall, 1981).

SELECTED CHEMICAL AND ELEMENTAL CONTAMINANTS

In this section, 16 different chemical, elemental, and elemental-compound contaminants are discussed. Many of these contaminants are well known, others not so well. The contaminants described here are representative samples of hundreds of potential substances that may impact aquatic ecosystems via sediment adsorption, absorption, sequestration, or seclusion.

Mercury/Methylmercury

As early as the 1950s, it was established that emissions of mercury to the environment could have serious effects on human health. These early studies demonstrated that fish and other wildlife from various ecosystems commonly attain mercury levels of toxicological concern when directly affected by mercury-containing emissions from human-related activities. Human health concerns arise when fish and wildlife from these ecosystems are consumed by humans. During the 1990s, a new trend emerged with regard to mercury pollution. Investigations found that fish in the northern-tier states of the United States, Canada, and Nordic countries, mainly from nutrient-poor lakes and often in very remote areas, commonly have high levels of mercury. More recent fish sampling surveys in other regions of the United States have shown widespread mercury contamination in streams, wetlands, reservoirs, and lakes.

Mercury becomes a toxicological problem because, like many environmental contaminants, mercury undergoes bioaccumulation. Bioaccumulation is the process by which organisms (including humans) can take up contaminants more rapidly than their bodies can eliminate them, thus the amount of mercury in their body accumulates over time. If for a period of time an organism does not ingest mercury, its body burden of mercury will decline. If, however, an organism continually ingests mercury, its body burden can reach toxic levels. The rate of increase or decline in body burden is specific to each organism. For humans, about half the body burden of mercury can be eliminated in 70 days if no mercury is ingested during that time. Biomagnification is the incremental increase in concentration of a contaminant at each level of a food chain. This phenomenon occurs because the food source for organisms higher on the food chain is progressively more concentrated in mercury and other contaminants, thus magnifying bioaccumulation rates at the top of the food chain. The bioaccumulation effect is generally compounded the longer an organism lives, so that larger predatory game fish will likely have the highest mercury levels. Adding to this problem is the fact that mercury concentrates in the muscle tissue of fish. So, unlike organic contaminants (e.g., PCBs, dioxins) that concentrate in the skin and fat, mercury cannot be filleted or cooked out of consumable game fish.

Human uptake of mercury occurs in two ways: (1) as methylmercury (CH_3Hg^+) from fish consumption, or (2) by breathing vaporous mercury (Hg^0) emitted from various sources such as metallic mercury dental amalgams and ambient air. Our bodies are much more adapted for reducing the potential toxicity effects from vaporous mercury, so health effects from this source are relatively rare. Methylmercury, on the other hand, affects the central nervous system and in severe cases irreversibly damages areas of the brain. In milder cases of mercury poisoning, adults complain of reductions in motor skills and dulled senses of touch, taste, and sight. These milder effects are generally reversible if exposure to mercury is halted. Unborn children are at greatest risk from low-level exposure to methylmercury. Recent research suggests that prenatal effects occur at intake levels 5 to 10 times lower than those of adults. If these results are confirmed, a substantial fraction of unborn children would be at risk.

Mercury cycles in the environment. With the exception of isolated cases of known point sources, the ultimate source of mercury to most aquatic ecosystems is deposition from the atmosphere, primarily associated with rainfall. Once in

DID YOU KNOW?

The conversion of inorganic mercury to methylmercury is important for two reasons: (1) methylmercury is much more toxic than inorganic mercury, and (2) organisms require considerably longer to eliminate methylmercury. Methylmercury-containing bacteria may be consumed by the next higher level in the food change, or the bacteria may release the methylmercury to the water where it can quickly adsorb to plankton, which are also consumed by the next level in the food chain (Krabbenhoft and Rickert, 2009).

surface water, mercury enters a complex cycle in which one form can be converted to another. It can be brought to the sediments by particle settling and then later released by diffusion or resuspension. It can enter the food chain, or it can be released back to the atmosphere by volatilization. The concentration of dissolved organic carbon (DOC) and pH have a strong effect on the ultimate fate of mercury in an ecosystem. Studies have shown that for the same species of fish taken from the same region, increasing the acidity of the water (decreasing pH) and/or the DOC content generally results in higher body burdens in fish. Many scientists currently think that higher acidity and DOC levels enhance the mobility of mercury in the environment, thus making it more likely to enter the food chain. Many of the details of the aquatic mercury cycle are still unknown, however, and remain areas of active research.

Cadmium

Cadmium, in its purest form, is a soft silver-white metal that is found naturally in the Earth's crust; however, the most common forms of cadmium found in the environment exist in combinations with other elements. Cadmium can enter the bloodstream of humans when they drink cadmium-contaminated water. The common source of exposure to cadmium is mainly through eating food, especially shellfish. Plants absorb or take up cadmium from soil, and the fish we eat take up cadmium from the water they live in. Exposure to cadmium can cause a number of harmful health effects. Eating food or drinking water with high levels of cadmium can severely irritate or bother a person's stomach and cause vomiting and diarrhea. However, the greatest concern is from exposure to lower doses of cadmium over a long period of time. Lower and long-term exposure to cadmium through drinking cadmium-contaminated water can cause kidney damage. The damage is not life threatening (although it has been determined that cadmium is a probable carcinogen), but it can lead to the formation of kidney stones and can affect the skeleton, which can be painful and debilitating. Lung damage has also been observed. The results of some animal studies show that animals given cadmium-contaminated water show high blood pressure, iron-poor blood, liver disease, nerve damage, or brain damage. These effects have not been observed in humans (ATSDR, 2012; Reigart and Roberts, 1999).

Lead

At the time of this writing, two major water lead-contamination events have been the focus of media reports and various federal, state, and local investigations. One of these events, lead contamination of the drinking water supply in Flint, Michigan, has little or nothing to do with lead contamination of the surface water body (Flint River) of record and its bottom sediments within. The lead in the Flint, Michigan, drinking water supply did not emanate from lead-contaminated sediments in the Flint River. Instead, the lead in Flint's drinking water was leached from lead-bearing water distribution piping within the system. How did this occur? Consider that Flint River water is very corrosive—corrosive to the point that when it entered Flint's water distribution system untreated and not properly pH adjusted to avoid its corrosive effects the lead in the lead-bearing piping was easily leached and seeped into the water flow. This could have been avoided, of course, simply by sampling, testing, and treating (with pH-adjusting chemicals) the corrosive water at the waterworks. Why the corrosive water was allowed to circulate in the lead-bearing water distribution piping in Flint, Michigan, is not the focus of this book and is currently being investigated by the appropriate authorities. But, the intense media scrutiny and regulatory and possible criminal investigations have shined a revealing light on the potential impact of this contamination event. Simply and tragically, lead intake by humans can be and is a very hazardous situation. High levels of lead in tapwater can cause health effects if the lead in the water enters the bloodstream and causes an elevated blood lead level. Adults and children exposed to large amounts of lead can experience brain and kidney damage.

The other recent surface water lead contamination event was discussed earlier—contamination of the Animas River from acid mine waste from the Gold King Mine. The lead-contaminated water that resulted from the Gold King Mine spill is just as serious as the Flint River event because in Cement Creek, Animas River, and downstream San Juan River the lead (including arsenic, cadmium, aluminum, and copper) cannot be readily removed by the self-purification process of the moving water bodies involved. Lead is persistent and resides long term in bottom sediments. When these bottom sediments are disturbed by humans (e.g., dredging) or natural processes (e.g., flooding), the lead contaminants are able to be transported elsewhere, including into plants via irrigation or into drinking water via water intake systems for potable water uses.

Polychlorinated Biphenyls or Aroclors

Polychlorinated biphenyls (PCBs) belong to a broad family of human-made organic chemicals known as chlorinated hydrocarbons. PCBs were domestically manufactured from 1929 until their manufacture was banned in 1979. They have a range of toxicity and vary in consistency from thin, light-colored liquids to yellow or black waxy solids. Due to their nonflammability, chemical stability, high boiling point, and electrical insulating properties, PCBs were used in hundreds of industrial and commercial applications, including electrical, heat transfer, and hydraulic equipment;

DID YOU KNOW?

A PCB congener is any single, unique, well-defined chemical compound in the PCB category. The name of a congener specifies the total number of chlorine substituents and the position of each chlorine. For example, 4,4'-dichlorobiphenyl is a congener having a biphenyl structure with two rings. In 1980, a numbering system was developed that assigns a sequential number to each of the 209 PCB congeners.

as plasticizers in paints, plastics, and rubber products; in pigments, dyes, and carbonless copy paper; and in many other industrial applications. Although no longer commercially produced in the United States, PCBs may still be present in products and materials produced before the 1979 PCB ban. Products that may contain PCBs include the following:

- Transformers and capacitors
- Other electrical equipment including voltage regulators, switches, reclosers, bushings, and electromagnets
- Oil used in motors and hydraulic systems
- Old electrical devices or appliances containing PCB capacitors
- Fluorescent light ballasts (see Sidebar 11.1)
- Cable insulation
- Thermal insulation material including fiberglass, felt, foam, and cork
- Adhesives and tapes
- Oil-based paint
- Caulking (see Sidebar 11.2)
- Plastics
- Carbonless copy paper
- Floor finish

The PCBs used in these products were chemical mixtures made up of a variety of individual chlorinated biphenyl components known as congeners. Most commercial PCB mixtures are known in the United States by their industrial trade names, the most common being Aroclor. The following sidebars are included here because it is not uncommon to find fluorescent light ballasts containing PCBs and used and deteriorated caulking products in or on the bottom sediments of rivers and lakes.

SIDEBAR 11.1. PCBs AND FLUORESCENT LIGHT BALLASTS

Polychlorinated biphenyls are contained within fluorescent light ballast (FLB) capacitors and in the interior potting material of old magnetic T12 lighting fixtures. The capacitor regulates the amount of electricity coming into the light fixture, and the potting material serves to insulate the FLB and reduce the "humming" noise. Because all PCB-containing FLBs currently in use have exceeded their designed life span, it makes them susceptible to leaking or rupturing at any time which may lead to increased exposures of building occupants. Residues from these sources are difficult and costly to clean up. Additionally, intact PCB-containing FLBs may emit small amounts of PCBs into the air during normal use of the lighting fixture (USEPA, 2016a).

SIDEBAR 11.2. PCBs IN CAULK

Polychlorinated biphenyls were widely used in caulking and elastic sealant materials, particularly from about the 1950s through the 1970s. These materials were primarily used in or around windows, door frames, stairways, building joints, masonry columns, and other masonry building materials. PCBs were used in these building materials because of their properties as a plasticizer. PCBs have been detected in caulk in buildings, including schools, with concentrations ranging from below 50 parts per million (ppm) to greater than 440,000 ppm. If caulk contains PCBs, the PCBs may be released into the air through off-gassing. This may occur when the caulk is intact and undisturbed or if it is deteriorating. PCBs in the air originating from caulk can then be absorbed into other building materials, creating secondary sources which can then re-emit PCBs into the air. When building materials containing PCBs are disposed of in rivers or lakes, PCBs leak off into the water and often end up in bottom sediments (USEPA, 2016b).

DID YOU KNOW?

For 30 years, ending in the later 1970s, the General Electric Company (GE) discharged as much as 1.3 million pounds of polychlorinated biphenyls into the Hudson River from its capacitor manufacturing plants in Hudson Falls and Fort Edward, New York. In 2002, the federal government ordered GE to conduct targeted environmental dredging of PCB-contaminated sediment in a 40-mile stretch of the Upper Hudson, and, after many years of study, the dredging was finally completed in 2015. The dredging was ordered because PCBs in the sediment are not safely buried. River sediment is continually redistributed across the bottom by erosion and river flows. This movement exposes PCB-contaminated sediment, making it available to fish. PCBs degrade naturally over time, but the process, called *natural dechlorination*, does not make them harmless. The USEPA considers all PCBs, regardless of the level of chlorination, to be hazardous.

Health Effects of PCBs

Polychlorinated biphenyls have been demonstrated to cause a variety of adverse health effects. PCBs have been shown to cause cancer in animals, as well as a number of serious non-cancer health effects, including effects on the immune system, reproductive system, nervous system, and endocrine system. Studies in humans provide supportive evidence for the potential carcinogenic and noncarcinogenic effects of PCBs. The different health effects of PCBs may be interrelated, as alterations in one system may have significant implications for the other systems of the body.

Copper

As discussed earlier with regard to concentrated animal feeding operations (CAFOs) and yellow boy, a major contaminant, copper, can enter surface waters and affect bottom sediments in the water bodies. Copper is a metal that occurs naturally in the environment, as well as in plants and animals. Low levels of copper are essential for maintaining good health. High levels can cause harmful effects such as irritation of the nose, mouth, and eyes; vomiting; diarrhea; stomach cramps; nausea; and even death. Copper has been found in at least 906 of the 1647 National Priority Sites identified by the USEPA. Copper is released into the environment by farming and mining but is also released by manufacturing operations and through wastewater (used water) releases into rivers and lakes. Copper is also released from natural sources such as volcanoes, windblown dusts, decaying vegetation, and forest fires. Copper released into the environment usually attaches to particles and sediments made of organic matter, clay, soil, or sand (ATSDR, 2004).

Arsenic

Arsenic is a naturally occurring element that is widely distributed in the Earth's crust. Arsenic is classified chemically as a metalloid, having both properties of a metal and a nonmetal; however, it is frequently referred to as a metal. Arsenic occurs naturally in soil and minerals; therefore, it may enter the air, water, and land from wind-blown dust or get into water from runoff and leaching. It may also enter the water environment via agricultural applications, mining, and smelting operations. Arsenic is usually found in the environment combined with other elements such as oxygen, chlorine, and sulfur. Because arsenic is found naturally in the environment, humans are exposed to some arsenic by eating food, drinking water, or breathing air. The predominant dietary source of arsenic is from fish. In most drinking water sources, the inorganic form of arsenic tends to be more predominant than organic forms. Inorganic arsenic in drinking water can exert toxic effects after acute (short-term) or chronic (long-term) exposure. Studies link inorganic arsenic ingestion to a number of health effects:

- *Cancerous effects*—Skin, bladder, lung, kidney, nasal passages, liver, and prostate cancer
- *Noncancerous effects*—Cardiovascular, pulmonary, immunological, neurological, and endocrine (e.g., diabetes) disorders (ATSDR, 2007a)

Acenaphthene

Acenaphthene is a polycyclic aromatic hydrocarbon (PAH) that is used on a large scale to make dyes, plastics, pesticides, and optical brighteners. Anyone who drinks water contaminated with PAHs is exposed. A number of PAHs have caused tumors in laboratory animals that were exposed to PAHs through their food and water. Effects include damage to skin, body fluids, and the immune system, which helps the body fight disease. As of yet, these effects have not been seen in humans (ATSDR, 1995a).

Disulfoton

Disulfoton is a manufactured substance used as a pesticide to control a variety of harmful pests that attack many field and vegetable crops. Disulfoton does not occur naturally. Pure disulfoton is a colorless oil with no identifiable odor or taste. The technical product is dark yellowish and has an aromatic odor. Disulfoton enters the environment principally when it is applied as a spray or as granules on field crops, vegetables, potted plants, and home gardens. Disulfoton also can enter the environment when it accidentally spills or leaks during storage and transport. Disulfoton may also enter the environmental from hazardous waste sites. Environmental contamination by disulfoton mainly affects soil and water. Natural chemical reactions and bacterial attack remove disulfoton from soil and water. Such reactions develop some byproducts that are more toxic than disulfoton. Fish accumulate disulfoton in their bodies; the levels of disulfoton in fish can be hundreds of times higher than the level in water. The estimated amount of time required for the concentration of disulfoton in river water to decrease to half of its initial level (half-life) is 7 days. People may be exposed to disulfoton by drinking contaminated water, eating contaminated food, breathing it in, or coming into contact with it. In people, disulfoton mainly causes harmful effects to the nervous system. Depending on the amount of disulfoton that enters the body, neurological effects such as inhibited cholinesterase activity, narrowing of the pupils, vomiting, diarrhea, drooling, difficulty in breathing, tremors, convulsions, or even death may occur (ATSDR, 1995b).

Fluoranthene

Fluoranthene is a polycyclic aromatic hydrocarbon (PAH). Although samples are often pale yellow, the compound is colorless. Fluoranthene is obtained from the high boiling fraction of coal tar and is commonly found in many combustion products. Fluoranthene has been classified by the International Agency for Research on Cancer as a Group 3 carcinogen (Griesbaum et al., 2012).

Heptachlor

Heptachlor is not found naturally in the earth. It is a human-made compound that looks like a white powder and smells like mothballs. Pure forms of heptachlor are white, but less pure forms of this substance appear tan. Between the 1960s and 1970s, heptachlor was used to kill termites found in the home, and farmers used it to kill insects found on farm crops, especially corn crops. In the later 1970s, the use of heptachlor was phased out. By 1988, the commercial sale of heptachlor was banned in the United States. Today, the use of heptachlor is restricted to controlling fire ants in power transformers. Heptachlor can enter the body through the consumption of food, water, or even milk that is contaminated with heptachlor. The health effects from exposure to heptachlor will vary depending on the exposure and the length of time. Animal studies show that heptachlor is very toxic to humans and animals; long-term exposure can affect the liver (ATSDR, 2007b).

Diazinon

Diazinon does not occur naturally in the environment. The pure chemical is an oil that is colorless and practically odorless. Commercial diazinon is a pale to dark brown liquid. Diazinon is the common name of an organophosphorus insecticide used to control pest insects in soil, on ornamental plants, and on fruit and vegetable field crops. Diazinon may enter the environment from agricultural and household application of the chemical to control insects. After application, it may be present in surface waters (such as rivers, lakes, and ponds). Short exposures to high levels of diazinon can affect the nervous system (ATSDR, 2008).

Dicofol

Dicofol, or kelthane, is a white crystalline, wettable powder dissolved in a liquid carrier (water). The primary hazard is the threat to the environment, and immediate steps should be taken to limit its spread to the environment. Because it is a liquid it can easily penetrate the soil and contaminate groundwater and nearby streams. It can cause illness by inhalation, skin absorption, or ingestion. It is used as a pesticide (NIH, 2015).

Pyrene

Pyrene is a polycyclic aromatic hydrocarbon (PAH). It is a colorless crystal-like solid but can also look yellow. Pyrene is used to make dyes, plastics, and pesticides. Exposure to pyrene can occur through the consumption of water that is contaminated. Health effects may include kidney disease. Other effects may include damage to skin, body fluids, and the immune system (ATSDR, 1995a).

Anthracene

Anthracene is a polycyclic aromatic hydrocarbon (PAH). It can vary in appearance from colorless to a pale yellow crystal-like solid. PAHS are created when products such as coal, oil, gas, and garbage are burned but the burning process is not complete. Anthracene is used to make dyes, plastics, and pesticides. It has also been used to make smoke screens and scintillation counter crystals. Exposure to anthracene can occur through the consumption of water that is contaminated. Once inside the body, anthracene appears to target the skin, blood, stomach, and intestines, as well as the lymph system (ATSDR, 1995a).

Benzo(*g,h,i*)Perylene

Benzo(*g*,*h*,*i*)perylene, a colorless crystal-like solid, is a polycyclic aromatic hydrocarbon (PAH). It is created when products such as coal, oil, gas, and garbage are burned but the burning process is not complete. Benzo(*g*,*h*,*i*)perylene is used to make dyes, plastics, pesticides, explosives, and drugs. It has also been used to make bile acids, cholesterol, and steroids. Exposure to benzo(*g*,*h*,*i*)perylene can occur through the consumption of water that is contaminated. Once inside the body, benzo(*g*,*h*,*i*) perylene can spread and target fat tissues. Target organs include the kidneys and liver (ATSDR, 1995a).

Chlorpyrifos

Chlorpyrifos is an insecticide that is a white crystal-like solid with a strong odor. Chlorpyrifos has been widely used in homes and on farms. In the home, it is used to control cockroaches, fleas, and termites; it is also used in some pet flea and tick collars. On the farm, it is used to control ticks on cattle and as a spray to control crop pests. Chlorpyrifos sticks tightly to soil sediments, which becomes an issue in water bodies. Alone, it does not mix will with water, so it rarely enters local water systems; however, because it sticks to sediments it can enter local water systems piggybacked on sediments that end up in the water. Exposure to high levels of chlorpyrifos may cause severe sweating, loss of bowel control, severe muscle tremors, seizures, loss of consciousness (coma), and death (ATSDR, 1997).

PHARMACEUTICALS AND PERSONAL CARE PRODUCTS

The acronym PPCPs was coined in a 1999 critical review published in *Environmental Health Perspectives* to refer to pharmaceuticals and personal care products. Although they are sometimes termed "emerging pollutants," it is important to point out that PPCPs are not truly emerging; rather, it is an understanding of the significance of their occurrence and their fate in the environment that is beginning to develop. PPCPs are a very broad, diverse collection of thousands of chemical substances, including prescription, veterinary, and over-the-counter (OTC) therapeutic drugs; fragrances; cosmetics; sunscreen agents; diagnostic agents; nutraceuticals; biopharmaceuticals; growth-enhancing chemicals used in livestock operations; and many others (Spellman, 2015). This broad collection of substances refers, in general, to any product used by individuals for personal health or cosmetic reasons (e.g., anti-aging cleansers, toners, exfoliators, facial masks, serums, lip balm). Sources of PPCPs include the following:

- Human bathing, shaving, and swimming
- Illicit drugs
- Veterinary drug use, especially antibiotics and steroids
- Agribusiness
- Residues from pharmaceutical manufacturing
- Residues from hospitals

People contribute PPCPs to the environment when

- Medication residues pass out of the body and into sewer lines.
- Externally applied drugs and personal care products they use wash down the shower drain.
- Unused or expired medications are placed in the trash.

The problem with PPCPs is that we do not know what we do not know about them; the jury is still out on their exact environmental impact. Through personal observation and after studying this issue for years, the author has found that most water professionals and practitioners in the field give little credence and even less attention to the possible environmental impacts of PPCPs. It must be pointed out, however, that this trend is slowly changing because of the increasing use and disposal of PPCPs. For the purpose of this discussion, it is important to point out that the statement, credited to the mythical hero Hercules, that "the solution to pollution is dilution" is reasonable and sensible and practicable *only to a point.* Many PPCPs are designed to adhere to skin tissue, for example. Consider that these same PPCPs can attach to surface water body sediments and sequester there for some time. What is the result, then, of PPCP-contaminated sediments in rivers, streams, lakes, and ponds? We really do not know the answer to this question, but it is difficult to imagine that aquatic organisms would have any use for facial creams and lipsticks. It is not difficult to assume, however, that they could ingest them. To what effect? Again, we do not know for sure. Then there are the drugs or pharmaceuticals. What happens when aquatic life is exposed to antibiotics, pain killers, aspirin, and illicit drugs? Thus far, studies have shown that pharmaceuticals are definitely present in our nation's water bodies. Further research suggests that certain drugs may cause ecological harm. But, the fact is more research is necessary to determine the extent of ecological harm and any role it may have in potential human health effects. To date, scientists have found no evidence of adverse human health effects from PPCPs in the environment; however, research has not been extensive and more information is needed (Spellman, 2015).

DID YOU KNOW?

Although PPCPs are used in large quantities, the concentrations of PPCPs currently being found in water suppliers are very small. The laboratory tests for these compounds do not report concentrations in parts per million (ppm) or parts per billion (ppb) but instead report concentrations in parts per trillion (ppt), which is the same as nanograms per liter. One part per million is equivalent to a shot glass full of water dipped from an Olympic swimming that is 2 meters deep. One part per billion is equivalent to one drop from an eye dropper filled from the same Olympic pool. One part per trillion is equivalent to 1 drop from 20 Olympic pools that are 2 meters deep, or 1 second in 31,700 years (Spellman, 2015).

REFERENCES AND RECOMMENDED READING

Adams, R.S. (1973). Factors influencing soil adsorption and bioactivity of pesticides. *Residue Reviews*, 47: 1–45.

Aiken, G.R., McKnight, D.M., Wershaw, R.L., and MacCarthy, P. (1985). *Humic Substances in Soil, Sediment, and Water: Geochemistry, Isolation, and Characterization*. New York: John Wiley & Sons.

Akhavein, A.A. and Linscott, D.L. (1968). The bipyridylium pesticides: paraquat and diquat. *Residue Reviews*, 23: 98–140.

ATSDR. (1995a). *Public Health Statement for Polycyclic Aromatic Hydrocarbons (PAHs)*. Atlanta, GA: U.S. Department of Health and Human Services, Agency for Toxic Substances and Disease Registry.

ATSDR. (1995b). *Public Health Statement for Disulfoton*. Atlanta, GA: U.S. Department of Health and Human Services, Agency for Toxic Substances and Disease Registry.

ATSDR. (1997). *Public Health Statement for Chlorpyrifos*. Atlanta, GA: U.S. Department of Health and Human Services, Agency for Toxic Substances and Disease Registry.

ATSDR. (2004). *Toxicological Profile for Copper*. Atlanta, GA: U.S. Department of Health and Human Services, Agency for Toxic Substances and Disease Registry.

ATSDR. (2007a). *Toxicological Profile for Arsenic*. Atlanta, GA: U.S. Department of Health and Human Services, Agency for Toxic Substances and Disease Registry.

ATSDR. (2007b). *Public Health Statement for Heptachlor and Heptachlor Epoxide*. Atlanta, GA: U.S. Department of Health and Human Services, Agency for Toxic Substances and Disease Registry.

ATSDR. (2008). *Toxicological Profile for Diazinon*. Atlanta, GA: U.S. Department of Health and Human Services, Agency for Toxic Substances and Disease Registry.

ATSDR. (2012). *Public Health Statement for Cadmium*. Atlanta, GA: U.S. Department of Health and Human Services, Agency for Toxic Substances and Disease Registry.

Bailey, G.W. and White, J.L. (1965). Herbicides: a compilation of their physical, chemical and biological properties. *Residue Reviews*, 10: 97–121.

Bailey, G.W., White, J.L., and Rothberg, T. (1968). Adsorption of organic herbicides by montmorillonite: role of pH and chemical character of adsorbate. *Soil Science Society of America Proceedings*, 32(2): 222–234.

Baker, E.T. and Feely, R.A. (1978). Chemistry of oceanic particulate matter and sediments: implications for bottom sediment resuspension: *Science*, 200(4341): 533–535.

Bilby, R.R. and Likens, G.E. (1979). Effect of hydrologic fluctuations on the transport of fine particulate organic carbon in a small stream. *Limnology and Oceanography*, 24(1): 69–75.

Bothner, M.H., Rendigs, R.R., Campbell, E. et al. (1986). *The Georges Bank Monitoring Program 1985: Analysis of Trace Metals in Bottom Sediments During the Third Year of Monitoring*, USGS Circular 988. Reston, VA: U.S. Geological Survey.

Bums, N.M. and Rosa, F. (1980). *In situ* measurement of the settling velocity of organic carbon particles and 10 species of phytoplankton. *Limnology and Oceanography*, 25(5): 855–864.

Carter, C.W. and Suffet, L.H. (1982). Binding of DDT to dissolved humic materials. *Environmental Science & Technology*, 16(11): 735–740.

Chase, R.R.P. (1979). Settling behavior of natural aquatic particulates. *Limnology and Oceanography*, 24(3): 417–426.

Chiou, C.T. (1981). Partition coefficient and water solubility in environment chemistry. In: *Hazard Assessment of Chemicals: Current Developments*, Vol. 1 (Saxena, J. and Fisher, F., Eds.), pp. 117–153. New York, Academic Press.

Chiou, C.T. and Shoup, T.D. (1985). Soil sorption of organic vapors and effects of humidity on sorption mechanism and capacity. *Environmental Science and Technology*, 19: 1196–1200.

Chiou, C.T., Peters, I.J., and Freed, B.H. (1979). A physical concept of soil–water equilibria for nonionic organic compounds. *Science*, 206(4420): 831–832.
Chiou, C.T., Porter, P.E., and Schmedding, D.W. (1983). Partition equilibriums of nonionic organic compounds between soil organic matter and water. *Environmental Science & Technology*, 17: 227–231.
Chiou, C.T., Porter, P.E., and Shoup, T.D. (1984). Reply to comment by W.G. MacIntyre and C.L. Smith. *Environmental Science & Technology*, 18(4): 295–297.
Chiou, C.T., Shoup, T.D., and Porter, P.E. (1985). Mechanistic roles of soil humus and minerals in the sorption of nonionic organic compounds from aqueous and organic solutions. *Organic Geochemistry*, 8: 9–14.
Chiou, C.T., Malcoim, R.L., Brinton, T.I., and Kile, D.E. (1986). Water solubility enhancement of some organic pollutants and pesticides by dissolved humic and fulvic acids. *Environmental Science & Technology*, 20(5): 502–507.
Christman, B.R. and Gjessing, E.T., Eds. (1983a). *Aquatic and Terrestrial Humic Materials.* Ann Arbor, MI: Ann Arbor Science.
Christman, B.R. and Gjessing, E.T. (1983b). Priorities in humic research. In: *Aquatic and Terrestrial Humic Materials* (Christman, B.R. and Gjessing, E.T., Eds.), pp. 517–528. Ann Arbor, MI: Ann Arbor Science.
Colby, R.R. (1963). *Fluvial Sediments: Summary of Source, Transportation, Deposition, and Measurement of Sediment Discharge*, USGS Bulletin 1181-A. Reston, VA: U.S. Geological Survey.
Davis, J.A. (1982). Adsorption of natural dissolved organic matter at the oxide/water interface. *Geochimica et Cosmochimica Acta*, 46: 2381–2393.
Duinker, J.C. (1980). Suspended matter in estuaries: adsorption and desorption processes. In: *Chemistry and Biochemistry of Estuaries* (Olausson, E., and Bato, I,. Eds.), pp. 121–145. New York: John Wiley & Sons
Edzwald, H.J.K., Upchurch, J.B., and O'Melia, C.R. (1974). Coagulation in estuaries. *Environmental Science & Technology*, 8(1): 58–63.
Eisma, D. (1981). Suspended matter as a carrier for pollutants in estuaries and the sea. In: *Maine Environmental Pollution*, Vol. 2 (Geyer, R., Ed.), pp. 281–295. Amsterdam: Elsevier.
Ertel, J.R. and Hedges, J.L. (1983). Bulk chemical and spectroscopic properties of marine and terrestrial humic acids, melanoidins and catechol-based synthetic polymers. In: *Aquatic and Terrestrial Humic Materials* (Christman, R.R. and Gjessing, E.T., Eds.), pp. 143–163. Ann Arbor, MI: Ann Arbor Science.
Faust, S.D. and Suffet, L.H. (1966). Recovery, separation, and identification of organic pesticides from natural and portable waters. *Residue Reviews*, 15: 45–111.
Garten, C.T. and Trabalka, J.R. (1983). Evaluation of models for predicting terrestrial food chain behavior ofxenobiotics. *Environmental Science & Technology*, 17(10): 590–595.
Griesbaum, K., Behr, A., Biedenkapp, D., Heinz-Werner, V., Garbe, D., Paetz, C., Collin, G., Mayer, D., and Hoke, H. (2012). Hydrocarbons. In: *Ullmann's Encyclopedia of Industrial Chemistry*, Vol. 18, pp. 134–189. Weinheim: Wiley-VCH.
Guy, H.P. (1969). Laboratory theory and methods for sediment analysis. In: *Techniques of Water-Resources Investigations*, Book 5, Chapter C1. Reston, VA: U.S. Geological Survey.
Guy, H.P. (1970). Fluvial sediment concepts. In: *Techniques of Water-Resources Investigations*, Book 3, Chapter C1. Reston, VA: U.S. Geological Survey.
Haque, R. and Freed, V.H. (1974). Behavior of pesticides in the environment: environmental chemodynamics. *Residue Reviews*, 52: 89–111.
Hassett, J.J., Banwart, W.L., and Griffin, R.A. (1983). Correlation of compound properties with sorption characteristics of nonpolar compounds by soils and sediments: concepts and limitations. In: *Environment and Solid Wastes* (Francis, C.W. and Auerbach, S.I., Eds.), pp. 161–178. Ann Arbor, MI: Ann Arbor Science.

Hayes, M.H.B., Pick, M.E., and Toms, B.A. (1975). Interactions between clay minerals and bipyridylium herbicides. *Residue Reviews*, 57: 1–27.
Helling, C.S. (1976). Dinitroaniline herbicides in soils. *Journal of Environmental Quality*, 5(1): 1–15.
Hunt, J.R. (1982). Particle dynamics in seawater: implications for predicting the fate of discharged particles. *Environmental Science & Technology*, 16(6): 303–309.
Hunter, K.A. and Liss, P.S. (1982). Organic matter and the surface charge of suspended particles in estuarine waters. *Limnology and Oceanography*, 27(2): 322–335.
International Union of Pure and Applied Chemistry. (1993). Glossary for chemists of terms used in toxicology. *Pure and Applied Chemistry*, 65(9): 2003–2122.
Jackson, T.A. (1975). Humic matter in natural waters and sediments. *Soil Science*, 119(1): 56–64.
Josephsen, J. (1982). Humic substances. *Environmental Science & Technology*, 16(1): 20A–24A.
Karickhoff, S.W. (1980). Sorption kinetics of hydrophobic pollutants in natural sediments. In: *Contaminants and Sediments* (Baker, R., Ed.), pp. 193–205. Ann Arbor, MI: Ann Arbor Science.
Karickhoff, S.W. (1981). Semi-empirical estimation of sorption of hydrophobic pollutants on natural sediments and soils. *Chemosphere*, 10(8): 833–846.
Karickhoff, S.W. (1984). Organic pollutant sorption in aquatic systems. *Journal of Hydraulic Engineering*, 110 (6): 707–735.
Karickhoff, S.W. and Brown, D.S. (1978). Paraquat sorption as a function of particle size in natural sediments. *Journal of Environmental Quality*, 7(2): 246–252.
Karickhoff, S.W. and Brown, D.S. (1979). *Determination of Octanol/Water Distribution Coefficients, Water Solubilities, and Sediment/Water Partition Coefficients for Hydrophobic Organic Pollutants*, EPA-600/4-79-032. Washington, DC: U.S. Environmental Protection Agency.
Karickhoff, S.W., Brown, D.S., and Scott, T.A. (1979). Sorption of hydrophobic organic pollutants on natural sediments. *Water Research*, 13: 241–248.
Kenaga, E.E. (1975a). Partitioning and uptake of pesticides in biological systems. In: *Environment Dynamics of Pesticides* (Haque, R. and Freed, V.H., Eds.), pp. 217–273. New York: Plenum Press.
Kenaga, E.E. (1975b). Guidelines for environmental study of pesticides: determination of bioconcentration potential. *Residue Review*, 44: 73–114.
Khan, S.U. (1972). Adsorption of pesticide by humic substances: a review. *Environmental Letters*, 3(1): 1–12.
Khan, S.U. (1980). Role of humic substances in predicting fate and transport of pollutants in the environment. In: *Dynamics, Exposure and Hazard Assessment of Toxic Chemicals* (Haque, R., Ed.), pp. 215–230. Ann Arbor, MI: Ann Arbor Science.
Khan, S.U. and Schnitizer, M. (1972). The retention of hydrophobic organic compounds by humic acid. *Geochimica et Cosmochimica Acta*, 36: 745–754.
Kononova, M.M. (1966). *Soil Organic Matter*, 2nd ed. Oxford: Pergamon Press.
Krabbenhoft, D.P. and Rickert, D.A. (2009). *Mercury Contamination of Aquatic Ecosystems*, USGS Fact Sheet FS-216-95. Reston, VA: U.S. Geological Survey.
Krom, M.D. and Sholkovitz, E.R. (1977). Nature and reactions of dissolved organic matter in the interstitial waters of marine sediments. *Geochimica et Cosmoshmica Acta*, 41: 1565–1573.
Lal, D. (1977). The oceanic microcosm of particles. *Science*, 198(4321): 997–1009.
Leo, A., Hansch, C., and Elkins, D. (1971). Partition coefficients and their uses. *Chemical Reviews*, 71(6): 525–616.
Lorenzen, C.J., Welschmeyer, N.A., Copping, A.E., and Vernet, M. (1983). Sinking rates of organic particles. *Limnology and Oceanography*, 28(4): 766–769.
MacIntryre, F. (1974). The top millimeter of the ocean. *Scientific American*, 203(1): 62–77.

Mackay, D., Bobra, A., Shiu, W.Y., and Yalkowsky, S.H. (1980). Relationships between aqueous solubility and octanol–water partition coefficients. *Chemosphere*, 9(4): 701–711.

Maier-Bode, H. and Hartel, K. (1981). Linuron and monolinuron. *Residue Reviews*, 77: 1–294.

Mathur, S.P. (1974). Phthalate esters in the environment: pollutants of natural products? *Journal of Environmental Quality*, 3(3): 189–197.

Means, J.C. and Wijayaratne, R. (1982). Role of natural colloids in the transport of hydrophobic pollutants. *Science*, 215(4535): 968–970.

Metcalf, R.L. (1972). Agricultural chemicals in relation to environmental quality: insecticides today and tomorrow. *Journal of Environmental Quality*, 1(1): 10–17.

Metcalf, R.L. and Pitts, J.N. (1969). Outline of environmental sciences. *Advances in Environmental Sciences*, 1(1): 1–23.

Meyers, P.A., Leenheer, M.J., Eadie, B.J., and Maule, S.J. (1984). Organic geochemistry of suspended and settling particulate matter in Lake Michigan. *Geochimica et Cosmochimica Acta*, 48: 443–452.

Mingelgrin, U. and Gerstl, Z. (1983). Reevaluation of partitioning as a mechanism of nonionic chemicals adsorption in soils. *Journal of Environmental Quality*, 12(1): 1–11.

Mulla, M.S. and Mian, L.S. (1981). Biological and environmental impacts of the insecticides malathion and parathion on non-target biota in aquatic ecosystems. *Residue Reviews*, 78: 101–137.

Mulla, M.S., Mian, L.S., and Kawecki, J.A. (1981). Distribution, transport and fate of the insecticides malathion and parathion in the environment. *Residue Reviews*, 81: 1–136.

Neely, W.B., Branson, D.R., and Blau, G.E. (1974). Partition coefficient to measure bioconcentration potential of organic chemicals in fish. *Environmental Science & Technology*, 8(13): 1113–1115.

NIH. (2015). *Dicofol*. Bethesda, MD: National Institutes of Health (https://pubchem.ncbi.nlm.nih.gov/compound/91566).

Norris, L.A. (1981). The movement, persistence, and fate of the phenoxy herbicides and TCDD in the forest. *Residue Reviews*, 80: 66–126.

O'Melia, C.R. (1980). Aquasols: the behavior of small particles in aquatic systems. *Environmental Science & Technology*, 14(9): 1052–1060.

Olsen, C.R., Cutshall, N.H., and Larsen, I.L. (1982). Pollutant-particle associations and dynamics in coastal marine environments: a review. *Marine Chemistry*, 11(3): 501–533.

Orians, G.H. (1995). Ecology, aggregate variables. In: *Encyclopedia of Environmental Biology* (Neiremberg, W.A., Ed.), pp. 581–588. San Diego, CA: Academic Press.

Oschwald, W.R. (1972). Sediment-water interactions. *Journal of Environmental Quality*, 1(4): 360–366.

Paris, D.F. and Lewis, D.C. (1973). Chemical and microbial degradation of 10 selected pesticides in aquatic systems. *Residue Reviews*, 45: 96–114.

Pavlov, S.T. and Dexter, R.N. (1979). Distribution of PCBs in estuarine ecosystems: testing the concept of equilibrium partitioning in the marine environment. *Environmental Science & Technology*, 13(1): 65–71.

Pionke, H.R. and Chesters, G. (1973). Pesticide–sediment–water interactions. *Journal of Environmental Quality*, 2(1): 29–45.

Plechanov, N., Josefsson, B., Dyrssen, D., and Lundquist, K. (1983). Investigations of humic substances in natural waters. In: *Aquatic and Terrestrial Humic Materials* (Christman, R.F. and Gjessing, E.T., Eds.), pp. 387–405. Ann Arbor, MI: Ann Arbor Science.

Poirrier, M.A., Bordelon, B.R., and Laseter, J. (1972). Adsorption and concentration of dissolved carbon-14 DDT by coloring colloids in surface-waters. *Environmental Science & Technology*, 6(12): 1033–1035.

Presley, B.J. and Trefry, J.H. (1980). Sediment–water interactions and the geochemistry of interstitial waters. In: *Chemistry and Biogeochemistry of Estuaries* (Olausson, E. and Cato, I., Eds.), pp. 187–226. New York: John Wiley & Sons.

Reigart, R.J. and Roberts, J.R. (1999). *Recognition and Management of Pesticide Poisonings*, 5th ed. Washington, DC: U.S. Environmental Protection Agency.
Rice, C.Pl, Eadie, B.J., and Erstfeld, K.M. (1982). Enrichment of PCBs in Lake Michigan surface films. *Journal of Great Lakes Research*, 8(2): 265–270.
Saar, R.A. and Weber, J.H. (1982). Fulvic acid: modifier of metal-ion chemistry. *Environmental Science & Technology*, 16(9): 510a–517a.
Schlagbauer, G.G.L. and Schlagbauer, A.W.J. (1972). The metabolism of carbamate pesticides: a literature analysis. *Residue Reviews*, 42: 1–91.
Schnitzer, M. (1978). Humic substances: chemistry and reactions. In: *Soil Organic Matter* (Schnitzer, M. and Khan, S.U., Eds.), pp. 1–64. Amsterdam: Elsevier.
Schnitzer, M. and Khan, S.U. (1972). *Humic Substances in the Environment*. New York: Marcel Dekker.
Schnitzer, M. and Khan, S.U., Eds. (1978). *Soil Organic Matter*. Amsterdam: Elsevier.
Schwarzenbach, R.R. and Westall, J. (1981). Transport of nonpolar organic compounds from surface-water to groundwater: laboratory sorption studies. *Environmental Science & Technology*, 15(11): 1360–1367.
Sheldon, R.W., Prakash, A., and Sutcliffe, W.H. (1972). The size distribution of particles in the ocean. *Limnology and Oceanography*, 17(2): 327–340.
Sholkovitz, E.R. (1976). Flocculation of dissolved organic and inorganic matter during the mixing of river water and seawater. *Geochimica et Cosmochimica Acta*, 40: 831–845.
Sigleo, A.C. and Helz, G.R. (1981). Composition of estuarine colloidal material: major and trace elements. *Geochimica et Cosmochimica Acta*, 45: 2501–2509.
Spellman, F.R. (2015). *The Science of Water: Concepts and Applications*, 3rd ed. Boca Raton, FL: CRC Press.
Stojanovic, B.J., Kennedy, M.V., and Shuman, F.L. (1972). Edaphic aspects of the disposal of unused pesticides, pesticide wastes and pesticide containers. *Journal of Environmental Quality*, 1(1): 54–62.
Stumm, W. (1977). Chemical interaction in particle separation. *Environmental Science & Technology*, 11(12): 1066–1070.
USEPA. (1978). *Environmental Pathways of Selected Chemicals in Freshwater Systems*. Part II. *Laboratory Studies*, EPA-600/7-78-074. Washington, DC: U.S. Environmental Protection Agency.
USEPA. (1979a). *Water Related Fate of 129 Priority Pollutants*, EPA-440/4-79-029a,b. Washington, DC: U.S. Environmental Protection Agency.
USEPA. (1979b). *Status Assessment of Toxic Chemicals*, EPA-600/2-79-210a–n. Washington, DC: U.S. Environmental Protection Agency.
USEPA. (1981). *Aquatic Fate Process Data for Organic Priority Pollutants*, EPA-440/4-81-014. Washington, DC: U.S. Environmental Protection Agency.
USEPA. (1982). *Treatability Manual*. Vol. 1. *Treatability Data*, EPA-600/2-82-001a. Washington, DC: U.S. Environmental Protection Agency.
USEPA. (2001). *Methods for Collection, Storage and Manipulation of Sediments for Chemical and Toxicological Analyses: Technical Manual*, EPA-823-B-01-002. Washington, DC: U.S. Environmental Protection Agency, Office of Water.
USEPA. (2010). *Solid Waste and Emergency Response Glossary*. Washington, DC: U.S. Environmental Protection Agency, Office of Water.
USEPA. (2016a). *PCB-Containing Fluorescent Light Ballasts (FLBs) in School Buildings*. Washington, DC: U.S. Environmental Protection Agency (https://www.epa.gov/pcbs/polychlorinated-biphenyl-pcb-containing-fluorescent-light-ballasts-flbs-school-buildings).
USEPA. (2016b). *Questions and Answers about Polychlorinated Biphenyls (PCBs) in Building Materials*. Washington, DC: U.S. Environmental Protection Agency (https://www.epa.gov/pcbs/questions-and-answers-about-polychlorinated-biphenyls-pcbs-building-materials).

USEPA. (2016c). *Table of Polychlorinated Biphenyl (PCB) Congeners*. Washington, DC: U.S. Environmental Protection Agency (https://www.epa.gov/pcbs/table-polychlorinated-biphenyl-pcb-congeners).

USGS. (2000). *Contaminant Sorption by Soil and Bed Sediment*, Fact Sheet 087-00. Reston, VA: U.S. Geological Survey.

USGS. (2007). *Glossary*. Reston, VA: U.S. Geological Survey.

Van Middelem, C.H. (1963). Parathion residues on leafy crops. *Residue Review*, 2: 77–95.

Verschueren, K. (1983). *Handbook of Environmental Data on Organic Chemicals*, 2nd ed. New York: Van Nostrand Reinhold.

Wauchope, R.D. (1978). The pesticide content of surface-waters draining from agricultural fields: a review. *Journal of Environmental Quality*, 7(4): 459–472.

Wershaw, R.L., Burcar, P.J., and Goldberg, M.C. (1969). Interaction of pesticides with natural organic material. *Environmental Science & Technology*, 3(3): 271–273.

Wikowski, P.J., Smith, J.A., Fusillo, T.V., and Chiou, C.T. (1987). *A Review of Surface-Water Sediment Fractions and Their Interactions with Persistent Manmade Organic Compounds*, USGS Circular 993. Reston, VA: U.S. Geological Survey.

Section III

Sediment Sampling

12 Sampling Sediments

In January, we take our nets to a no-name stream in the foothills of the Blue Ridge Mountains of Virginia to do a special kind of macroinvertebrate monitoring—looking for winter stoneflies (*Allocapnia*). Winter stoneflies have an unusual life cycle. Soon after hatching in early spring, the larvae bury themselves in the streambed. They spend the summer lying dormant in the mud, thereby avoiding problems such as overheated streams, low oxygen concentrations, fluctuating flows, and heavy predation. In later November, they emerge, grow quickly for a couple of months, and then lay their eggs in January. January monitoring of winter stoneflies helps in interpreting the results of spring and fall macroinvertebrate surveys. In spring and fall, a thorough benthic survey is conducted based on Protocol II of the U.S. Environmental Protection Agency's *Rapid Bioassessment Protocols for Use in Streams and Rivers* (Plafkin et al., 1989). Some sites on various rural streams have poor diversity and sensitive families. Is the lack of macroinvertebrate diversity because of specific warm-weather conditions, high water temperature, low oxygen, or fluctuating flows, or is some toxic contamination present? In the January screening, if winter stoneflies are plentiful, seasonal conditions could probably be blamed for the earlier results; if winter stoneflies are absent, the site probably suffers from toxic contamination (based on the rural location, probably emanating from nonpoint sources) that is present all year. Though different genera of winter stoneflies are found in the region of southwestern Virginia, *Allocapnia* is sought because it is present even in the smallest streams (Figure 12.1).

THE HOUSE OF *E. COLI*

When humans participate in and enjoy recreation in local streams and rivers and when natural disasters such as flooding occur, the disturbance of streambed sediments can bother, annoy, and disrupt the house of *Escherichia coli*. If *E. coli* are in residence in the affected streams or river sediments, their life cycle can be disrupted. Disturbing the residence of *E. coli*, of course, has ramifications. For example, disturbance of bottom sediments allows *E. coli* to become elevated in the water column. Bacterial counts, including that of *E. coli*, rise and fall in streams, rivers, and lakes according to the amount of runoff occurring from the surrounding land; this is how surface freshwater bodies initially become contaminated. Sources of bacterial contamination include runoff from animals, sewage overflow, and septic tanks. Bacterial contaminants within the runoff lead to elevated colonies of microbes in the water body. When the water becomes contaminated, the microbes settle into their new residence—the bottom sediments. These microbes are bad news in the making, as they can impact water quality at a later time. It is important to note that other pathogenic microbes (e.g., viruses, protozoans) take up residence in stream, lake, and river bottom sediments (sediments are literally a sink for such organisms), but *E. coli* is of particular interest to water quality practitioners because *E. coli* is a

FIGURE 12.1 Sediment sampling for *Allocapnia* in the Blue Ridge Parkway region of Virginia. (Photograph by author.)

bacterium that is commonly used as an indicator of sewage contamination. *E. coli* is easily sampled and tested in the water and sediment (Spellman, 2015; Whitman and Nevers, 2003). The following text discusses sediment sampling methodologies; however, to present a holistic view of freshwater sediment sampling, it is important to present a brief overview of stream and lake water sampling first.

BIOMONITORING

The life in and physical characteristics of a stream ecosystem provide insight into the historical and current status of its quality. The assessment of a water body ecosystem based on organisms living in it is called *biomonitoring*. The assessment of a system based on its physical characteristics is called a *habitat assessment*. Biomonitoring and habitat assessments are two tools that water practitioners and stream ecologists use to assess the water quality of a stream. Biological monitoring involves the use of various organisms, such as periphytons, fish, and macroinvertebrates (combinations of which are referred to as *assemblages*), to assess environmental condition. Biological observation is more representative as it reveals cumulative effects as opposed to chemical observation, which is representative only at the actual time of sampling. Again, identifying the presence of various assemblages of organisms is key to conducting biological assessments and biosurveys. When selecting the appropriate assemblages for a particular biomonitoring situation, the advantages of using each assemblage must be considered along with the objectives of the program. Some of the advantages of using periphytons (algae), benthic macroinvertebrates, and fish in a biomonitoring program are presented in this section.

DID YOU KNOW?

Periphytons are a complex matrix of benthic algae, cyanobacteria, heterotrophic microbes, and detritus attached to submerged surfaces in most aquatic ecosystems.

Advantages of Using Periphytons

1. Algae generally have rapid reproduction rates and very short life cycles, making them valuable indicators of short-term impacts.
2. As primary producers, algae are most directly affected by physical and chemical factors.
3. Sampling is simple and inexpensive, requires few people, and has minimal impact on resident biota.
4. Relatively standard methods exist for the evaluation of functional and non-taxonomic structural (biomass, chlorophyll measurements) characteristics of algal communities.
5. Algal assemblages are sensitive to some pollutants that may not visibly affect other aquatic assemblages or may only affect other organisms at higher concentrations (e.g., herbicides) (APHA, 1971; Carins and Dickson, 1971; Karr, 1981; Patrick, 1973; Rodgers et al., 1979; USEPA, 1983; Weitzel, 1979).

Advantages of Using Fish

1. Fish are good indicators of long-term (several years) effects and broad habitat conditions because they are relatively long-lived and mobile (Karr et al., 1986).
2. Fish assemblages include a range of species that represent a variety of trophic levels (omnivores, herbivores, insectivores, planktivores, piscivores). They tend to integrate effects of lower trophic levels; thus, fish assemblage structure is reflective of integrated environmental health.
3. Fish are at the top of the aquatic food web and are consumed by humans, making them important for assessing contamination.
4. Fish are relatively easy to collect and identify at the species level. Most specimens can be sorted and identified in the field by experienced fishery professionals and subsequently released unharmed.
5. Environmental requirements of most fish are comparatively well known. Life history information is extensive for many species, and information on fish distributions is commonly available.
6. Aquatic life uses (water quality standards) are typically characterized in terms of fisheries (coldwater, coolwater, warmwater, sport, forage). Monitoring fish allows direct evaluation of fishability and fish propagation, thus recognizing the importance of fish to anglers and commercial fishermen.
7. Fish account for nearly half of the endangered vertebrate species and subspecies in the United States (Warren and Burr, 1994).

ADVANTAGES OF USING MACROINVERTEBRATES

Benthic macroinvertebrates are the larger organisms such as aquatic insects, insect larvae, and crustaceans that live in the bottom portions of a waterway for part of their life cycle. They are ideal for use in biomonitoring, as they are ubiquitous, relatively sedentary, and long lived. They provide a cross-section of the situation, as some species are extremely sensitive to pollution while others are more tolerant. Just as for toxicity testing, however, biomonitoring does not tell us *why* animals are present or absent. Benthic macroinvertebrates are excellent indicators for several reasons:

1. Biological communities reflect overall ecological integrity (i.e., chemical, physical, and biological integrity); therefore, biosurvey results directly assess the status of a water body relative to the primary goal of the Clean Water Act (CWA).
2. Biological communities integrate the effects of different stressors and thus provide a broad measure of their aggregate impact.
3. Because they are ubiquitous, communities integrate the stressors over time and provide an ecological measure of fluctuating environmental conditions.
4. Routine monitoring of biological communities can be relatively inexpensive, because they are easy to collect and identify.
5. The status of biological communities is of direct interest to the public as a measure of a particular environment.
6. Where criteria for specific ambient impacts do not exist (e.g., nonpoint sources that degrade habitats), biological communities may be the only practical means of evaluation.

Benthic macroinvertebrates act as continuous monitors of the water they live in. Unlike chemical monitoring, which provides information about water quality at the time of measurement (a snapshot), biological monitoring can provide information about past or episodic pollution (a videotape). This concept is analogous to miners who took canaries into deep mines with them to test for air quality. If the canary died, the miners knew the air was bad and they had to leave the mine. Biomonitoring a water body ecosystem uses the same theoretical approach. Aquatic macroinvertebrates are subject to pollutants in the water body; consequently, the health of the organisms reflects the quality of the water they live in. If the pollution levels reach a critical concentration, certain organisms will migrate away, fail to reproduce, or die, eventually leading to the disappearance of those species at the polluted site. Normally, these organisms will return if conditions improve in the system (Bly and Smith, 1994).

Biomonitoring (and the related concept bioassessment) surveys are conducted before and after an anticipated impact to determine the effect of the activity on the water body habitat. Moreover, surveys are performed periodically to monitor water body habitats and watch for unanticipated impacts. Finally, biomonitoring surveys are designed to reference conditions or to set biocriteria for determining that an impact has occurred; that is, they establish monitoring thresholds that signal future impacts or necessary regulatory actions (Camann, 1996).

DID YOU KNOW?

The primary justification for bioassessment and monitoring is that degradation of water body habitats affects the biota using those habitats; therefore, the living organisms themselves provide the most direct means of assessing real environmental impacts.

SEDIMENT SAMPLING IN FRESHWATER SYSTEMS

A few years ago, my sampling partner and I were preparing to perform sediment sampling protocols in a wadeable section in one of the countless reaches of the Yellowstone River in Wyoming. It was autumn, windy and cold. Before we stepped into the slow-moving frigid waters, we stood for a moment at the bank and took in the surroundings. The pallet of autumn is austere in Yellowstone. The coniferous forests east of the Mississippi lack the bronzes, the coppers, the peach-tinted yellows, and the livid scarlets that set the mixed stands of the East aflame. All we could see in that line were the quaking aspens and their gold. This autumnal gold, which provides the closest thing to eastern autumn in the West, is mined from the narrow, rounded crowns of *Populus tremuloides*. The aspen trunks stand stark white and antithetical against the darkness of the firs and pines, the shiny pale gold leaves sensitive to the slightest rumor of wind. Agitated by the slightest hint of breeze, the gleaming upper surfaces bounced the sun into our eyes. Each tree scintillated, like a show of gold coins in free fall. The aspens' bright, metallic flash seemed, in all their glittering motion, a valiant dying attempt to fill the spectrum of fall.

Because they were bright and glorious, we did not care that they could not approach the colors of an eastern autumn, although nothing is comparable to experiencing leaf-fall in autumn along the Appalachian Trail. This spirited display of gold against dark green lightened our hearts and eased the task that was before us, warming the thought of entering the bone-chilling water. With the aspens gleaming gold against the pines and firs, it simply did not seem to matter. Notwithstanding the glories of nature, one should not be deceived. Conducting sampling in a water body is not only the nuts and bolts of sediment sampling but is also very hard and important work.

SAMPLING PURPOSE AND PROJECTS

Sediment samples are collected by water practitioners for a variety of reasons, including chemical, physical, toxicological, and biological analysis. Because of the inherent variability of sediments, collection techniques should be evaluated and chosen for each sampling site and each sampling purpose. Choosing the most appropriate sampling device and technique depends on (1) the purpose of the sampling, (2) the location of the sediment, and (3) the characteristics of the sediment. When the sampling site and collection technique have been selected, then the specific methodologies for the actual collection of the samples should be closely followed. The experience and judgment of the sample collector should be relied upon as much as possible to obtain a representative sample of the sediment environment compatible with the objectives of the sampling. Whatever sampling technique and device is used, the specific rationale and collection methodologies should be stated in each evaluation and report of the data. Note that the purpose of the sediment sampling should be well defined before any sediment sampling plan is developed. Below are brief descriptions of sediment sampling projects that have been used in environmental studies (Ohio EPA, 2001):

- *Bioassays*—Sediment bioassay samples are used to determine if there is toxicity to representative aquatic organisms from contaminated bulk sediments. Sediment bioassay samples are usually collected within the top 10 centimeters of the sediment surface with equipment that causes the least disturbance to the sediment surface during collection.
- *Biosurvey sampling*—Macroinvertebrates are often collected for biosurveys from soft, fine-grained sediments.
- *Monitoring*—Chemical and physical analyses of sediments can be used as tools to monitor pollutant discharges to a river or lake system. In order to be able to make valid comparisons among stations or reference sites, consistent sampling techniques should be maintained.
- *Contaminant source identification*—Sediments can be used to help locate nonpoint, historical, or intermittent discharges that may not be readily apparent using samples collected from the water column. Sediments are used to identify the location of these sources by upstream incremental collection of samples from a contaminated site.
- *In situ measurements*—Sediment oxygen demand (SOD) is an *in situ* measure of the oxygen consumed by biochemical decomposition of organic matter in stream or lake sediment deposits. SOD can be used to evaluate pollutant source control performance or as a metric (input) for use in water quality models.
- *Dredging*—Sediment samples are often collected for use in dredging and dredge spoil management decisions. These samples should be collected within the vertical profile of the dredging project to account for probable stratification. Discrete sampling is preferred, and the use of composite samples for dredge management decisions should be made with caution.
- *Trends/historical contamination*—Sediment sampling is also used as a tool in the evaluation of the effectiveness of pollution source controls. This can be accomplished with discrete vertical sampling (assuming the sediments have not been mixed or otherwise disturbed) or by reproducing earlier sampling efforts.
- *Complaint investigation*—Sediment sampling to help address citizen complaints requires a great deal of assessment and judgment by the sample collector. The design of each complaint sampling investigation should be evaluated on a case-by-case basis. Because of cost and often long turnaround times, sediment sampling for the sole purpose of resolving citizen complaints should be used judiciously.

- *Sediment collection technique evaluation*—Comparison of samples using sediment collection techniques and devices can be made to determine the easiest and most effective sampling method. Evaluation of other techniques such as sediment traps can also be made to make sediment collection more reproducible.
- *Nonpoint pollution assessment*—Sediment samples can be collected for evaluation of nonpoint pollution. Selection of parameter coverage for analysis of the samples can sometimes be important in defining the source of sediment (e.g., high pesticide/herbicide contamination would indicate agricultural runoff).
- *Nutrient cycling*—Sediment samples can be collected in lake or river habitats to determine the potential release of nutrients (e.g., phosphorus) back into the water column.
- *Bedload/sediment dynamics*—Prediction of sediment resuspension, both modeling and measurement procedures, are still experimental. The dynamics of the movement, transport, and fate of contaminants adsorbed to sediment are not thoroughly understood and are beyond the scope of this book.

SAMPLING PLAN

A proper, representative, and accurate sampling and analysis procedure for freshwater and freshwater sediments requires planning. Moreover, when planning a sampling outing, it is important to determine the precise objectives. One important consideration is to determine whether sampling will be accomplished at a single point or at isolated points. Additionally, frequency of sampling must be determined. That is, will sampling be accomplished at hourly, daily, weekly, monthly, or even longer intervals? Whatever sampling frequency of sampling is chosen, the entire process will probably continue over a protracted period (i.e., preparing for water and sediment sampling in the field might take several months from the initial planning stages to the time when actual sampling occurs). An experienced freshwater practitioner or ecologist should be centrally involved in all aspects of planning. The sampling plan should be written and approved by the project manager prior to the collection of sediment samples to maximize resource allocation.

In *Monitoring Water Quality: Intensive Stream Bioassay* (USEPA, 2000), the USEPA recommended that the following issues should be considered when planning the sampling program (this applies to sediment sampling as well):

- Availability of reference conditions for the chosen area
- Appropriate dates for sampling in each season
- Appropriate sampling gear
- Availability of laboratory facilities
- Sample storage
- Data management
- Appropriate taxonomic keys, metrics, or measurement for macroinvertebrate analysis

- Habitat assessment consistency
- Availability of a U.S. Geological Survey topographical map
- Familiarity with safety procedures

With regard to specifically sampling sediment only, keep in mind that sediment sampling usually entails relatively higher expenses for the personnel required, collection effort, and analysis of the samples than the collection and analysis of water samples. Also, whether sampling water or sediment, the sampling plan should incorporate a statement as to the purpose and the data quality objectives of the proposed sampling. But, keep in mind that sample collection is often governed by logistical and resource constraints rather than specific project objectives. As a result, the data from such studies are often incomplete, and the benefits from the collection of that data are reduced if not eliminated as a result of the constraints. If resources are unavailable to perform an adequate study to meet the data quality objectives, then the sampling project should be reevaluated. In addition to the USEPA sampling issues listed above, a brief description of the sampling project should be included in the sampling plan. A description of how the sediment sampling will be integrated with other planned studies and an explanation of how the sediment sampling information will be used should be stated.

Data Quality Objectives

The data quality objectives section of the sampling plan should state what type of information must be collected in order to meet the objectives of the sampling project. In addition to the USEPA issues above, this information should also include the following:

- Purpose of the sampling
- How the data from the sampling will be used
- What actions will be taken as a result of the sampling
- Identification of the laboratory performing the analyses
- The parameters for analysis including method detection
- Number and type of quality control samples such as field blanks, equipment rinses, field duplicates, station replicates, reference and back samples
- Statistical analysis and criteria (allowable errors) used to evaluate the data
- Standards, background, or benchmark criteria that will be used to compare the analytical results
- Number and location of samples to be collected to meet the purposes of the project
- How the information will be reported
- Whether the data will be entered into an electronic database and, if so, the structure and file type of the database

When the initial objectives (issues) have been determined and the plan devised, then the sampler can move on to other important aspects of the sampling procedure. Along with the items just mentioned, it is imperative for the sampler to understand what sediment sampling is all about.

Water and sediment sampling is one of the most basic and important aspects of water quality management (Tchobanoglous and Schroeder, 1985). Water and sediment sampling go hand in hand in determining the quality of both. With regard to water sampling, it allows for rapid and general water quality classification. Rapid classification is possible because quick and easy cross-checking between stream biota and a standard stream biotic index is possible. Biological sampling is typically used for general water quality classification in the field because sophisticated laboratory apparatus is usually not available. Additionally, stream communities often show a great deal of variation in basic water quality parameters such as dissolved oxygen (DO), biochemical oxygen demand (BOD), suspended solids, and coliform bacteria. Such variation can be observed in eutrophic lakes, where oxygen saturation levels may vary throughout the day, dropping to levels less than 0.5 mg/L. Also, the concentration of suspended solids may double immediately after a heavy rain. Moreover, the sampling method chosen must take into account the differences in the habits and habitats of the aquatic organisms.

The first step toward accurate measurement of the water and sediment quality of a stream is to make sure that representative samples are collected. Laboratory analysis is meaningless if the sample collected is not representative of the aquatic environment being analyzed. As a rule, samples should be taken at many locations, as often as possible. If, for example, you are studying the effects of sewage discharge into a stream, you should first take at least six samples upstream of the discharge, six samples at the discharge, and at least six samples at several points below the discharge for two to three days (the *six six six sampling rule*). If these samples show wide variability, then the number of samples should be increased. On the other hand, if the initial samples exhibit little variation, then a reduction in the number of samples may be appropriate (Kittrell, 1969).

Sampling History

A thorough review and assessment of existing data and information for the sampling areas should be performed to assist in the planning process. A brief summary of that information and an assessment should be included in the written sampling plan. When reviewing existing information, attention should be given to the purpose of the collection of the historical data and what sampling techniques, analytical procedures, and laboratories were used in performing the analyses. This information is important in order to determine the usefulness of the historical data for the proposed project.

Dates of Collection

The general time of year when the samples will be collected should be considered during the planning of the sampling activities. In general, sediment sampling in the low-flow conditions of summer and fall is the most practical. Seasonal variations in sediment deposits and quality can occur due to high flows and ice scour on rivers, leaf litter in the fall, land use practices (e.g., agricultural pesticide applications), or seasonal variations in benthic populations. Winter may be a convenient time to sample some inland lakes through the ice, whereas ice cover may be a significant safety concern in the collection of river sediment samples. The analytical laboratory should be contacted early in the planning process for proper coordination to ensure that all needs are met.

SAMPLING STATION SELECTION

After determining the number of samples to be taken, sampling stations (locations) must be determined. Several factors determine where the sampling stations should be set up. These factors include stream habitat types, the position of the wastewater effluent outfalls, the stream characteristics, stream developments (dams, bridges, navigation locks, and other manmade structures), the self-purification characteristics of the stream, and the nature of the objectives of the study (Velz, 1970). The chemical and physical nature of sediments is strongly influenced by the size of the individual particles of sediment. Sediments composed of sands (0.06 to 2.0 mm) and larger sized particles are often stable inorganic silicate minerals. These larger particles from non-consolidated deposits have a relatively lower specific capacity (amount of interstitial water) and a more neutral surface electrical charge. These types of materials are usually not associated with contaminants and are not recommended for analysis. Fine-grained silts and clays (<0.06 mm), however, have a much larger specific capacity, unbalanced electrical charges, and much larger ratio of surface area to volume. These properties make the finer grained sediments much more chemically, physically, and biologically interactive. These are the types of sediments that should be submitted for analysis, and most of the sediment sampling locations should be biased toward collecting these types of sediments. The stream habitat types used in this discussion are as follows:

- *Cobble (hard substrate)*—Cobble is prevalent in the riffles (and runs) that are a common feature throughout most mountain and piedmont streams. In many high-gradient streams, this habitat type will be dominant; however, riffles are not a common feature of most coastal or other low-gradient streams.
- *Snags*—Snags and other woody debris that have been submerged for a relatively long period (not recent deadfall) provide excellent sampling habitats.
- *Vegetated banks*—When lower banks are submerged and have roots and emergent plants associated with them, they are sampled in a fashion similar to snags. Submerged areas of undercut banks are good sediment areas to sample.
- *Sand* (and other fine sediment)—Usually the most productive sediment habitat in streams, this habitat may be the most prevalent in some streams.

It is usually impossible to go out and sample each and every type of sediment present in a waterway. This would be comparable to counting different sizes of grains of sand on the beach. Instead, in a sediment sampling program, the most common sampling methods are the *transect* and the *grid*. Transect sampling involves taking

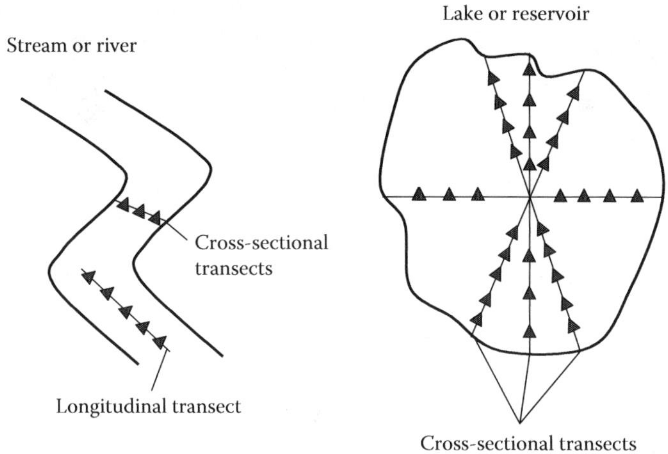

FIGURE 12.2 Transect sampling.

samples along a straight line at either uniform or random intervals (see Figure 12.2). The transect approach samples a cross-section of a lake or stream or the longitudinal section of a river or stream. The transect sampling method allows for a more complete analysis by including variations in habitat.

In grid sampling, an imaginary grid system is placed over the study area. The grids may be numbered, and random numbers are generated to determine which grids should be sampled (see Figure 12.3). This type of sampling method allows for quantitative analysis because the grids are all of a certain size; for example, to sample a stream for assorted sediments, grids that are 0.25 m^2 may be used. The weight or composition of sediments per square meter can then be determined.

Random sampling requires that each possible sampling location have an equal chance of being selected. Numbering all sampling locations and then using a computer, calculator, or a random numbers table to create a series of random numbers can accomplish this. An illustration of how to put the random numbers to work is provided in the following example. Given a pond that has 300 grid units, find eight random sampling locations using the following sequence of random numbers taken from a standard random numbers table: 101, 209, 007, 018, 099, 100, 017, 069, 096, 033, 041, 011. The first eight numbers of the sequence could be selected, and only those grids would be sampled to obtain a random sample.

Estimating Particle Size Percentage

A goal of sediment collection is >30% silt and clays in the sediment sample. If these sediment types are not found, then it should be noted on the laboratory submission sheets and field collection form. The percentage of silts and clays in a sample can be estimated in the field by marking a line on a clear jar and then marking 30% of the way up to that line on the jar with another line. Fill the jar to the top line with

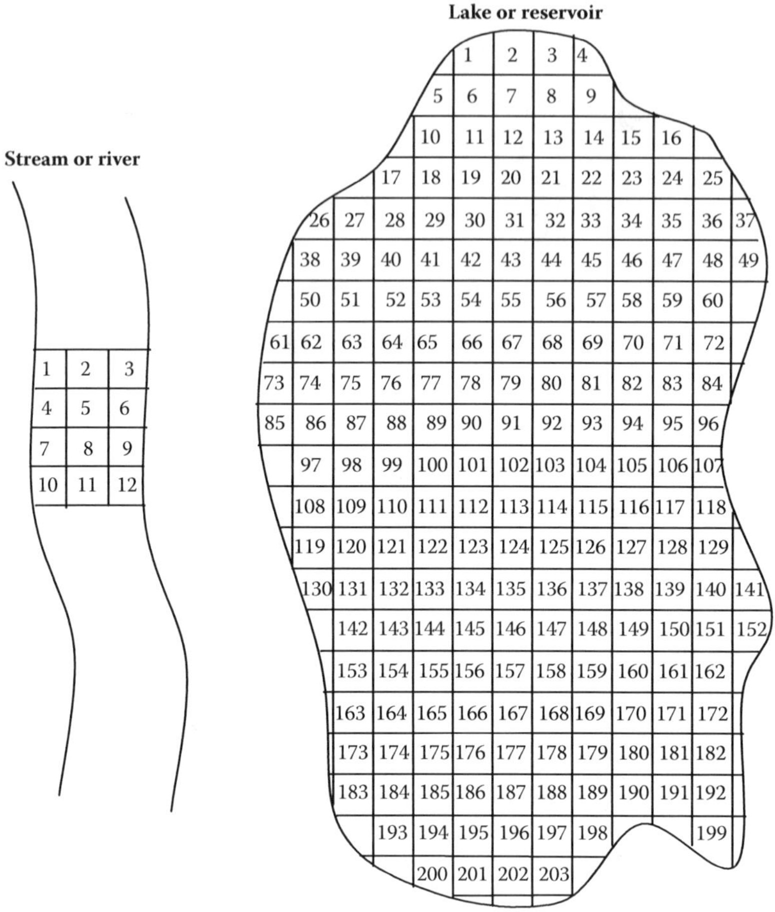

FIGURE 12.3 Grid sampling.

sediment, vigorously shake the jar, and set it aside to settle. A 1-inch headspace in the jar allows for easier mixing. After settling for 10 minutes, an estimate of the particle size distribution can be made with a visual inspection of the sediment stratification in the jar. If the lines stop below the 30% line, then the silt/clay fraction is likely to be <30%. It is assumed that the finer grained sediments are located in still waters of the sample area in deep water, at stream margins, behind boulders and other obstructions, or at the inside bends of river meanders. An initial reconnaissance of the sample area should be performed, if possible, prior to completion of the sampling plan. This reconnaissance can often identify field limitations in the study design that can be addressed prior to sample collection. An initial reconnaissance should include a cursory bathymetric survey using a wading staff in shallow streams and rivers or an echo-sounding (sonar) depth finder for deeper waters. Local knowledge or recent navigation charts can often provide information similar to that of an echo-sounding survey.

Sampling Frequency and Notes

After establishing the sampling methodology and the sampling locations, the frequency of sampling must be determined. The more samples collected, the more reliable the data will be. A frequency of once a week or once a month will be adequate for most aquatic studies. Usually, the sampling period covers an entire year so yearly variations may be included. The details of sample collection will depend on the type of problem that is being solved and will vary with each study. When a sample is collected, it must be carefully identified with the following information:

1. Location (name of water body and place of study; longitude and latitude)
2. Date and time
3. Site (sampling location)
4. Name of collector
5. Weather (e.g., temperature, precipitation, humidity, wind)
6. Miscellaneous (any other important information, such as observations)
7. Field notebook

With regard to the last item, on each sampling day notes on field conditions should be taken; for example, miscellaneous observations and weather conditions can be entered. Additionally, notes that describe the condition of the water are also helpful (e.g., color, turbidity, odor, algae). All unusual findings and conditions should also be entered.

SAMPLING EQUIPMENT AND SAMPLE TYPES

Sampling Equipment

Basic and standard water and sediment sampling equipment includes the following:

1. Jars (two), at least quart size, plastic, wide-mouth with tight cap; one should be empty and the other filled about two thirds with 70% ethyl alcohol
2. Hand lens, magnifying glass, or field microscope
3. Fine-point forceps
4. Heavy-duty rubber gloves
5. Plastic sugar scoop or ice-cream scoop
6. Kink net (rocky-bottom stream) or dip net (muddy-bottom stream)
7. Buckets (two)
8. String or twine (50 yards) and a tape measure
9. Stakes (four)
10. Orange to measure velocity (a stick, an apple, or a fish float may also be used in place of an orange)
11. Reference maps providing general information relevant to the sampling area, including the surrounding roadways, as well as a hand-drawn station map
12. Station ID tags
13. Spray water bottle
14. Pencils (at least two)

Sample equipment specifically designed and intended for grab, core, and *in situ* sediment sampling operations in freshwater surface waters include the following:

1. *Spoon or scoop sampler*—For grab samples taken in zero to slight current conditions in all types of substrates.
   - These samplers are used only in relatively calm and shallow water.
   - Relatively little sample disturbance occurs.
   - They are simple and inexpensive.
   - Fines may wash out when retrieved through the water column.
2. *Birge–Ekman bottom grab sampler*—For grab samples in zero to very slight current conditions in clay and silt substrates.
   - This sampler is used in relatively calm water.
   - Jaw shape and cut are excellent.
   - Relatively little sample disturbance occurs.
   - Pebbles and branches may interfere with the jaw closure.
   - Stability is poor. Its light weight allows for a tendency to "swim" in a current and sometimes causes missed triggers.
   - Sample area is 0.02 m^2.
   - Weight with sample is 10 kg (22 lb).
3. *Petite Ponar® and Petersen samplers*—For grab samples in zero to very slight current conditions in clay and fine gravel.
   - Relatively calm/sheltered waters are necessary.
   - Stability is good.
   - Jaw shape and cut are poor.
   - Sample disturbance occurs.
   - Less washout occurs if extra weights are used.
   - These samplers can be cumbersome, and they require a winch.
   - Sample area is 0.1 to 0.2 m^2.
   - Weight with sample is 30 to 50 kg (66 to 110 lb).
4. *Box (rectangular) sampler*—For core samples in zero to moderate current conditions in clay to sand substrates.
   - This sampler accommodates a large sample size.
   - Sampler is difficult to handle.
   - A boat or barge with a winch is required.
5. *Shipek sampler*—For grab samples in zero to strong current conditions in clay to gravel substrate.
   - A boat or barge with a winch is required (although smaller Shipek samplers can be used manually).
   - This is one of the most reliable in terms of triggering, stability, washout, and leaching.
   - Jaw shape and cut are excellent with an extremely clean cutting action.
   - Sample area is 0.04 m^2.
   - Weight with sample is 60 to 70 kg (132 to 154 lb); smaller Shipek samplers with sample weigh 20 to 30 kg (44 to 66 lb).

6. *Manual core sampler*—For core samples in zero to strong current conditions in clay to sand.
   - Inserts are required for sandy samples.
   - This sampler is recommended for use in shallow water.
   - It is deployed by hand or by driver (hammer).
   - Extension handles can be used for deeper waters.
7. *Coring tube sampler*—For core samples in zero to moderate current conditions in clay to sand.
   - Inserts are required for sandy samples.
   - Sampling is quick and easy.
   - Sample is relatively undisturbed.
   - Sample volume is small.
   - Samples are sometimes compressed.
8. *Split spoon sampler*—For core samples in zero to moderate current conditions in flay to sand.
   - Inserts are required for sandy samples.
   - This sampler is recommended for use in shallow water.
   - It is deployed by hand or by driver (hammer).
   - Vertical profile remains intact and is visible.
   - Point design can reduce sample compaction.
   - Stones can interfere with collection.
   - Equipment is heavy.
9. *Gravity core sampler*—For core samples in zero to moderate current conditions in silts and clays.
   - This sampler is recommended for rivers.
   - It requires careful handling to avoid sediment spillage.
   - It can be used at depths up to 10 m (32.8 ft).
10. *Phleger sampler*—For core samples in zero to moderate current conditions in silts.
    - This sampler is good for short cores in soft sediments.
    - Sampling is quick and easy.
    - Sample is relatively undisturbed.
    - Sample volume is small.
11. *Kajak–Brinkhurst core sampler*—For core samples in zero to moderate current conditions in clay to sand.
    - Inserts are required for sand samples.
    - Sampling is quick and easy.
    - Sample is relatively undisturbed.
    - Sample volume is small.
    - Samples are sometimes compressed.
12. *Sediment oxygen demand (SOD) sampler*—For *in situ* SOD sampling in zero to moderate current conditions in clay to gravel.
    - This sampler is used to determine sediment oxygen demand.
    - It is not intended for the collection of sediment samples.

DID YOU KNOW?

It is important to note that some of the sampling equipment discussed and described thus far comes with a cost and sometimes a design that are not suitable for the sediment sampling at hand. Sometimes it may be desirable to design and fashion a homemade sampler for a specific stream or lake. Figure 12.4 shows a rough drawing of a homemade sampling dredge that has been successfully used by the author for sampling streams in Wyoming, Washington, and Montana.

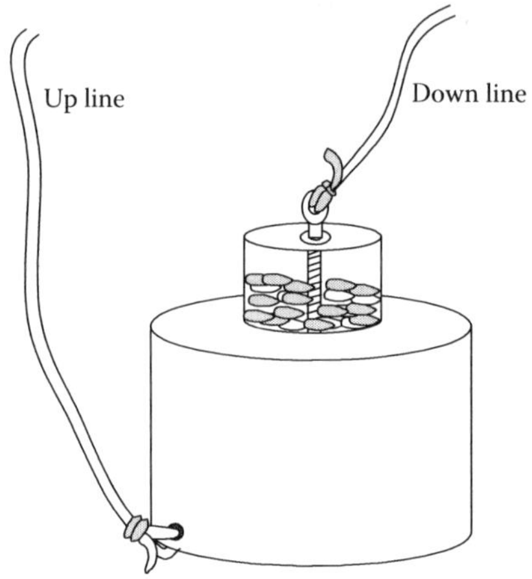

FIGURE 12.4 Homemade sampling dredge.

Sample Types

A description and rationale for the types of samples to be collected should be included in the written sampling plan.

- *Cores for vertical discrete grab samples*—Most appropriate for historical contamination information or dredging decisions at heavily contaminated areas.
- *Cores for depth integrated composite samples*—Most appropriate for reference.
- *Scoops and dredges for surface (top 2 to 4 cm) sediment grab samples*—Most appropriate for benthic, sediment oxygen demand (*in situ*), recent ambient conditions, and recent contaminant investigation.
- *Scoops and dredges for surface sediment composite samples*—May be used to reduce costs for specific conditions or situations such as ambient or specific historical data. In general, however, discrete sampling is preferred if resources are available. An example of a discrete sample would be taking a 1-cm section of sediment from a core sample that was originally 1 m long.

Types of Sediment Samplers

As discussed previously, the three main types of sediment sampling devices are *grab samplers*, *core samplers*, and *dredge samplers*. Grab samplers are used to collect surficial sediments for assessment of the horizontal distribution of sediment characteristics. Core samplers are typically used to sample thick sediment deposits, to collect sediment profiles for the determination of the vertical distribution of sediment characteristics, or to characterize the entire sediment column. Dredge samplers are used primarily to collect benthos. Dredges cause disruption of sediment and porewater integrity, as well as the loss of fine-grained sediments. For these reasons, only grab and core samplers are recommended for sediment physicochemistry or toxicity evaluations. Because many grab samplers are appropriate for collecting benthos as well (ASTM, 2000c; Klemm et al., 1990), grab samplers are likely to be more useful than dredges in sediment quality assessments. Therefore, dredges are not considered further in this discussion. Figure 12.5 provides recommendations regarding the type of sampler that would be appropriate given different study objectives. For many study objectives, either cores or grab samplers can be used; however, in practice, one will often be preferred over the other depending on other constraints such as amount of sample required for the analyses and equipment availability (USEPA, 2001).

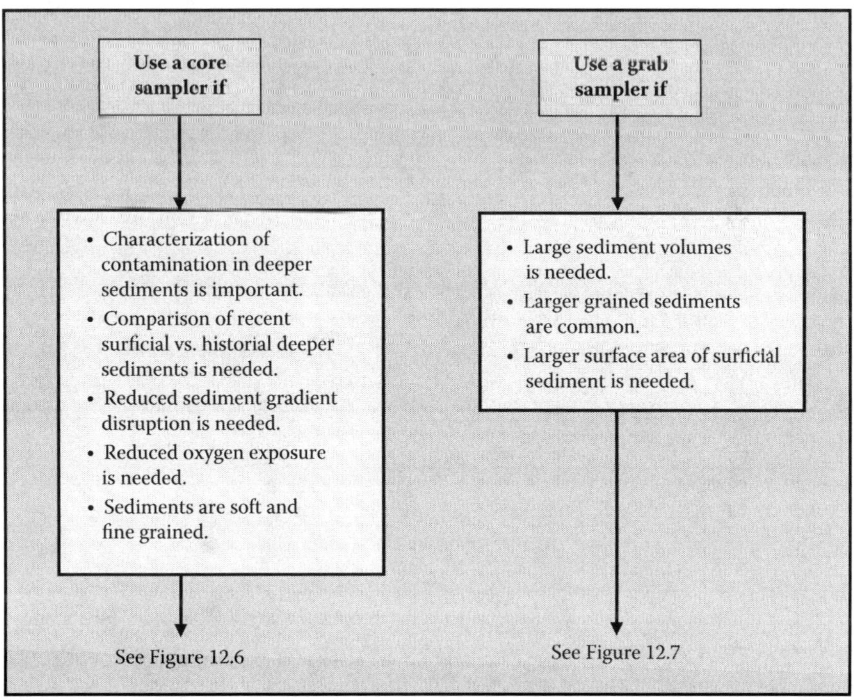

FIGURE 12.5 General types of considerations or objectives that are appropriate for grab or core sampling devices. (From USEPA, *Methods for Collection, Storage and Manipulation of Sediments for Chemical and Toxicological Analyses: Technical Manual*, EPA-823-B-01-002, U.S. Environmental Protection Agency, Office of Water, Washington, DC, 2001.)

Grab Samplers

Grab samplers consist either of a set of jaws that shut when lowered into the surface of the bottom sediment or a bucket that rotates into the sediment when it reaches the bottom. Grab samplers have the advantages of being relatively easy to handle and operate, readily available, moderately priced, and versatile in terms of the range of substrate types they can effectively sample. Of the grab samplers, the *Van Veen*, *Ponar*, and *Petersen* are the most commonly used. These samplers are effective in most types of surface sediments and in a variety of environments (e.g., lakes, rivers, estuaries, marine waters). In shallow, quiescent water, the Birge–Ekman sampler also provides acceptable samples and allows for relatively nondisruptive sampling; however, this sampler is typically limited to soft sediments. The Van Veen sampler, or the modified Van Veen, is used in several national and regional estuarine monitoring programs, including the National Oceanic and Atmospheric Administration National Status and Trends Program, the USEPA Environmental Monitoring and Assessment Program (EMAP), and the USEPA National Estuary Program because it can sample most types of sediment, is less subject to blockage and loss of sample than the Petersen or Ponar samplers, is less susceptible to forming a bow wave during descent, and provides generally high sample integrity (Klemm et al., 1990). The support frame further enhances the versatility of the Van Veen sampler by allowing the addition of either weights (to increase penetration in compact sediments) or pads (to provide added bearing support in extremely soft sediments). However, this sampler is relatively heavy and requires a power switch to operate safely (GLNPO, 1994).

Grab sampler capacities range from approximately 0.5 to 75 L. If a sampler does not have sufficient capacity to meet the study plan requirements, additional samples can be collected and composited to obtain the requisite sample size. Grab samplers penetrate to different depths depending on their size, weight, and the bottom substrate. Heavy, large-volume samplers such as the Smith–McIntyre, large Birge–Ekman, Van Veen, and Petersen devices can effectively sample to a depth of 30 cm. These samplers might actually sample sediments that are too deep for certain study objectives (i.e., they are not reflective of recently deposited sediments). Smaller samplers such as the small Birge–Ekman, standard and petite Ponar, and standard Shipek devices can effectively collect sediments in a maximum depth of 10 cm. The mini-Shipek can sample to a depth of 3 cm.

Another consideration in choosing a grab sampler is how well it protects the sample from disturbance and washout. Grab samples are prone to washout, which results in the loss of surficial, fine-grained sediments that are often important from

DID YOU KNOW?

Ponar, Van Veen, or Ekman samplers are commonly used and generally preferred for grab sampling. Ekman samplers, however, are less efficient in deep waters.

a biological and contaminant standpoint. The Ponar, Ted Young modified grab, and Van Veen samplers are equipped with mesh screens and rubber flaps to cover the jaws. This design allows water to pass through the samplers during descent, reducing disturbance from bow waves at the sediment–water interface. The rubber flaps also serve to protect the sediment from washout during ascent.

The use of small or lightweight samplers, such as the small Birge–Ekman, petite Ponar, and mini-Shipek, can be advantageous because of easy handling, particularly from a small vessel or when using only a hand line; however, these samplers are not recommended for use in strong currents or high waves. This is particularly true for the Birge–Ekman sampler, which requires relatively calm conditions for proper performance. Lightweight samplers generally have the disadvantage of being less stable during sediment penetration. They tend to fall to one side due to inadequate or incomplete penetration, resulting in unacceptable samples.

In certain very shallow water applications, such as a stream assessment at a Superfund site, it might be difficult to use even a lightweight sampler to collect a sample. In these cases, it might be acceptable to collect sediment from depositional areas, using a shovel or other hand implement; however, such sampling procedures are discouraged as a general rule, and the use of a hand corer or similar device is preferred.

Figure 12.6 summarizes grab samplers based on two important site factors: depth and sediment article size. This figure also shows how appropriate grab samplers are chosen depending on certain common study constraints such as sample depth and volume desired, as well as the ability to subsample directly from the sampler (ASTM, 2000c). Based on all of these factors, the Ponar or Van Veen samplers are perhaps the most versatile of the grab samplers, thus their common usage in sediment studies.

Careful use of grab samplers is required to avoid problems such as loss of fine-grained surface sediments from the bow wave during descent, mixing of sediment layers upon impact, lack of sediment penetration, and loss of sediment from tilting or washout upon ascent (ASTM, 2000a; Baudo, 1990; Environment Canada, 1994; Golterman et al., 1983; Plumb, 1981). When deploying a grab sampler, the speed of descent should be controlled, with no free fall allowed. In deep waters, use of a winching system is recommended to control both the rate of descent and the rate of ascent. A ball-bearing swivel should be used to attach the grab sampler to the cable to minimize twisting during descent. After the sample is collected, the sampling device should be lifted slowly off the bottom, then steadily raised to the surface at a speed of about 30 cm/s (Environment Canada, 1994).

DID YOU KNOW?

In sand, gravel, firm clay, or till sediments, grab samplers might be preferred over core samplers (when only surface material needs to be collected and sampling at depth is not necessary) because the latter are often less efficient in these sediment types.

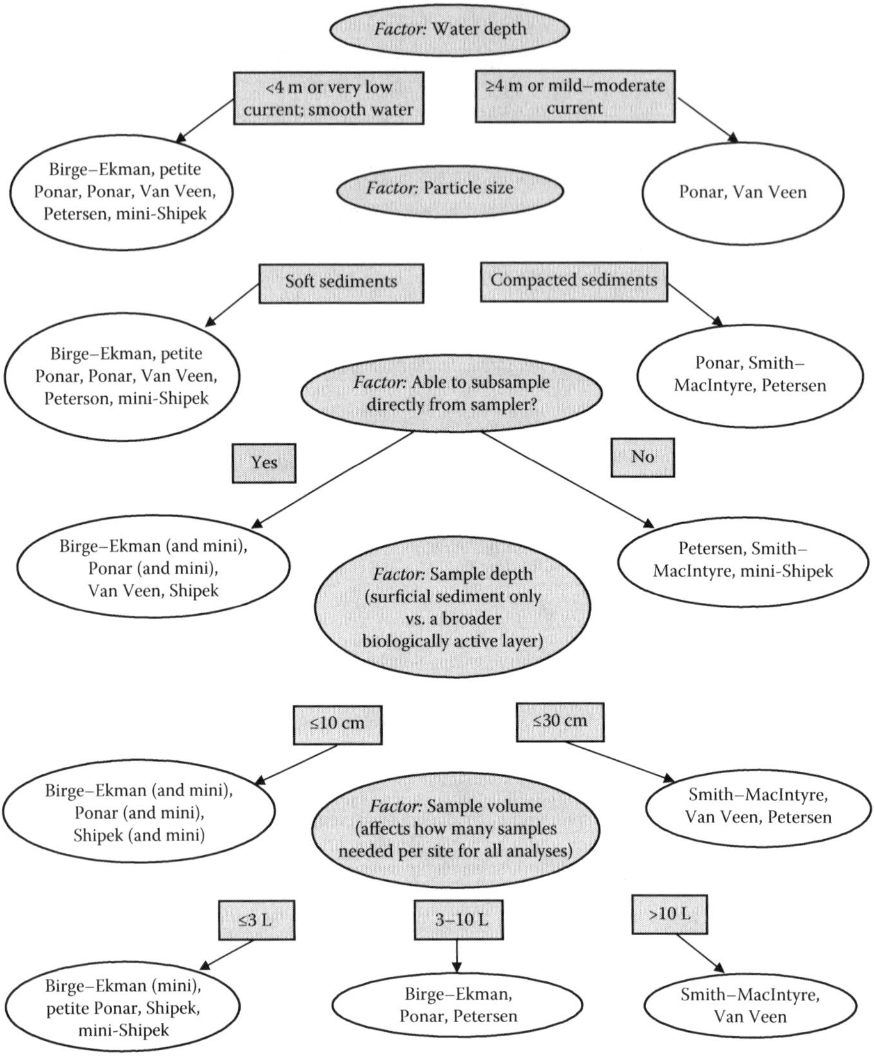

FIGURE 12.6 Flowchart for selecting appropriate grab samplers based on site-specific or design factors.

Core Samplers

Core samplers (corers) are used (1) to obtain sediment samples for geological characterizations and dating, (2) to investigate the historical input of contaminants to aquatic systems, and (3) to characterize the depth of contamination at a site. Corers are an essential tool in sediments in which three-dimensional maps of sediment contamination are necessary. Corers may consist of the following components (Mudroch and Azcue, 1995):

- A hollow metal (or plastic) pipe that serves as the core barrel
- Easily removed plastic liners or core tubes that fit into the core barrel and retain the sediment sample
- A valve or piston mounted on top of the core barrel that is open and allows water to flow through the barrel during descent but shuts upon penetration of the corer into the sediment to prevent the sediment from sliding through the corer during the ascent
- A core catcher to retain the sediment sample
- A core cutter for penetration of the sediment
- Removable metal weights (usually lead coated with plastic) or piston-driven impact or vibration to increase penetration of the corer into the sediment
- Stabilizing fins to ensure vertical descent of the corer

Core devices are recommended for projects in which it is critical to maintain the integrity of the sediment profile, because they are considered to be less disruptive than dredge or grab samplers. Additionally, core samplers should be used where it is important to maintain an oxygen-free environment because they limit oxygen exchange with the air more effectively than grab samplers. Core samplers should also be used where thick sediment deposits must be representatively sampled (e.g., for dredging projects).

One limitation of core samplers is that the volume of any given depth horizon within the profile sample is relatively small. Thus, depending on the number and type of analyses needed, repetitive sampling at a site might be required to obtain the desired quantity of material from a given depth. Some core samplers are prone to plugging or rodding, where the friction of the sediment within the core tube prevents it from passing freely and the core sample is compressed or does not sample to the depth required. This limitation is more likely with smaller diameter core tubes and heavy clay sediments. Except for piston corers and vibracorers, there are a few core devices that function efficiently in substrates with significant proportions of sand, gravel, clay, or till (USEPA, 2001).

Coring devices are available in various designs, lengths, and diameters. With the obvious exception of hand corers (Figure 12.7), only a few corers can be operated without a mechanical winch. The more common of these include the standard

FIGURE 12.7 Coring sampler.

DID YOU KNOW?

Vibracorers are recommended for studies requiring deep cores (>1 m) or where the sediment consists of very compacted or large grained material (e.g., gravel).

Kajak–Brinkhurst corer, suitable for sampling soft, fine-grained sediments, and the *Phleger corer*, suitable for a wider variety of sediment types ranging from soft to sandy, semi-compacted material, as well as peat and plant roots in shallow lakes or marshes (Mudroch and Azcue, 1995). The Kajak–Brinkhurst corer uses a larger core tube and therefore recovers a greater quantity of sediment than the Phleger corer. Both corers can be used with different liner materials, including stainless steel or polyvinyl chloride (PVC). Stainless steel liners should not be used if trace metal contamination is an issue.

Gravity corers are appropriate for recovering up to 3-m-long cores from soft, fine-grained sediments. Several models include stabilizing fins on the upper part of the corer to promote vertical penetration into the sediment and weights that can be mounted externally to enhance penetration (Mudroch and Azcue, 1995). A variety of liner materials are available, including stainless steel; Lexan®, and PVC. For studies in which metals are a concern, stainless steel liners should not be used (USEPA, 2001).

Vibracorers are perhaps the most commonly used coring device in sampling programs in the United States because they collect deep cores in most types of sediments, yielding excellent sample integrity. Vibracorers are one of the only sampling devices that can reliably collect thick sediment samples (up to 10 m or more). Vibracorers have an electric-powered, mechanical vibrator located at the head end of the corer which applies thousands of vertical vibrations per minute to help penetrate the sediment. A core tube and rigid liner (preferably of relatively inert material such as cellulose acetate butyrate) of varying diameter depending on the specific vibrator head used are inserted into the head and the entire assembly is lowered in the water. Depending on the horsepower of the vibrating head and its weight, a vibracorer can penetrate very compact sediments and collect cores up to 6 m long. Use of a heavy vibracorer requires a large vessel to maintain balance and provide adequate lift to break the corer out of the sediment and retrieve it (GLNPO, 1994; PSEP, 1997).

When deployed properly, *box corers* can obtain undisturbed sediment samples of excellent quality. The basic box corer consists of a stainless steel box equipped with a frame to add stability and facilitate vertical penetration on low slopes. Box corers are recommended particularly for studies of the sediment–water interface or when there

DID YOU KNOW?

Box corers are especially recommended for (1) studies of the sediment–water interface; (2) collecting larger volumes of sediment from a given depth (generally less than 1 meter in depth, though); (3) for *in situ* studies involving interstitial water characterization; and (4) collecting subsamples for different analyses from the same station.

DID YOU KNOW?

The recommended depth of sediment sampling is dependent on the study objectives. Issues that determine the appropriate depth of sampling include regulatory objectives (e.g., depth of dredging for sediment remediation), need to characterize sediments at depth (e.g., materials to be dredged vs. shallow depositional areas in some Superfund sites), historical comparisons, sediment deposition rates, and time period of contamination.

is a need to collect larger volumes of sediment from the depth profile. Because of the heavy weight and larger size of almost all box corers, they can be operated only from a vessel with a large lifting capacity and sufficient deck space. Sediment inside a box corer can be subsampled by inserting narrow core tubes into the sediment; thus, they are an ideal sampler for obtaining acceptable subsamples for different analyses at a given station. Carlton and Wetzel (1985) described a box corer that permits the sediment and overlying water to be held intact as a laboratory microcosm under either the original *in situ* conditions or other laboratory controlled conditions. A box corer was developed that enables horizontal subsampling of the entire sediment volume recovered by the device (Mudroch and Azcue, 1995). Figure 12.8 summarizes the core samplers that are appropriate for given site factors such as depth and particle size and other study constraints, such as sample depth and volume required and the lifting capacity necessary to use the sampling device. Given the factors examined for general monitoring studies, the Phleger, Alpine, and Kajak–Brinkhurst corers might be the most versatile. For dredged material evaluations and projects requiring sediment profile characterizations >3 m in sediment depth, the vibracorer and piston corers are the samplers of choice.

Collection of core samples with hand-coring devices should be executed with care to minimize disturbance or compression of sediment during collection. To minimize disruption of the sediment, core samples should be kept as stationary and vibration free as possible during transport. These cautions are particularly applicable to cores collected by divers.

The speed of descent of coring devices should be controlled, especially during the initial penetration of the sediment, to avoid disturbance of the surface and to minimize compression due to frictional drag from the sides of the core liner (ASTM, 2000d). In deep waters, winches should be used where necessary to minimize twisting and tilting and to control the rate of both descent and ascent. With the exception of piston corers or vibracorers, which are equipped with their own mechanical impact features, only the weight or piston mechanism of the sampler should be used to force it into the sediment. The sampler should be raised to the surface at a steady rate, similar to that described for grab samplers. Where core caps are required, it is essential to quickly and securely cap the core samples when the samples are retrieved. The liner from the core sampler should be carefully removed and kept in a stable position until the samples are processed. If there is little to no overlying water in the tube and the sediments are relatively consolidated, it is not necessary to keep the core sample tubes vertical. Core sample tubes should be quickly capped and

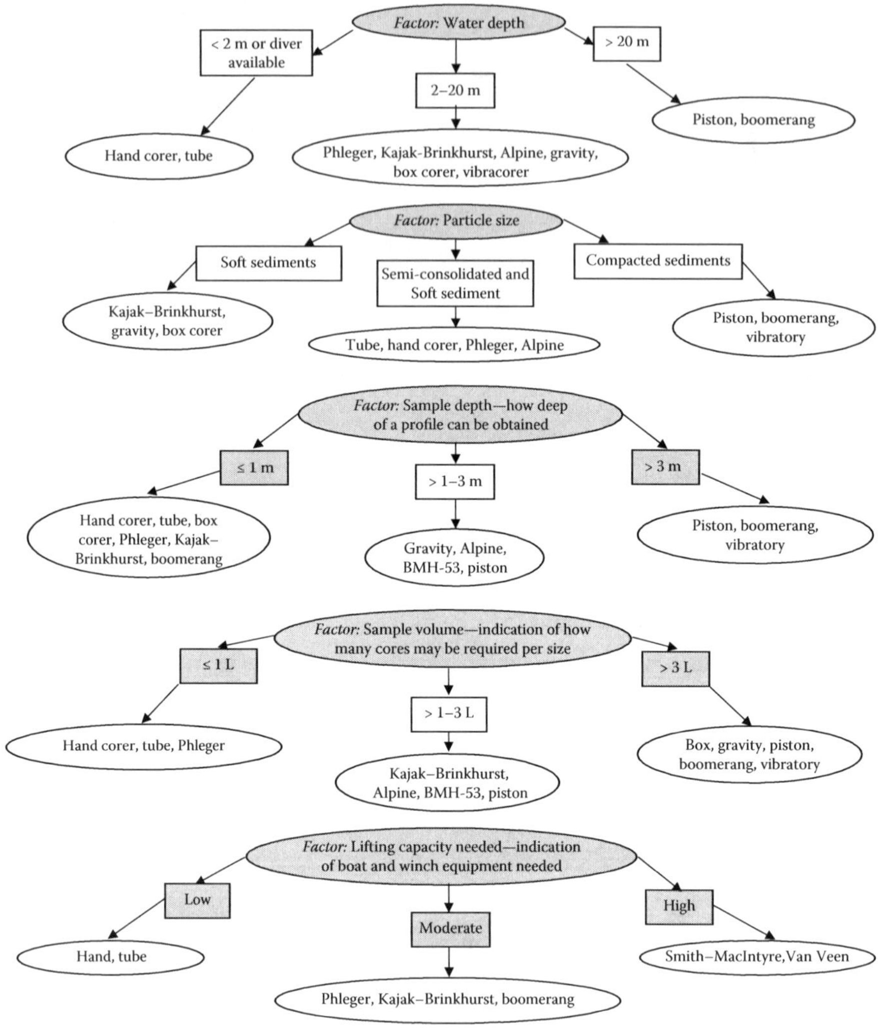

FIGURE 12.8 Flowchart for selecting appropriate grab samplers based on site-specific or design factors.

taped to secure the sample. If sediment oxidation is a concern (e.g., due to potential changes in metal bioavailability or volatile substances), then the head space of the core tube should be purged with an inert gas such as nitrogen or argon.

Dredge Samplers

A dredge sampler is designed to obtain a sample of the bottom material in a slow-moving stream and the organisms in it. The simple homemade dredge shown in Figure 12.4 works well in water too deep to sample effectively with hand-held tools. The homemade dredge is fashioned from a #3 coffee can and a smaller can with a tight-fitting plastic lid (peanut cans work well). To use the homemade dredge, first invert it

DID YOU KNOW?

To reduce the probability of cross-contamination of samples, it is useful to sample reference or relatively clean sites first and then suspected contaminated sites.

under water so the can fills with water and no air is trapped. Then, lower the dredge as quickly as possible with the "down" line. The idea is to bury the open end of the coffee can in the bottom. Then, quickly pull the "up" line to bring the can to the surface with a minimum loss of material. Dump the contents into a sieve or observation pan to sort. This approach works best in bottoms composed of sediment, mud, sand, and small gravel. The bottom sampling dredge can also be used for a number of different analyses. For example, because the bottom sediments represent a good area in which to find macroinvertebrates and benthic algae, the communities of organisms living on or in the bottom can be easily studied quantitatively and qualitatively. A chemical analysis of the bottom sediment can be conducted to determine what chemicals are available to organisms living in the bottom habitat (USEPA, 2001).

SAMPLE SUITABILITY

Only sediments that are correctly collected with grab or core sampling devices should be used for subsequent physicochemical, biological, or toxicity testing. The key words in this statement are "correctly collected." For example, the suitability of correctly collected grabs can be ascertained by noting that the samplers were closed when retrieved, are relatively full of sediment (but not overfilled), and do not appear to have lost surficial fines. Core samples are suitable if the core as inserted vertically in the sediment and an adequate depth was sampled.

A sediment sample should be inspected as soon as it is secured. If a collected sample fails to meet any of the suitability conditions listed below for the respective sampling device, then the sample might have to be rejected and another sample collected at the site. The location of consecutive attempts should be as close to the original attempt as possible and located in the upstream direction of any existing current. Rejected sediment samples should be discarded in a manner that will not affect subsequent samples at the station or other possible sampling stations.

SAMPLE CONTAINERS

The way in which sediment samples are contained might alter contaminant bioavailability and concentration by introducing contaminants to the sample or by changing the physical, chemical, or biological characteristics of the sample. The point here is that any material that is in contact with a field sample has the potential to contaminate the sample or adsorb components from the sample; for example, samples can be contaminated by zinc from glassware, metals from metallic containers, and organic compounds from rubber or plastic materials. The use of appropriate materials, along with appropriate cleaning procedures, can minimize or mitigate interferences from sample containers.

TABLE 12.1
Recommended Sampling Containers, Holding Times, and Storage Conditions for Common Types of Sediment Analyses

Contaminant	Container	Holding Time	Storage Conditions
Ammonia	P, G	28 days	R; F
Sulfate	P, G	28 days	R; F
Sulfide	P, G	28 days	R or NaOH; pH > 9
Oil and grease	G	28 days	HCl; pH < 2
Mercury	P, G	6 weeks	H_2SO_4; pH < 2; R
Metals (except Cr or Hg)	P, G	6 months	HNO_3; pH < 2; F
Extractable organics, including phthalates, airosamines, organochlorine pesticides, PCBs, aromatics, isophrone, PAHs, haloethers, chlorinated hydrocarbons, and TCDD	G, PTFE-lined cap	7 days (until extraction) 30 days (after extraction)	R; F
Purgables (halocarbons and aromatics)	G, PTFE-lined septum	14 days	R; F
Pesticides	G, PTFE-lined cap	7 days (until extraction) 30 days (after extraction)	R; F
Sediment toxicity (acute and chronic)	P, PTFE	2 weeks[a]	R; dark
Bioaccumulation testing	P, PTFE	2 weeks[a]	R; dark

Sources: ASTM (2000a); USEPA (1983, 1993).

Abbreviations: P, plastic; G, glass; PTFE, polytetrafluoroethylene; R, refrigerate; F, freeze.

[a] Holding time might be longer, depending on the magnitude and type of contaminants present.

With regard to container materials, borosilicate glass and high-density polyethylene, polycarbonate, and fluorocarbon plastics should be used whenever possible to minimize leaching, dissolution, and sorption (APHA, 1995; ASTM, 2000a). Direct contact between sediment samples and the following substances should be avoided: PVC, natural or neoprene rubber, nylon, talcum powder, polystyrene, galvanized metal, brass, copper, lead, other metal materials, soda glass, paper tissues, and painted surfaces. Table 12.1 summarizes the appropriate types of sampling containers and allowable holding times for various types of contaminants associated with sediments.

In general, sediments and porewaters with multiple or unknown chemical types should be stored in containers made from high-density polyethylene plastic or polytetrafluoroethylene (PTFE; Teflon®) as these materials are least likely to add chemical artifacts or interferences and they are much less fragile than glass.

Samples for organic contaminant analysis should be stored in brown borosilicate glass containers with PTFE lid liners. If volatile compounds will be analyzed, containers should have a septum to minimize escape of volatile gases during storage and analysis. Extra containers should be provided for these analyses in the event that reanalysis of the sample is required. If samples are contaminated with photoreactive compounds such as PAHs, exposure to light should be minimized by using brown glass containers or clear containers wrapped tightly with an opaque material (e.g., clean aluminum foil). Plastic or acid-rinsed glass containers are recommended when the chemicals of concern are heavy metals.

With regard to container preparation, many vendors have commercially available precleaned containers for a variety of applications. For chemical and toxicological analyses, certified precleaned containers are often a cost-effective way to limit the potential for container contamination of samples. For this reason, manufacturer-supplied precleaned containers are often a prerequisite in quality assurance project plans (QAPPs).

If new containers are used, new glassware and plasticware should be soaked in 1:1 concentrated acid prior to use (Environment Canada, 1994). Soaking overnight is adequate for glassware. For plasticware, the recommended procedure involves soaking for 7 days in hydrochloric acid (HCl) followed by 7 days in nitric acid (HNO_3), followed by 7 days in deionized water. Shorter soaking times might be satisfactory in most instances (ASTM, 2000a). Used sample containers should be washed following these steps: (1) non-phosphate detergent wash, (2) triple water rinse, (3) water-miscible organic solvent wash (acetone followed by pesticide-grade hexane), (4) water rinse, (5) acid wash (e.g., 5% concentrated HCl), and (6) triple rinse with deionized distilled water. A dichromate-sulfuric acid cleaning solution can generally be used in place of both the organic solvent and the acid (steps 3 through 5), but it might attack any silicone adhesive present in the container.

If a sample is to be refrigerated, the container should be filled to the brim to reduce oxygen exposure. This is particularly critical for volatile compounds, such as acid-volatile sulfides (AVSs). If a sample is to be frozen, the container should be filled to approximately 90% of its volume (i.e., 10% headspace) to allow for expansion of the sample during freezing. For studies in which it is critical to maintain the collected sediment under anoxic conditions (e.g., where metals are the pollutants of concern), the container should be purged with an inert gas (e.g., nitrogen) before filling and then again before capping tightly.

All sediment containers should be properly labeled with a waterproof marker prior to sampling. Containers should be labeled on their sides in addition to or instead of labeling the lids. Each label should include, at a minimum, the study title, station location, sample identification, date and time of collection, sample type, and name of collector. Blind sample labeling (i.e., a sample code) should be used, along with a sample log that identifies information about each sample to minimize potential analytical bias. Additional information such as required analyses and any preservative used might also be included on the label, although this information is typically recorded on the chain-of-custody form. Labeled containers should be stabilized in an upright position in the transport or storage container. Extra containers should be carried on each sample trip.

RECORDKEEPING: MEASUREMENTS AND OBSERVATIONS

After completing the sediment suitability assessment, it is important to ensure that all of the field data sheets have been completed properly and that the information is legible. In the field records taken, be sure they include the site's identifying name and the sampling date on each sheet. This information will function as a quality control element. Before leaving the stream location, make sure that all sampling equipment and devices have been collected and rinsed properly. Double-check to make sure that sample jars are tightly closed and properly identified. Keep a copy of the field data sheets for comparison with future monitoring trips and for personal records.

The next step is to prepare for laboratory work. This step includes all of the work necessary to set up a laboratory for processing samples into subsamples and identifying sediment contaminants. A professional geologist, biologist, entomologist, or freshwater ecologist or a professional advisor should supervise the identification procedure. During the identification and recordkeeping process, documentation should include the following:

- Type of vessel used (e.g., size, power, type of engine)
- Notation of the system used to define the position of the sampling site
- Notation of the system used to identify and track samples
- Name of personnel collecting the samples
- Level of personal protective equipment (PPE) worn
- Notation of any visitors to the site
- Sketch of sampling area with photographs, if possible
- Ambient weather conditions, including wind speed and direction, wave action, current, tide, vessel traffic, temperature of both the air and water, and thickness of ice, if present
- Type of sediment collection device and any modifications that were made during sampling
- Calibration data

EQUIPMENT DECONTAMINATION

Reused samplers, scoops, spatulas, mixing bowls, sample containers, glassware, and any other utensils that come in contact with sediment samples must be cleaned and rinsed before the first sampling run and after each run. The most suitable method of decontamination depends on the parameter being measured. At a minimum, use the following method when preparing all sampling equipment and sample containers. Wearing latex gloves,

1. Rinse each sampling device, sample bottle, or piece of glassware with site water and use a brush and phosphate-free detergent on glassware.
2. Rinse three times with clean water.
3. Collect and dispose of washwater from decontamination properly.

REFERENCES AND RECOMMENDED READING

APHA. (1971). *Standard Methods for the Examination of Water and Wastewater*, 17th ed. Washington, DC: American Public Health Association.

APHA. (1995). *Standard Methods for the Examination of Water and Wastewater*, 18th ed. Washington, DC: American Public Health Association.

APHA. (1998). *Standard Methods for the Examination of Water and Wastewater*, 20th ed. Washington, DC: American Public Health Association.

ASTM. (2000a). Standard guide for collection, storage, characterization, and manipulation of sediments for toxicological testing, E1391-94. In: *ASTM Standards on Environmental Sampling*. West Conshohocken, PA: American Society for Testing and Materials International.

ASTM. (2000b). Standard guide for general planning of waste sampling, D4687. In: *ASTM Standards on Environmental Sampling*. West Conshohocken, PA: American Society for Testing and Materials International.

ASTM. (2000c). Standard guide for selecting grab sampling devices for collecting benthic macroinvertebrates, D4387. In: *ASTM Standards on Environmental Sampling*. West Conshohocken, PA: American Society for Testing and Materials International.

ASTM. (2000d). Standard guide for core-sampling submerged, unconsolidated sediments, D4823-95. In: *ASTM Standards on Environmental Sampling*. West Conshohocken, PA: American Society for Testing and Materials International.

ASTM (2000e). Standard practice for preparation of sediment samples for chemical analysis, D3976-92. In: *ASTM Standards on Environmental Sampling*. West Conshohocken, PA: American Society for Testing and Materials International.

ASTM. (2000f). Standard guide for conducting acute toxicity tests with fishers, macroinvertebrates, and amphipods, E729-96. In: *ASTM Standards on Environmental Sampling*. West Conshohocken, PA: American Society for Testing and Materials International.

ASTM. (2000g). Standard test methods for ammonia nitrogen in water, D1426-93. In: *ASTM Standards on Environmental Sampling*. West Conshohocken, PA: American Society for Testing and Materials International.

AWRI. (2000). *Plankton Sampling*. Allendale, MI: Robert B. Annis Water Resource Institute, Grand Valley State University.

AWWA. (1995). *Water Treatment*, 2nd ed. Denver, CO: American Water Works Association.

Bahls, L.L. (1993). *Periphyton Bioassessment Methods for Montana Streams*. Helena: Montana Water Quality Bureau, Department of Health and Environmental Science.

Barbour, M.T., Garretson, J., Snyder, B.D., and Stabling, J.B. (1997). *Revision to Rapid Bioassessment Protocols for Use in Streams and Rivers: Periphytons, Benthic Macroinvertebrates, and Fish*. Washington, DC: U.S. Environmental Protection Agency.

Baudo, R. (1990). Sediment sampling, mapping and data analysis. In: *Sediments: Chemistry and Toxicity of In-Place Pollutants* (Giesy, J.P. and Muntau, H., Eds.), pp. 15–60. Chelsea, MI: Lewis Publishers.

Bly, T.D. and Smith, G.F. (1994). *Biomonitoring Our Streams: What's It All About*? Nashville, TN: U.S. Geological Survey.

Botkin, D.B. (1990). *Discordant Harmonies*. New York: Oxford University Press.

Camann, M. (1996). *Freshwater Aquatic Invertebrates: Biomonitoring*. Arcata, CA: Humboldt State University.

Carins, Jr., J. and Dickson, K.L. (1971). A simple method for the biological assessment of the effects of waste discharges on aquatic bottom-dwelling organisms. *Journal of Water Pollution Control Federation*, 43: 755–772.

Carlton, R.G. and Wetzel, R.G. (1985). A box corer for studying metabolism of epipelic microorganisms in sediment under *in situ* conditions. *Limnology and Oceanography*, 30: 422.

Environment Canada. (1994). *Guidance Document on Collection and Preparation of Sediments for Physicochemical Characterization and Biological Testing*, Report EPS I/RM/29. Gatineau, Quebec: Environment Canada, Environmental Protection Services.

GLNPO. (1994). *Assessment and Remediation of Contaminated Sediments (ARCS) Program*, EPA-905-B94-002. Chicago, IL: U.S. Environmental Protection Agency, Great Lakes National Program Office.

Golterman, H.L., Sly P.G., and Thomas, R.L. (1983). *Study of the Relationship between Water Quality and Sediment Transport*. Mayenne, France: UNESCO.

Hill, B.H. (1997). The use of periphyton assemblage data in an index of biotic integrity. *Bulletin of the North American Ethological Society*, 14: 158.

Huff, W.R. (1993). Biological indices define water quality standard. *Water Environment & Technology*, 5: 21–22.

Karr, J.R. (1981). Assessment of biotic integrity using fish communities. *Fisheries*, 66: 21–27.

Karr, J.R., Bausch, K.D., Hagermeier, P.L., Yant, P.R., and Schlosser, I.J. (1986). *Assessing Biological Integrity in Running Waters: A Method and Its Rationale*, Special Publ. 5. Champaign: Illinois Natural History Survey.

KDOW. (1993). *Methods for Assessing Biological Integrity of Surface Waters*. Frankfort: Kentucky Department of Environmental Protection, Division of Water.

Kittrell, F.W. (1969). *A Practical Guide to Water Quality Studies of Streams*. Washington, DC: U.S. Department of Interior.

Klemm, D.J., Lewis, P.A., Fulk, F., and Lazorchak, J.M. (1990). *Macroinvertebrate Field and Laboratory Methods for Evaluating the Biological Integrity of Surface Waters*, EPA-600-4-90-030. Cincinnati, OH: U.S. Environmental Protection Agency, Environmental Monitoring and Support Laboratory.

Mudroch, A. and Azcue, J.M. (1995). *Manual of Aquatic Sediment Sampling*. Boca Raton, FL: Lewis Publishers.

O'Toole, C., Ed. (1986). *The Encyclopedia of Insects*. New York: Facts on File.

Ohio EPA. (2001). *Ohio EPA Sediment Sampling Guide and Methodologies*. Columbus: State of Ohio Environmental Protection Agency.

Patrick, R. (1973). Use of algae, especially diatoms, in the assessment of water quality. In: *Biological Methods for the Assessment of Water Quality*, ASTM STP 528 (Carins, Jr., J. and Dickson, K.L., Eds.), pp. 76–95. Philadelphia, PA: American Society for Testing and Materials International.

Plafkin, J.L., Barbour, M.T., Porker, K.D., Gross, S.K., and Hughes, R.M. (1989). *Rapid Bioassessment Protocols for Use in Streams and Rivers: Benthic Macroinvertebrates and Fish*, EPA-440-4-89-001. Washington, DC: U.S. Environmental Protection Agency.

Plumb, R.H. (1981). *Procedures for Handling and Chemical Analysis of Sediment and Water Samples*. Buffalo, NY: State University College at Buffalo, Great Lakes Laboratory.

Puget Sound Estuary Program. (1996). *Recommended Guidelines for Measuring Selected Environmental Variables in Puget Sound*. Olympia, WA: U.S. Environmental Protection Agency; Olympia, WA: Puget Sound Water Quality Authority.

Puget Sound Estuary Program. (1997). *Recommended Guidelines for Sampling Marine Sediment, Water Column, and Tissue in Puget Sound*. Olympia, WA: Puget Sound Estuary Program, Puget Sound Water Quality Action Team.

Rodgers, Jr., J.H., Dickson, K.L., and Cairns, Jr., J. (1979). A review and analysis of some methods used to measure functional aspects of periphyton. In: *Methods and Measurements of Periphyton Communities: A Review*, ASTM STP 690 (Weitzel, R.L., Ed.), pp. 142–167. West Conshohocken, PA: American Society for Testing and Materials International.

Spellman, F.R. (2015). *The Science of Water: Concepts and Applications*, 3rd ed. Boca Raton, FL: CRC Press.

Stevenson, R.J. (1996). An introduction to algal ecology in freshwater benthic habitats. In: *Algal Ecology: Freshwater Benthic Ecosystems* (Stevenson, R.J., Bothwell, M., and Lowe, R.L., Eds.), pp. 3–30. San Diego, CA: Academic Press.

Stevenson, R.J. (1998). Diatom indicators of stream and wetland stressors in a risk management framework. *Environmental Monitoring and Assessment*, 51: 107–108.

Stevenson, R.J. and Pan, Y. (1999). Assessing ecological conditions in rivers and streams with diatoms. In: *The Diatoms: Application to the Environmental and Earth Sciences* (Stoermer, E.G. and Smol, J.P., Eds.), pp. 11–40. Cambridge, U.K.: Cambridge University Press.

Tchobanoglous, G. and Schroeder, E.D. (1985). *Water Quality*. Reading, MA: Addison-Wesley.

USEPA. (1983). *Technical Support Manual: Waterbody Surveys and Assessments for Conducting Use Attainability Analyses*, EPA 440486037. Washington, DC: U.S. Environmental Protection Agency.

USEPA. (1993). *Statement of Work for Organics Analysis, Multi-Media, Multi-Concentration*, Document No. ILMO1.0–ILMO-1.9. Washington, DC: U.S. Environmental Protection Agency.

USEPA. (2000). *Monitoring Water Quality: Intensive Stream Bioassay*. Washington, DC: U.S. Environmental Protection Agency.

USEPA. (2001). *Methods for Collection, Storage and Manipulation of Sediments for Chemical and Toxicological Analyses: Technical Manual*, EPA-823-B-01-002. Washington, DC: U.S. Environmental Protection Agency, Office of Water.

Velz, C.J. (1970). *Applied Stream Sanitation*. New York: Wiley Interscience.

Warren, Jr., M.L. and Burr, B.M. (1994). Status of freshwater fishes of the United States: overview of an imperiled fauna. *Fisheries*, 19(1): 6–18.

Weitzel, R.L. (1979). Periphyton measurements and applications. In: *Methods and Measurements of Periphyton Communities. A Review*, ASTM STP 690 (Weitzel, R.L., Ed.), pp. 3–33. West Conshohocken, PA: American Society for Testing and Materials International.

Whitman, R.L. and Nevers, M.B. (2003). Foreshore sand as a source of *Escherichia coli* in nearshore water of a Lake Michigan beach. *Applied and Environmental Microbiology*, 69: 5555–5562.

13 Collection of Porewater

Sediment porewater, or interstitial water as used in this discussion, is defined as the water occupying the spaces between sediment particles. Interstitial water might occupy about 50% (or more) of the volume of a depositional (silt–clay) sediment. The interstitial water is in contact with sediment surfaces for relatively long periods of time and therefore might become contaminated due to partitioning of the contaminants from the surrounding sediments. In addition, interstitial waters might reflect groundwater–surface water transition zones in upwelling or downwelling areas. In these areas, their chemistry might be more reflective of ground or surface waters at the site. Therefore, flow, residence time, and other physicochemical factors (e.g., pH, temperature, redox potential, organic carbon, sulfides, carbonates, mineralogy) might have varying roles in determining whether interstitial waters are contaminated. Because many interstitial waters are relatively static in many depositional sediments, the contaminants in the interstitial water and in the solid phase are considered to be at thermodynamic equilibrium. This makes interstitial waters useful for assessing contaminant levels and associated toxicity. Interstitial water is often isolated to provide either a matrix for toxicity testing or an indication of the concentration or partitioning of contaminants within the sediment matrix (USEPA, 2001).

INTERSTITIAL WATER COLLECTION AND SAMPLING

The collection of interstitial water has become increasingly important in sediment quality monitoring and remediation programs. Moreover, interstitial water sampling has become especially important in regulatory programs because interstitial water toxicity tests yield additional information not currently provided by solid-phase, elutriate (i.e., separated by washing), or sediment extract tests (Carr and Chapman, 1992; SETAC, 2001). Additionally, interstitial water toxicity tests have proved to be useful in sediment toxicity identification evaluation (TIE) studies (Burgess, 1996; Burton et al., 2003; Carr, 1998) as test procedures and sample manipulation techniques are generally less expensive, faster, and easier to conduct than solid-phase tests (SETAC, 2001). Sediment types ranging from sandy to uncompacted silt–clays are most suitable for interstitial water sampling (Sarda and Burton, 1995; SETAC, 2001). Such sampling is not typically performed on sediments with coarse particle size (such as gravel) or on hard, compacted clays, as the potential for interstitial water contamination in these sediment types is relatively low.

Note that for the purposes of discussion in this book, the principle aim is to describe sampling procedures that minimize changes in the *in situ* condition of the water. It is also important to recognize that most sediment collection and processing methods have been shown to alter interstitial water chemistry (e.g., Bufflap and Allen, 1995; Sarda and Burton, 1995; Schults et al., 1992), thereby potentially altering contamination bioavailability and toxicity.

DID YOU KNOW?

The Superfund program has initiated a project to develop media-specific benchmark values for those chemicals commonly found in surface water, sediment, or soil samples at sites. The values are referred to as *Ecotox Thresholds* (ETs), and they are defined as media-specific contaminant concentrations above which there is sufficient concern regarding adverse ecological effects to warrant further site investigations.

Centrifugation, pressurization, suction, and other laboratory-based methods are commonly used as alternatives to *in situ* interstitial water collection. Although these methods have been shown to alter interstitial water chemistry, they are sometimes necessary or preferred, especially when larger sample volumes are required (e.g., for toxicity testing).

As both *in situ* and laboratory-based or *ex situ* methods might be appropriate for many study objectives, it is critical to use the same procedures for all stations sampled in a study or program so appropriate sample comparisons can be made. Furthermore, the sediment depth at which interstitial water is sampled (using either *in situ* or *ex situ* extraction methods) should match the depth of interest in the study (SETAC, 2001). For example, samples for dredging remediation should be sampled to the depth to be disturbed by dredging activity, whereas samples for a status and trends survey should be collected at the biologically active depth (often <15 cm). Figure 13.1 summarizes the major considerations for selecting *in situ* or *ex situ* procedures in a given study.

The two major issues of concern regarding interstitial water sample integrity are (1) the ability of the sampling device to maintain physicochemical conditions in the natural state by minimizing adsorption or leaching of chemicals to and from the device, and (2) the ability to maintain the sample in the redox state existing at the site. Precautions required to reduce the likelihood of sample artifacts will vary with each study as indicated in the following sections.

IN SITU COLLECTION

In situ methods might be superior to *ex situ* methods for collecting interstitial water, as they are less subject to sampling/extraction related artifacts and therefore might be more likely to maintain the chemical integrity of the sample (ASTM, 2000a; Sarda and Burton, 1995; SETAC, 2001). However, *in situ* methods have generally produced relatively small volumes of interstitial water and are often limited to wadeable or diver-accessible water depths. These logistical constraints have limited their use and applicability in sediment monitoring studies. The principal methods for *in situ* collection of interstitial water involve either deploying peepers (Adams, 1991; Bottomly and Bayly, 1984; Brumbaugh et al., 1994; Bufflap and Allen, 1995; Carignan and Lean, 1991; Carignan et al., 1985) or suction techniques (Howes et al., 1985; Knezovich and Harrison, 1988; Watson and Frickers, 1990). A summary

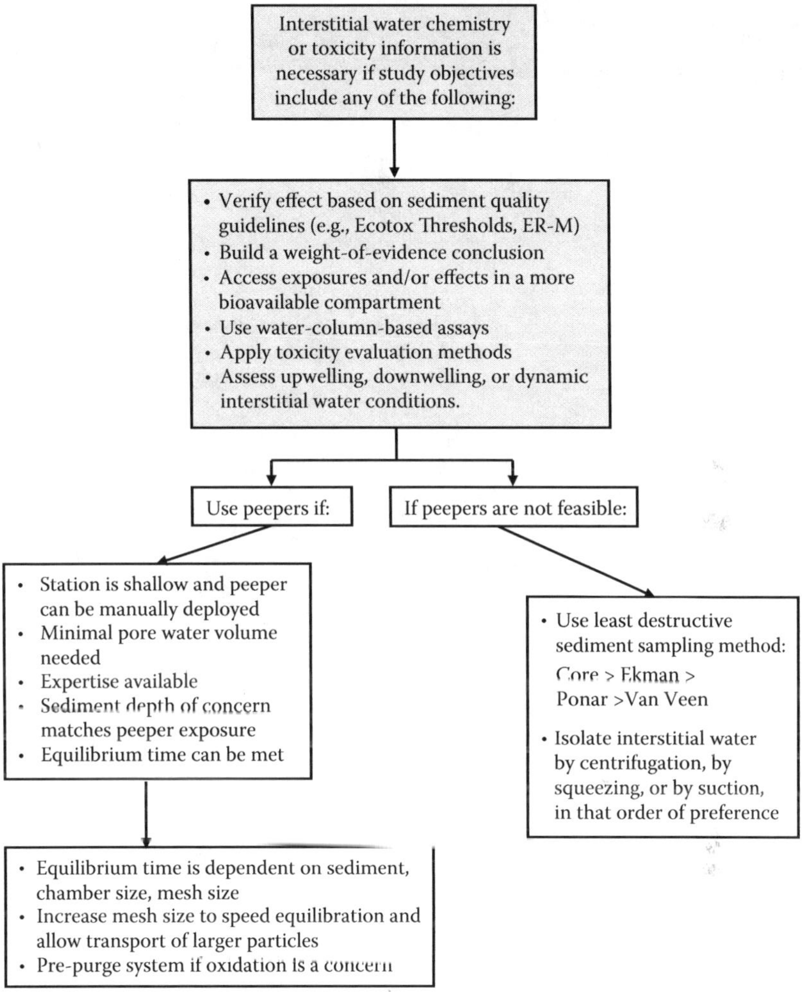

FIGURE 13.1 Considerations for selecting the appropriate type of interstitial water sampling method.

of these methods is provided in Table 13.1. Both methods have a high likelihood of maintaining *in situ* conditions. In cases where *in situ* deployment is impractical, peepers or suction device can be place in relatively undisturbed sediments collected by core or grab samplers.

Peeper Methods

Peepers are small chambers with membrane or mesh walls containing either distilled water or clean water of the appropriate salinity or hardness. Samples are collected by burying the devices in sediments and allowing surrounding interstitial waters to infiltrate. In principle, dissolved solutes will diffuse through the porous wall into

TABLE 13.1
***In Situ* Interstitial Water Collection Methods**

Device	Sediment Depth (cm)	Sample Volume (L^3)	Advantages	Disadvantages
Peeper	0.2–10	≤0.5	Most accurate method, reduced artifacts, no lab processing; relatively free of effects from temperature, oxidation, and pressure; inexpensive and easy to construct; some selectivity possible, depending on the nature of sample, via specific membranes; wide range of membrane/mesh pore sizes and internal solutes or substrates available	Requires deployment by hand, thus requiring diving in water >0.6 m in depth; requires hours to days for equilibration (varies with site and chamber); methods are not standardized and are used infrequently; some membranes, such as dialysis/cellulose, are subject to biofouling; must deoxygenate chamber and materials to prevent oxidation effects; some construction materials yield chemical artifacts; some chambers only allow small sample volumes; care must be used on collection to prevent sample oxidation
In situ suction	0.2–30	≤0.25	Reduced artifacts, gradient definition; rapid collection; no lab processing; closed system, which prevents contamination; methods include airstone, syringes, probes, and core-type samplers	Requires custom, nonstandard collection devices; small volumes; limited to softer sediments; core airstone method; difficult in some sediments and in deeper water (>1 m); method might require diving for deployment in deep waters; methods used infrequently and by limited number of laboratories

Sources: Adapted from Sarda and Burton (1995) and SETAC (2001).
Note: Incorporation of filtration into any collection method might result in loss of metal and organic compounds.

the peeper, and the contained water will reach equilibrium with the ambient interstitial water. The design concept for sediment peepers originated as modifications of the dialysis bag technique used by Mayer (1976) and Hesslein (1976), and it has been modified successfully for use in laboratory sediment toxicity tests (Doig and Liber, 2000). The initial designs consisted of either a flat base plate or a cylindrical dialysis probe (Bottomley and Bayly, 1984) with compartments covered by dialysis membranes and a manifold for collection of multiple samples at various depths in the sediment profile. Further modifications to these designs have incorporated sampling ports, large sample compartments, and various types of membranes with different port sizes. These modifications are usually required based on specific project objectives regarding sample volumes and contaminants of interest.

Various peeper devices have been used effectively to collect interstitial water. For example, a simplified design using a 1-μm polycarbonate membrane over the opening of a polyethylene vial was successful in capturing elevated levels of copper and zinc (Brumbaugh et al., 1994). Other designs have been used to collect nonpolar organic compounds in a variety of aquatic systems (Axelman et al., 1999; Bennett et al., 1996) and in overlying water (Huckins et al., 1990). Peepers have also been used to expose organism to sediments *in situ* (Burton et al., 2003). Burton et al. (1999) successfully introduced organisms to aerobic sediments using peepers; however, anoxic sediments are not amenable to *in situ* organism exposure.

The use of various materials might be advisable when constructing peepers, depending on the contaminants of concern; for example, for many contaminants, peepers constructed from acrylic materials appear to yield interstitial water samples with minimal chemical artifacts (Burton et al., 2003). Some polymer materials might be inappropriate for studies of certain nonpolar organic compounds. Cellulose membranes are also unsuitable, as they decompose too quickly. Plastic samplers can contaminate anoxic sediments with diffusible oxygen (Carignan et al., 1994).

In preparation for interstitial water collection, peeper chambers should be filled with deoxygenated water, which can be prepared by nitrogen purging for 24 hours prior to insertion. If sediment oxidation is a concern, the peepers should be transported to the deployment site in a sealed oxygen-free water bath to avoid potential changes to the sediment water equilibrium caused by dissolved oxygen interactions. However, during peeper equilibration periods, anoxic conditions are likely to be quickly reestablished. In addition, when samples are collected and processed, exposure to oxygen should be minimized. Following initial placement, the equilibrium time for peepers may range from hours to a month, but a deployment period of 1 to 2 weeks is most often used (Adams, 1991; Call et al., 1999; Steward and Malley, 1999). Equilibration time is a function of sediment type, study objectives, contaminants of concern, and temperature (e.g., Carr et al., 1989; Howes et al., 1985; Mayer, 1976; Simon et al., 1985; Skalski and Burton, 1991). Membrane pore size also affects equilibration time, with larger pore sizes being used to achieve reduced equilibration times (Sarda and Burton, 1995). For example, using a peeper with a 149-μm pore size, Adams (1991) reported equilibration of conductivity within hours of peeper insertion into the sediment. Thus, it appears that equilibration time is a function of the type of contaminant, sediment type, peeper volume, and mesh pore size.

Peepers with large-pored membranes shorten equilibration times and allow particulates to enter the chamber. The larger solids tend to settle to the bottom of the peeper chamber, and caution should be used to avoid collecting the solids when retrieving the water sample from the chamber. Colloidal particles will remain suspended in the sample and thereby present an artifact, but the concentration of such particles is typically lower than that found in laboratory-centrifuged samples (Chin and Gschwend, 1991).

Suction Methods

Various suction devices are available for collecting interstitial water. A typical suction device consists of a syringe or tube of varying length, with one or more ports located at the desired sampling positions (ASTM, 2000a). The device is inserted

DID YOU KNOW?

The potential for high variability in interstitial water chemical characteristics should be taken into account when developing the sampling design. This is the case because several studies and analyses of interstitial water from replicate peepers have demonstrated from low to high heterogeneity in water quality characteristics (Frazier et al., 1996; Sarda and Burton, 1995).

into the sediment to the desired depth and a manual, spring-operated, or vacuum gas suction is applied to directly retrieve the water sample. A variation on this approach employs a peeper-like porous cup or perforated tube with filters. The unit is inserted into the sediment for a period of time, allowing interstitial water to infiltrate the chamber before suction is applied. The samples are then retrieved by suction. Another variation that has been used successfully employs an airstone embedded into the sediment which forces interstitial water upward where it can be collected via syringe or tube. All of these suction methods generally yield smaller quantities of interstitial water than peepers, and chemical (toxicological) artifacts are more likely due to greater potential exposure of interstitial water to oxygen (ASTM, 2000a).

Processing of Field-Collected Interstitial Water Samples

Following sample retrieval, interstitial water might have to be recovered and stabilized quickly to prevent oxidative changes or volatilization (Carignan, 1984). Containers should be filled, with no headspace, to minimize changes in dissolved oxygen and contaminant bioavailability. Procedures for stabilization are dependent on the analyses to be performed. When nonvolatile compounds are the target analytes, acidification is often stipulated, whereas organic carbon and methane may be stabilized with saturated mercury chloride (Mudroch and MacKnight, 1994). Samples to be analyzed for toxicity are normally cooled to 4°C as soon as possible for transport to the laboratory. U.S. Environmental Protection Agency (USEPA) methods for toxicity testing of surface waters and effluents recommend that samples not be frozen in storage or transport (USEPA, 1991). However, later information suggests that freezing of interstitial water may not affect toxicity in some cases (Carr and Chapman, 1995; Ho et al., 1997; SETAC, 2001). Unless a demonstration of acceptability is made for the sites of interest, interstitial water samples should not be frozen prior to biological testing. Samples for chemical analyses should be preserved immediately, if appropriate, or cooled to 4°C as soon as possible.

EX SITU EXTRACTION OF INTERSTITIAL WATER

Ex-situ interstitial water collection methods are often necessary when relatively large volumes of interstitial water are required (such as for toxicity testing), when *in situ* collection is not viable, or when a brief sampling time is critical. Although these extraction methods can be done in the field or in the laboratory, extraction in

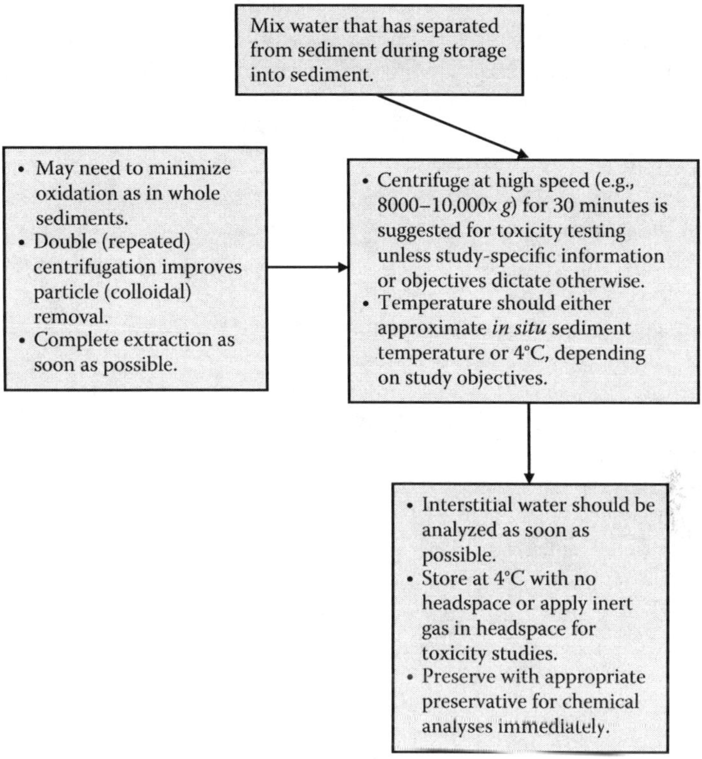

FIGURE 13.2 Summary of recommended procedures and considerations for laboratory isolation of interstitial water.

the laboratory just prior to analysis or testing is preferable so the sample is maintained as close to its original state as much as possible during transport and storage (SETAC, 2001). Guidance in this section reflects recommendations presented in several publications, including proceedings from workshops devoted entirely to interstitial water extraction methods, water handling, and their use in toxicity applications: (1) a dredged materials management program workshop on interstitial water extraction methods and sample storage in relation to tributyltin analysis (Hoffman, 1998), and (2) a Pellston workshop on interstitial water toxicity testing, including interstitial water extraction methods and applications (SETAC, 2001). Figure 13.2 summarizes many of the issues associated with laboratory isolation of interstitial water discussed in this section.

DID YOU KNOW?

Emphasis should be placed on minimizing the duration of all sample manipulations whenever possible.

Centrifugation and squeezing are the two most common techniques for collecting interstitial water, and they are generally preferred when large volumes are required. Other methods include pressurization (e.g., vacuum filtration) devices, which can be used to recover small volumes of interstitial water. Regardless of the method used, interstitial water should be preserved immediately for chemical analyses, if appropriate, or analyzed as soon as possible after sample collection if unpreserved (such as for toxicity testing) (Hoffman, 1998; SETAC, 2001). Significant chemical changes can occur even when interstitial water is stored for periods as short as 24 hours (Hulbert and Brindle, 1975; Kemble et al., 1999; Sarda and Burton, 1995; SETAC, 2001; Watson et al., 1985).

If sediments are anoxic, as most depositional sediments are, sample processing, including mixing of interstitial water that has separated from the sediment, should be conducted in an inert atmosphere or with minimal atmospheric contact. Exposure to air can result in oxidation of contaminants, thereby altering bioavailability (Bray et al., 1973; Howes et al., 1985; Lyons et al., 1979). Air exposure can also result in loss of volatile sulfides, which might increase the availability of sulfide-bound metals (Allen et al., 1993; Bufflap and Allen, 1995). In addition, iron and manganese oxyhydroxides are quickly formed upon exposure to air. These compounds readily complex with trace metals, thus altering metals-related toxicity (Bray et al., 1973; Bufflap and Allen, 1995; Burton, 1991; Troup et al., 1974). Maintaining anoxic processing conditions is not necessary when study objectives are concerned with exposures to aerobic sediments, or if target contaminants are unaffected by oxidation in short-term toxicity or bioaccumulation testing.

As just mentioned and repeated here for emphasis, interstitial water filtration should be avoided (SETAC, 2001). Numerous studies have shown that filters reduce toxicity and contaminant concentrations by retaining contaminant-associated particles and also by contaminant sorption onto the filter matrix (Bray et al., 1973; Sasson-Brickson and Burton, 1991; Schults et al., 1992; Troup et al., 1974). If filtration is stipulated by a test method, treated filters (e.g., pre-soaked in distilled, deionized water, or combusted at 400°C overnight for glass-fiber filters) should be used, and unfiltered samples should also be tested for toxicity and contaminant concentrations. The characteristics of filters and the filtering apparatus should also be carefully considered, as different filters have different sorptive capacities for different contaminants.

DID YOU KNOW?

Filtration should be avoided unless required by a test method because it might reduce interstitial water toxicity. Double (serial) centrifugation (low speed followed by high speed) should be used instead. If filtering is required by a test method, pretreated filters should be used to reduce potential contamination (SETAC, 2001).

CENTRIFUGATION

Centrifugation is the generally preferred laboratory method for collection of interstitial water (SETAC, 2001). It is a relatively simple procedure that allows rapid collection of large volumes of interstitial water. It also facilitates the maintenance of anoxic conditions (if required). However, centrifugation, like other *ex situ* procedures, might yield chemical or toxicology artifacts due to the extraction procedures themselves, which might alter the natural equilibrium between interstitial water and sediment.

Prior to centrifugation, the sediment sample is homogenized (i.e., complete mixing of the sediment to obtain a consistency of physicochemical properties throughout the sample) and partitioned among centrifuge bottles. If the homogenized sample is stored prior to centrifugation, interstitial water might accumulate on the surface of the sediment. This overlying water should be mixed into the sediment before subsampling for centrifugation. Samples are then partitioned among centrifuge bottles. In general, approximately 50% of sediment moisture content can be extracted as interstitial water. If interstitial water volume requirements are lower, smaller sediment subsamples may be used.

Based on research to date, both slower and faster centrifugation speeds (and associated differences in colloid/suspended solids removal) may be appropriate depending on the study objectives. For many programs that are interested in characterizing site toxicity, high-speed centrifugation may not be appropriate because one is interested in the toxicity potential of the interstitial water in its entirety (i.e., including colloidal material). However, if one is interested in comparing interstitial water contaminant concentration to specific sediment quality values or model exposure compartments, for example, then high-speed centrifugation might be necessary. As our knowledge is still limited in this area, it is perhaps most important to note that centrifugation speed often has a dramatic effect on observed sample toxicity and chemical characteristics. Therefore, in any sediment monitoring study, one centrifugation protocol (including speed and time) should be identified and used throughout for all samples.

Centrifugation has been performed at various temperatures. ASTM (2000a) recommended that the centrifugation temperature reflect the *in situ* sediment temperature to ensure that the equilibrium between the particulate and interstitial water is not altered. Alternatively, a temperature of 4°C may be preferred to minimize temperature-mediated chemical and biological processes (Environment Canada, 1994). When centrifuging coarse sand, it might be desirable to use a modified centrifuge bottle to aid interstitial water recovery (USEPA/USACE, 1998). The modified bottle is equipped with an internal filter that can recover 75% of the interstitial water, as compared to 25 to 30% recovery from squeezing (Saager et al., 1990).

As discussed in Chapter 12, all containers have limitations with regard to adsorption or leaching of chemicals, ease of use, and reliability. For example, polytetrafluorothylene (PTF) bottles have been used successfully up to 2500× *g* when filled to 80% of capacity but collapse at 3000× *g* (Burgess et al., 1993). Polycarbonate bottles have been used successfully for tributytin analyses in interstitial water (Hoffman, 1998). If small volumes of water are required for testing, higher speed centrifugation can be performed with glass tubes (up to 10,000× *g*) (Word et al., 1987). Larger glass tubes, however, cannot be centrifuged at such high speeds. If metal toxicity is not a concern, then high-speed centrifugation in larger stainless steel centrifuge tubes is suitable. If test samples are contaminated with photoreactive compounds such as polycyclic aromatic hydrocarbons (PAHs), exposure of the sample to light should be minimized to limit degradation or alteration of potentially toxic compounds. This can be accomplished by using reduced lighting.

Sediment Squeezing

Isolation of interstitial water by squeezing has been performed using a variety of procedures and devices (Adams, 1991; Carr, 1998; Carr and Chapman, 1992; Carr et al., 1989; Jahnke, 1988; Kalil and Goldhaker, 1973; Long et al., 1990; Reeburgh, 1967; Watson and Frickers, 1990). Inexpensive, low-pressure mechanical squeezers can be constructed that may provide specialized capacities such as collection of interstitial water profiles from core samples (Bender et al., 1987). In all cases, the interstitial water is passed through a filter that is a part of the squeezing apparatus.

Squeezing has been shown to produce a number of artifacts due to shifts in equilibrium from pressure, temperature, and gradient changes (e.g., Bollinger et al., 1992; Froelich et al, 1979; Kriukov and Manheim, 1982; Schults, 1992). Squeezing can affect the electrolyte concentration in the interstitial water particularly with a decrease in chemical concentration in the interstitial water near the end of the squeezing process. However, others have reported that squeezing did not produce artifacts in interstitial water toxicity studies (Carr, 1998; Carr and Chapman, 1995; SETAC, 2001). It is therefore recommended that, if squeezing is performed, moderate pressures be applied along with electrolyte (conductivity) monitoring during extraction (Kriukov and Manheim, 1982). Squeezing should also be performed at *in situ* ambient temperatures, as significant alterations to interstitial water composition can occur when squeezing is conducted at temperatures different from ambient conditions (e.g., Bischoff et al., 1970; Mangelsdorf et al., 1969; Sayles et al., 1973).

Other sources of interstitial water alteration during squeezing are contamination from overlying water, internal mixing of interstitial water during extrusion, and solid-solution reactions as interstitial water is expressed through the overlying sediment. As interstitial waters are displaced into upper sediment zones, they come in contact with solids with which they are not in equilibrium. This intermixing causes solid-solution reactions to occur. Most interstitial water chemical species are rapidly transformed, as observed with ammonia and trace metals (Rosenfield, 1979; Santschi et al., 1997). Bollinger et al. (1992) found elevated levels of several ions and dissolved organic carbon in squeezed samples as compared to samples collected by *in situ* peepers. The magnitude of the artifact will depend on the pollutant sediment characteristics and redox potential.

Pressurized and Vacuum Devices

Other methods for extraction of interstitial water from sediment samples can include vacuum filtration (Jenne and Zachara, 1987; Knezovich and Harrison, 1987; Winger and Lasier, 1991), gas pressurization (Reeburgh, 1967), and displacement (Adams,

1991). These methods typically recover only small volumes of interstitial water and are not commonly used. Use of a hand vacuum with an aquarium stone is an effective vacuum filtration method (Sarda and Burton, 1995; Winger and Lasier, 1991). The procedure typically involves attaching the air stone to a 50-mL syringe via plastic tubing, inserting it into the sediment to the desired depth, and then applying suction. This method can recover relatively large volumes of interstitial water. Santschi et al. (1997) used this procedure to extract up to 1500 mL from 4 L of sediment. Sarda and Burton (1995) found that ammonia concentrations in water obtained by this procedure were similar to those collected by *in situ* peepers. Drawbacks to this method include loss of equilibrium between the interstitial water and the solids, filter clogging, and oxidation (Brinkman et al., 1982).

REFERENCES AND RECOMMENDED READING

Adams, D.D. (1991). Sampling sediment pore water. In: *CRC Handbook of Techniques for Sediment Sampling* (Mudroch, A. and MacKnight, S.D., Eds.), pp. 117–202. Boca Raton, FL: CRC Press.

Allen, H.E., Fu, G., and Deng, B. (1993). Analysis of acid-volatile sulfide (AVS) and simultaneously extracted metals (SEM) for the estimation of potential toxicity in aquatic sediments. *Environmental Toxicology and Chemistry*, 12: 1441–1453.

APHA. (1971). *Standard Methods for the Examination of Water and Wastewater*, 17th ed. Washington, DC: American Public Health Association.

APHA. (1998). *Standard Methods for the Examination of Water and Wastewater*, 20th ed. Washington, DC: American Public Health Association.

ASTM. (2000a). Standard guide for collection, storage, characterization, and manipulation of sediments for toxicological testing, E1391-94. In: *ASTM Standards on Environmental Sampling* West Conshohocken, PA: American Society for Testing and Materials International.

ASTM. (2000b). Standard guide for general planning of waste sampling, D4687. In: *ASTM Standards on Environmental Sampling*. West Conshohocken, PA: American Society for Testing and Materials International.

ASTM. (2000c). Standard guide for selecting grab sampling devices for collecting benthic macroinvertebrates, D4387. In: *ASTM Standards on Environmental Sampling*. West Conshohocken, PA: American Society for Testing and Materials International.

ASTM. (2000d). Standard guide for core-sampling submerged, unconsolidated sediments, D4823-95. In: *ASTM Standards on Environmental Sampling*. West Conshohocken, PA: American Society for Testing and Materials International.

ASTM. (2000e). Standard practice for preparation of sediment samples for chemical analysis, D3976-92. In: *ASTM Standards on Environmental Sampling*. West Conshohocken, PA: American Society for Testing and Materials International.

ASTM. (2000f). Standard guide for conducting acute toxicity tests with fishers, macroinvertebrates, and amphipods, E729-96. In: *ASTM Standards on Environmental Sampling*. West Conshohocken, PA: American Society for Testing and Materials International.

ASTM. (2000g). Standard test methods for ammonia nitrogen in water, D1426-93. In: *ASTM Standards on Environmental Sampling*. West Conshohocken, PA: American Society for Testing and Materials International.

AWRI. (2000). *Plankton Sampling*. Allendale, MI: Robert B. Annis Water Resource Institute, Grand Valley State University.

AWWA. (1995). *Water Treatment*, 2nd ed. Denver, CO: American Water Works Association.

Axelman, J., Carina, N., Broman, D., and Kristoffer, N. (1999). Accumulation of polycyclic aromatic hydrocarbons in semipermeable membrane devices and caged mussels (*Mytilus edulis* I.) in relation to water column phase distribution. *Environmental Toxicology and Chemistry*, 18(11): 2454–2461.

Bahls, L.L. (1993). *Periphyton Bioassessment Methods for Montana Streams*. Helena: Montana Water Quality Bureau, Department of Health and Environmental Science.

Barbour, M.T., Garretson, J., Snyder, B.D., and Stabling, J.B. (1997). *Revision to Rapid Bioassessment Protocols for Use in Streams and Rivers: Periphytons, Benthic Macroinvertebrates, and Fish*. Washington, DC: U.S. Environmental Protection Agency.

Baudo, R. (1990). Sediment sampling, mapping and data analysis. In: *Sediments: Chemistry and Toxicity of In-Place Pollutants* (Giesy, J.P. and Muntau, H., Eds.), pp. 15–60. Chelsea, MI: Lewis Publishers.

Bender, M., Martin, W., Hess, J., Sayles, F., Ball, L., and Lambert, C. (1987). A whole-core squeezer for interfacial pore-water sampling. *Limnology and Oceanography*, 32: 1214–1225.

Bennett, E.R., Metcalfe, C.D., and Metcalfe, T.L. (1996). Semi-permeable membrane devices (SPMDs) for monitoring organic contaminants in the Otonabee River, Ontario. *Chemosphere*, 33: 363–375.

Bischoff, J.L., Greet, R.E., and Luistro, A.O. (1970). Composition of interstitial waters of marine sediments: temperature of squeezing effect. *Science*, 167: 1245–1246.

Bly, T.D. and Smith, G.F. (1994). *Biomonitoring Our Streams: What's It All About*? Nashville, TN: U.S. Geological Survey.

Bollinger, R., Bandl, H., Hohener, P., Hanselmann, K.W., and Bachofen, R. (1992). Squeeze-water analysis for the determination of microbial metabolites in lake sediments-comparison of methods. *Limnology and Oceanography*, 37: 448–455.

Botkin, D.B. (1990). *Discordant Harmonies*. New York: Oxford University Press.

Bottomly, E.Z. and Bayly, I.L. (1984). A sediment pore water sampler used in root zone studies of the submerged macrophyte, *Myriophyllum spicatum*. *Limnology and Oceanography*, 29: 671–673.

Bray, J.T., Bricker, O.P., and Troup, B.N. (1973). Phosphate in interstitial waters of anoxic sediments: oxidation effects during sampling procedure. *Science*, 180: 1362–1364.

Brinkman, A.G., van Raaphorst, W., and Lijklema, L. (1982). *In situ* sampling of interstitial water from lake sediments. *Hydrobiologia*, 92: 659–663.

Brumbaugh, W.G., Ingersoll, C.G., Kemble, N.E., May, T.W., and Zajicek, J.L. (1994). Chemical characterization of sediments and porewater from the Upper Clark Fork River and Milltown Reservoir, Montana. *Environmental Toxicology and Chemistry*, 13: 1971–1973.

Bufflap, W.E. and Allen, H.E. (1995). Sediment pore water collection methods: a review. *Water Research*, 29: 165–177.

Burgess, R.M. (1996). Enrichment of marine sediment colloids with polychlorinated biphenyls: trends resulting from PCB solubility and chlorination. *Environmental Science & Technology*, 30(8): 2556–2566.

Burgess, R.M., Schweitzer, K.A., McKinney R.A., and Phelps, D.K. (1993). Contaminated marine sediments: water column and interstitial toxic effects. *Environmental Toxicology and Chemistry*, 12: 127–138.

Burton, Jr., G.A. (1991). Assessment of freshwater sediment toxicity. *Environmental Toxicology and Chemistry*, 10: 1585–1627.

Burton, Jr., G.A., Rowland, C., Lavoie, D., and Nordstrom, N. (1999). *Assessment of In Situ Stressors and Sediment Toxicity in the Lower Housatonic River*, Final Report to R.F. Weston, Manchester, NH.

Burton, Jr., G.A., Rowland C.D., Greenberg M.S., Lavoie, D.R., Nordstrom, J.F., and Eggert, L.M. (2003). A tiered, weight-of-evidence approach for evaluating aquatic ecosystems. In: *Sediment Quality Assessment and Management: Insight and Progress* (Munawar, M., Ed.). The Netherlands: Backhuys Publishers.

Call, D.J., Christine, N., Polkinghorne, T.P., Markee, L.T., Brooke, D.L., Geiger, J.W., Gorsuch, K., and Robillard, N. (1999). Silver toxicity to *Chironomus tentans* in two freshwater sediments. *Environmental Toxicology and Chemistry*, 18(1): 30–39.

Camann, M. (1996). *Freshwater Aquatic Invertebrates: Biomonitoring*. Arcata, CA: Humboldt State University.

Carignan, R. (1984). Interstitial water sampling by dialysis: methodological notes. *Limnology and Oceanography*, 29: 667–670.

Carignan, R. and Lean, D.R.S. (1991). Regeneration of dissolved substances in a seasonally anoxic lake: the relative importance of processes occurring in the water column and in the sediments. *Limnology and Oceanography*, 36: 683–703.

Carignan, R., Rapin, F., and Tessier, A. (1985). Sediment porewater sampling for metal analysis: a comparison of techniques. *Geochimica et Cosmochimica Acta*, 49: 2493–2497.

Carignan, R., St. Pierre, S., and Gachter, R. (1994). Use of diffusion samples in oligotrophic lake sediments: effects of free oxygen in sampler material. *Limnology and Oceanography*, 39: 468–474.

Carins, Jr., J. and Dickson, K.L. (1971). A simple method for the biological assessment of the effects of waste discharges on aquatic bottom-dwelling organisms. *Journal of Water Pollution Control Federation*, 43: 755–772.

Carlton, R.G. and Wetzel, R.G. (1985). A box corer for studying metabolism of epipelic microorganisms in sediment under *in situ* conditions. *Limnology and Oceanography*, 30: 422.

Carr, R.S. (1998). Marine and estuarine porewater toxicity testing. In: *Microscale Testing in Aquatic Toxicology: Advances, Techniques, and Practice* (Wells, P.G., Lee, K., and Blaise, C., Eds.), pp. 523–538. Boca Raton, FL: CRC Press.

Carr, R.S. and Chapman, D.C. (1992). Comparison of solid-phase and pore-water approaches for assessing the quality of marine and estuarine sediments. *Chemical Ecology*, 7: 19–30.

Carr, R.S. and Chapman, D.C. (1995). Comparison of methods for conducting marine and estuarine sediment porewater toxicity tests: extraction, storage, and handling techniques. *Archives of Environmental Contamination and Toxicology*, 28: 69–77.

Carr, R.S., Williams, J.W., and Fragata, C.T.B. (1989). Development and evaluation of a novel marine sediment porewater toxicity test with the polychaete *Dinophilus gyrociliatus*. *Environmental Toxicology and Chemistry*, 8: 533–543.

Chin, Y. and Gschwend, P.M. (1991). The abundance, distribution, and configuration of pore water organic colloids in recent sediment. *Geochimica et Cosmochimica Acta*, 55: 1309–1317.

Doig, L. and Liber, K. (2000). Dialysis minipeeper for measuring porewater metal concentrations in laboratory sediment toxicity and bioavailability tests. *Environmental Toxicology and Chemistry*, 19: 2882–2889.

Environment Canada. (1994). *Guidance Document on Collection and Preparation of Sediments for Physicochemical Characterization and Biological Testing*, Report EPS I/RM/29. Gatineau, Quebec: Environment Canada, Environmental Protection Services.

Frazier, B.E., Naimo, T.J., and Sandheinrich, M.B. (1996). Temporal and vertical distribution of total ammonia nitrogen and un-ionized ammonia nitrogen in sediment porewater from the upper Mississippi River. *Environmental Toxicology and Chemistry*, 15: 92–99.

Froelich, P.M., Klinkhammer, G.P., Bender, M.L., Luedtke, N.A., Heath, G.R. et al. (1979). Early oxidation of organic matter in pelagic sediments of the eastern equatorial Atlantic: suboxic diagenesis. *Geochimica et Cosmochimica Acta*, 43: 1075–1090.

GLNPO. (1994). *Assessment and Remediation of Contaminated Sediments (ARCS) Program*, EPA 905-B94-002. Chicago, IL: U.S. Environmental Protection Agency, Great Lakes National Program Office.

Golterman, H.L., Sly P.G., and Thomas, R.L. (1983). *Study of the Relationship between Water Quality and Sediment Transport*. Mayenne, France: UNESCO.

Hesslein, R.H. (1976). An *in-situ* sampler for close internal pore water studies. *Limnology and Oceanography*, 21: 912–914.

Hill, B.H. (1997). The use of periphyton assemblage data in an index of biotic integrity. *Bulletin of the North American Ethological Society*, 14: 158.

Ho, K.T., McKinney, R., Kuhn, A., Pelletier, M., and Burgess, R. (1997). Identification of acute toxicants in new Bedford Harbor sediments. *Environmental Toxicology and Chemistry*, 16(3): 551–558.

Hoffman, E. (1998). *Tributyltin Analysis: Clarification of Interstitial Water Extraction and Analysis Methods—Interim*, EPA 206-553-0038. Seattle, WA: U.S. Environmental Protection Agency, Dredged Material Management Program.

Howes, B.L., Dacey, J.W.H., and Wakeham, S.G. (1985). Effects of sampling technique on measurements of porewater constituents in salt marsh sediments. *Limnology and Oceanography*, 30: 221–227.

Huckins, J.N., Ruvwefwn, M.Q., and Mnuqwwe, F.K. (1990). Semipermeable membrane devices containing model lipid: a new approach to monitoring the bioavailability of lipophilic contaminants and estimating their bioconcentration potential. *Chemosphere*, 20: 533–552.

Huff, W.R. (1993). Biological indices define water quality standard. *Water Environment & Technology*, 5: 21–22.

Hulbert, M.H. and Brindle, M.P. (1975). Effects of sample handling on the composition of marine sedimentary pore water. *Geological Society of America Bulletin*, 86(1): 109–110.

Jahnke, R.A. (1988). A simple, reliable, and inexpensive pore-water sample. *Limnology and Oceanography*, 33: 483–487.

Jenne, E.A. and Zachara, J.M. (1984). *Factors Influencing the Sorption of Metals: Fate and Effects of Sediment-Bound Chemicals in Aquatic Systems*. New York: Pergamon Press.

Kalil, E.K. and Goldhaker, M. (1973). A sediment squeezer for removal of pore waters without air contact. *Journal of Sediment Petrology*, 43: 554–557.

Karr, J.R. (1981). Assessment of biotic integrity using fish communities. *Fisheries*, 66: 21–27.

Karr, J.R., Bausch, K.D., Hagermeier, P.L., Yant, P.R., and Schlosser, I.J. (1986). *Assessing Biological Integrity in Running Waters: A Method and Its Rationale*, Special Publ. 5. Champaign: Illinois Natural History Survey.

KDOW. (1993). *Methods for Assessing Biological Integrity of Surface Waters*. Frankfort: Kentucky Department of Environmental Protection, Division of Water.

Kemble, N.E., Dwyer, F.J., Ingersoll, C.G., Dawson, T.D., and Norberg-King, T.J. (1999). Tolerance of freshwater test organisms to formulated sediments for use as control materials in whole-sediment toxicity tests. *Environmental Toxicology and Chemistry*, 18: 222–230.

Kittrell, F.W. (1969). *A Practical Guide to Water Quality Studies of Streams*. Washington, DC: U.S. Department of Interior.

Klemm, D.J., Lewis, P.A., Fulk, F., and Lazorchak, J.M. (1990). *Macroinvertebrate Field and Laboratory Methods for Evaluating the Biological Integrity of Surface Waters*, EPA-600-4-90-030. Cincinnati, OH: U.S. Environmental Protection Agency, Environmental Monitoring and Support Laboratory.

Knezovich, J.P. and Harrison, F.L. (1987). A new method for determine the concentration of volatile organic compounds in sediment interstitial water. *Bulletin of Environmental Contamination and Toxicology*, 38: 937–940.

Knezovich, J.P. and Harrison, F.L. (1988). The bioavailability of sediment-sorbed chlorobenzenes to larvae of the midge *Chironomus decorus*. *Ecotoxicology and Environmental Safety*, 15: 226–241.

Kriukov, P.A. and Manheim, F.T. (1982). Extraction and investigative techniques for study of interstitial waters of unconsolidated sediments: a review. In: *The Dynamic Environment of the Ocean Floor* (Fanning, K.A. and Manheim, F.T., Eds.), pp. 3–26. Washington, DC: Lexington Books.

Long, E.R., Buchman, M.F., Bay, S.M., Breteler, R.H., Carr, R.S. et al. (1990). Comparative evaluation of five toxicity tests with sediments from San Francisco Bay and Tomales Bay, California. *Environmental Toxicology and Chemistry*, 9: 1193–1214.

Lyons, W.B., Gaudette, J., and Smith, G. (1979). Porewater sampling in anoxic carbonate sediments: oxidation artifacts. *Nature*, 277: 48–49.

Mangelsdorf, P.C. and Wilson, T.R.S. (1969). Potassium enrichments in interstitial waters of recent marine sediments. *Science*, 165: 171.

Mayer, L.M. (1976). Chemical water sampling in lakes and sediments with dialysis bags. *Limnology and Oceanography*, 21: 909.

Mudroch, A. and Azcue, J.M. (1995). *Manual of Aquatic Sediment Sampling*. Boca Raton, FL: Lewis Publishers.

Mudroch, A. and MacKnight, S.D. (1994). *CRC Handbook of Techniques for Aquatic Sediment Sampling*, 2nd ed. Boca Raton, FL: CRC Press.

O'Toole, C., Ed. (1986). *The Encyclopedia of Insects*. New York: Facts on File.

Ohio EPA. (2001). *Ohio EPA Sediment Sampling Guide and Methodologies*. Columbus: State of Ohio Environmental Protection Agency.

Patrick, R. (1973). Use of algae, especially diatoms, in the assessment of water quality. In: *Biological Methods for the Assessment of Water Quality*, ASTM STP 528 (Cairns, Jr., J. and Dickso, K.L., Eds.), pp. 76–95. Philadelphia, PA: American Society for Testing and Materials International.

Plumb, R.H. (1981). *Procedures for Handling and Chemical Analysis of Sediment and Water Samples*. Buffalo, NY: State University College at Buffalo, Great Lakes Laboratory.

Puget Sound Estuary Program. (1997). *Recommended Guidelines for Sampling Marine Sediment, Water Column, and Tissue in Puget Sound*. Olympia, WA: Puget Sound Estuary Program, Puget Sound Water Quality Action Team.

Reeburgh, W.S. (1967). An improved interstitial water sampler. *Limnology and Oceanography*, 12: 163–165.

Rodgers, Jr., J.H., Dickson, K.L., and Cairns, Jr., J. (1979). A review and analysis of some methods used to measure functional aspects of periphyton. In: *Methods and Measurements of Periphyton Communities: A Review*, ASTM STP 690 (Weitzel, R.L., Ed.), pp. 142–167. West Conshohocken, PA: American Society for Testing and Materials International.

Rosenfeld, J.K. (1979). Ammonia absorption in nearshore anoxic sediments. *Limnology and Oceanography*, 24: 356–364.

Saager, P.M., Sweerts, J.-P., and Ellermeijer, H.J. (1990). A simple pore-water sampler for coarse, sandy sediments of low porosity. *Limnology and Oceanography*, 35: 747–751.

Santschi, P.H., Lenhart, J., and Honeyman, B.D. (1997). Heterogeneous processes affecting trace contaminant distribution in estuaries: the role of natural organic matter. *Marine Chemistry*, 58: 99–125.

Sarda, N. and Burton, Jr., G.A. (1995). Ammonia variation in sediments: spatial, temporal and method-related effects. *Environmental Toxicology and Chemistry*, 14: 1499–1506.

Sasson-Brickson, G. and Burton, Jr., G.A. (1991). *In situ* and laboratory sediment toxicity testing with *Ceriodaphnia dubia*. *Environmental Toxicology and Chemistry*, 10: 201–207.

Sayles, F.L., Wilson, T.R.S., Hume, D.N., and Mangelsdorf, Jr., P.C. (1973). *In situ* sampler for marine sedimentary porewaters: evidence for potassium depletion and calcium enrichment. *Science*, 180: 154–156.

Schults, D.W., Ferraro, S.P., Smith, L.M., Roberts F.A., and Poindexter, C.K. (1992). A comparison of methods for collection interstitial water for trace organic compounds and metals analyses. *Water Research*, 26: 989–995.

SETAC. (2001). *Summary of a SETAC Technical Workshop: Porewater Toxicity Testing: Biological, Chemical, and Ecological Considerations with a Review of Methods and Applications, and Recommendations for Future Areas of Research*. Pensacola, FL: Society of Environmental Toxicology and Chemistry.

Simon, N.S., Kennedy, M.M., and Massoni, C.S. (1985). Evaluation and use of a diffusion controlled sampler for determining chemical and dissolved oxygen gradients at the sediment water interface. *Hydrobiologia*, 126: 135–141.

Skalski, C. and Burton, G.A. (1991). Laboratory and *In Situ* Sediment Toxicity Evaluations Using Early Life Stages of *Pimephales promelas*, master's thesis, Wright State University, Dayton, OH.

Spellman, F.R. (2015). *The Science of Water: Concepts and Applications*, 3rd ed. Boca Raton, FL: CRC Press.

Stevenson, R.J. (1996). An introduction to algal ecology in freshwater benthic habitats. In: *Algal Ecology: Freshwater Benthic Ecosystems* (Stevenson, R.J., Bothwell, M., and Lowe, R.L., Eds.), pp. 3–30. San Diego, CA: Academic Press.

Stevenson, R.J. (1998). Diatom indicators of stream and wetland stressors in a risk management framework. *Environmental Monitoring and Assessment*, 51: 107–108.

Stevenson, R.J. and Pan, Y. (1999). Assessing ecological conditions in rivers and streams with diatoms. In: *The Diatoms: Application to the Environmental and Earth Sciences* (Stoermer, E.G. and Smol, J.P., Eds.), pp. 11–40. Cambridge, U.K.: Cambridge University Press.

Steward, A.R. and Malley, D.F. (1999). Effect of metal mixture (Cu, Zn, Pb, and Ni) on cadmium partitioning in littoral sediments and its accumulation by the freshwater macrophyte *Eriocailon septangilare. Environmental Toxicology and Chemistry*, 18(3): 436–447.

Tchobanoglous, G. and Schroeder, E.D. (1985). *Water Quality*. Reading, MA: Addison-Wesley.

Troup, B.N., Bricker, O.P., and Bray, J.T. (1974). Oxidation effect on the analysis of iron in the interstitial water of recent anoxic sediments. *Nature*, 249: 237–239.

USEPA. (1983). *Technical Support Manual: Waterbody Surveys and Assessments for Conducting Use Attainability Analyses*, EPA 440486037. Washington, DC: U.S. Environmental Protection Agency.

USEPA. (1991). *Methods for Measuring the Acute Toxicity of Effluents and Receiving Waters to Freshwater and Marine Organisms*, 4th ed., EPA-600/4-90/027F. Cincinnati, OH: U.S. Environmental Protection Agency.

USEPA. (2001). *Methods for Collection, Storage and Manipulation of Sediments for Chemical and Toxicological Analyses: Technical Manual*, EPA-823-B-01-002. Washington, DC: U.S. Environmental Protection Agency, Office of Water.

USEPA/USACE. (1998). *Evaluation of Dredged Material Proposed for Discharge in Waters of the U.S.: Testing Manual*, EPA-823-B-98-004. Washington, DC: U.S. Environmental Protection Agency, Office of Water.

Velz, C.J. (1970). *Applied Stream Sanitation*. New York: Wiley Interscience.

Warren, Jr., M.L. and Burr, B.M. (1994). Status of freshwater fishes of the United States: overview of an imperiled fauna. *Fisheries* 19(1): 6–18.

Watson, P.G. and Frickers, T.E. (1990). A multilevel, *in situ* pore-water sample for use in intertidal sediments and laboratory microcosms. *Limnology and Oceanography*, 35: 1381–1389.

Watson, P.G., Frickers, P., and Goodchild, C. (1985). Spatial and seasonal variations in the chemistry of sediment interstitial waters in the Tamar estuary. *Estuarine, Coastal and Shelf Science*, 21: 105–119.

Weitzel, R.L. (1979). Periphyton measurements and applications. In: *Methods and Measurements of Periphyton Communities: A Review*, ASTM STP 690 (Weitzel, R.L., Ed.), pp. 3–33. West Conshohocken, PA: American Society for Testing and Materials International.

Whitman, R.L. and Nevers, M.B. (2003). Foreshore sand as a source of *Escherichia coli* in nearshore water of a Lake Michigan beach. *Applied and Environmental Microbiology*, 69: 5555–5562.

Winger, P.V. and Lassier, P.J. (1991). A vacuum-operated porewater extractor for estuarine and freshwater sediments. *Archives of Environmental Contamination and Toxicology*, 21: 321–324.

Word, J.Q., Ward, J.A., Franklin, L.M., Cullinan, V.I., and Kiesser, S.L. (1987). *Evaluation of the Equilibrium Partitioning Theory for Estimating the Toxicity of the Nonpolar Organic Compound DDT to the Sediment Dwelling Amphipod Rheopoxynius abronius*. Sequim, WA: Battelle Marine Research Laboratory.

14 Sediment Physicochemical Characteristics

INTRODUCTION

After an environmental practitioner devises a sediment collection plan, determines the sediment collection location, collects sediments, properly packages the collected samples, and labels the sample package contents, the samples are carefully transported to an examination and testing laboratory. Such a laboratory is typically tasked with determining certain physicochemical characteristics of the sediments, in conjunction with toxicity testing or chemical analysis or inorganic or organic contaminants. This characterization should include measurement of certain parameters known to mediate the availability of contaminants in sediment (ASTM, 2000f). Bulk chemical concentrations alone should not be used to evaluate bioavailability (USEPA, 1998). The following parameters are generally measured:

- pH (porewater)
- Ammonia (porewater)
- Total organic carbon
- Particle size distribution (e.g., percent sand, silt, and clay)
- Percent water content
- Salinity and hardness of porewater
- Conductivity of porewater

Depending on the experimental design or study objectives, more extensive characterization may be necessary. Several additional characteristics that may assist in study implementation, data interpretation, or quality assurance/quality control (QA/QC) (e.g., assessing sediment integrity, artifact production, optimal extraction, test procedures) include the following:

- Sediment biochemical oxygen demand (BOD)
- Sediment chemical oxygen demand (COD)
- Sediment oxygen demand (SOD)
- Cation exchange capacity (CEC)
- Redox (Eh) or oxidation-reduction potential (ORP)
- Total inorganic carbon

- Total volatile solids
- Acid-volatile sulfides (AVSs)
- Simultaneously extracted metals (SEMs)
- Metals
- Synthetic organic compounds (pesticides, PCBs, PAHs, TCDD-dioxin)
- Oil and grease
- Petroleum hydrocarbons
- Dissolved organic carbon (DOC) in the porewater

Note that many of the measurements of many sediment physicochemical characteristics use analytical techniques originally developed for soils and waters, and the literature should be consulted for details regarding the recommended methodology (Guy, 1969; Page et al., 1982; Plumb, 1981). The following sections provide the rationale for making each type of sediment physicochemical measurement, as well as a brief description of the measurement techniques for specific procedures.

SEDIMENT PHYSICOCHEMICAL PARAMETERS AND MEASUREMENT

pH

Sediment pH is often one of the single most important factors controlling speciation and equilibria for many chemicals, including sulfides, ammonia, cyanide, and metals, all of which ionize under the influence of pH. The U.S. Environmental Protection Agency (USEPA) ammonia water-quality criterion, for example, is dependent in part on pH because ammonia toxicity is largely governed by the un-ionized ammonia fraction, which is pH dependent (USEPA, 1999). Metal (Cd, Cu, Ni, Pb, and Zn) speciation and bioavailability are also known to be affected by pH (Ho et al., 1999; Schubauer-Berigan and Ankley et al., 1991).

Generally, pH is measured using a pH meter consisting of a potentiometer, a glass electrode, a reference electrode, and a temperature compensating device. A circuit is completed through the potentiometer when the electrodes are submersed. General-purpose process pH electrodes are available in a wide variety of configurations for inline and submersion applications. Generally, electrodes with gel-filled references require less maintenance than electrodes with liquid-filled references. The latest instruments have microprocessors that automatically calculate and display the slope. Some older instruments have a percent-slope readout or millivolt readout. For instruments with a millivolt readout, the measured electrode potential is calculated as the difference between millivolts measured at the known pH of two buffers.

Plumb (1981) and Gonzalez (1996) described a method for measuring pH in sediment using a pH probe and meter. The probe was inserted into the sediment and pH directly measured after at least a 5-minute equilibration time. Electrodes have also been used for direct measurements of pH in sediment porewater or in a 1:1 mixture of sediment to water (Jackson, 1958). Direct measurement of sediment pH is also possible using electrodes with spear-up designs allowing for greater penetration into the sample (USEPA, 2001).

Ammonia in Porewater

Nitrogen, a nutrient associated with over-enrichment of aquatic environments, exists in several forms, including ammonia. Ammonia is highly soluble in water, where it is found in an un-ionized form as NH_3 and in an ionized form as NH_4^+. The extent of ionization is dependent on pH, temperature, and salinity (in seawater). Ammonia in sediments and porewater is generally the result of microbial degradation of nitrogenous organic material such as amino acids (Ankely et al., 1990). Porewater concentrations of ammonia as high as 50 mg/L have been measured in otherwise uncontaminated sediments (Kristensen and Blackburn, 1987; Murray et al., 1978), while ammonia in porewaters from contaminated sediments can range from 50 to more than 200 mg/L (Ankley et al., 1990; Schubauer-Berigan and Ankley, 1991).

The toxic effects of ammonia are generally considered to be associated with the un-ionized fraction (NH_3) rather than the ionic components (HN_4^+ and $NH_4SO_4^-$), which coexist in equilibria. This equilibrium is highly dependent on pH, temperature, pressure, salinity, and ionic concentrations of ammonia. The toxic un-ionized ammonia fraction can be calculated using known total ammonia values and measurements of pH, pressure, salinity, and temperature (Thurston et al., 1981; Whitfield, 1978).

Methods available to measure ammonia in the porewater include the following (APHA, 1995; USEPA, 1983):

1. Titrimetric method
2. Ammonia-selective electrode method
3. Ammonia-selective electrode method using known addition
4. Phenate method
5. Automated phenate method

A preliminary distillation step may be required if interferences (e.g., sample constituents that interact with procedural reagents) are present (APHA, 1995; ASTM, 2000g). Once distilled, the sample can be analyzed using any of the methods listed above.

The distillation and titration methods are frequently used when ammonia concentrations are greater than 5.0 mg/L. The ammonia-selective electrode method is appropriate when concentrations range between 0.03 and 1400 mg NH_3-N/L. Ammonia readings are calibrated against ammonia standards. To verify meter readings, confirmatory subsamples can be preserved and analyzed for ammonia using the standard Nessler technique described in APHA (1995). For the phenate method, distillation with sulfuric acid is recommended when interferences are present (APHA, 1995). The automated phenate method is suitable for porewaters with ammonia concentrations in the range of 0.02 and 2.0 mg NH_3-N/L.

Total Organic Carbon Content

The total organic carbon (TOC) content of sediment is a measure of the total amount of oxidizable organic material. TOC is the sum of dissolved organic carbon (DOC), particulate organic carbon (POC) or suspended organic carbon (SOC), and colloids.

TOC is an important parameter to measure in sediments because it is a major determinant of non-ionic organic chemical bioavailability (DiToro et al., 1991). Metal bioavailability is also affected by the amount of TOC present in sediments. TOC is usually expressed as a percentage of the bulk sediment and is used to normalize the dry-weight sediment concentration of a chemical to the organic carbon content of the sediment. The USEPA equilibrium partitioning sediment guidelines estimate bioavailability as a function of contaminant concentration sorbed to sediment organic carbon and contaminant concentration in the porewater under equilibrium conditions (USEPA, 1998). Recently, the presence of soot carbon from the combustion of organic carbon (e.g., fossil fuels) has been recognized as a fraction of the TOC in sediment. Soot carbon may alter the geochemistry and bioavailability of some organic contaminants (Gustuffson et al., 1997).

The organic carbon content of sediments has been measured using several methods, including wet oxidation titration, modified titration, and combustion after removal of carbonate by the addition of HCL and subsequent drying. USEPA methods (USEPA, 1986b, 1987), including SW-846 and 430/0-86-004, are often used to measure TOC. One of two methods may be used to separate organic from inorganic carbon before analyzing for TOC (Plumb, 1981): (1) ignition and using HCl as the acid for pretreating sediment, or (2) differential combustion, which uses thermal combustion to separate the two forms of carbon. Total organic carbon analyses should be based on high-temperature combustion rather than on chemical oxidation (USEPA/USACE, 1998), because some classes of organic compounds are not fully degraded by combined chemical and ultraviolet oxidation techniques. Inorganic carbon (e.g., carbonates and bicarbonates) can be a significant proportion of the total carbon in some sediments. Therefore, samples could be treated with acid to remove the inorganic carbon prior to TOC analysis. The procedure described by the Puget Sound Estuary Program (1987a) is recommended for TOC analysis because this method uses high-temperature combustion using an induction furnace. The USEPA recommends a similar method using catalytic combustion and non-dispersive infrared detection (Leonard, 1991) for quantifying TOC.

Particle Size Distribution (Percent Sand, Silt, and Clay)

Particle size is used to characterize the physical characteristics of sediments. Because particle size influences both chemical and biological characteristics, it can be used to normalize chemical concentrations and account for some of the variability found in biological assemblages (USEPA, 1998) or in laboratory toxicity testing (Hoss et al., 1999; USEPA, 2000). Particle size can be characterized in varying detail. The broadest divisions that generally are considered useful for characterizing particle size distributions are percentages of gravel, sand, silt, and clay; however, each of these size fractions can be subdivided further so that additional characteristics of the size distribution are determined (Puget Sound Estuary Program, 1996). Particle size determinations can either include or exclude organic material. If organic material is removed prior to analysis, the *true* (i.e., primarily inorganic) particle size distribution is determined. If organic material is included in the analysis, the *apparent* (i.e., organic plus inorganic) particle size distribution is determined. Because true and apparent distributions may differ, detailed comparisons between samples analyzed by these different methods are questionable. Therefore, if comparisons among samples between studies is desired, sediment particle size should be measured using constituent methods (Puget Sound Estuary Program, 1996).

Sediment particle size can be measured by a number of different methods (Allen, 1975; ASTM, 2000a; Plumb, 1981; Puget Sound Estuary Program, 1996). The best method will depend on the particle properties of the sample (Singer et al., 1988). Particle size distribution is often determined by wet sieving the sample (Plumb, 1981; Puget Sound Estuary Program, 1996; Singer et al., 1988; USEPA, 1979), the hydrometer method (Day, 1965; Patrick, 1958), the pipet method (Guy, 1969; Rukavina and Duncan, 1970), settling techniques (Sanford and Swift, 1971), or x-ray absorption (Duncan and Lattaie, 1979; Rukavina and Duncan, 1970). The pipet method may be superior to the hydrometer method (Sternberg and Creager, 1961). Combinations of multiple methods may provide refined measurements of particle size distribution. Gee and Bauder (1986) used sieving and pipetting after soluble salts were removed. Gonzalez (1996) used a combination of sieve and hydrometer methods.

Recommended methods for measuring sediment particle size distribution are those of the Puget Sound Estuary Program (1996) and the USEPA (1995). Percent gravel, sand, silt, and clay are determined as apparent distribution using a minimum sediment sample size of 100 g taken from a homogenized sediment sample. Organic matter should be removed prior to analysis by oxidation using hydrogen peroxide. Wet sieving followed by dry sieving (mechanical shaking) separates the two coarse particle size groups. The silt–clay fraction is subdivided using a pipet technique that depends on the differential settling rates of the two different particle size fractions. All fractions are dried to a constant weight. Cooled samples are stored in a desiccator and weighed.

To obtain an accurate determination of particle sizes for the fine fraction, the Coulter (particle size) counter method may be employed (McCabe and Jarvis, 1973; Vanderpleog, 1981). This method gives the fraction of particles with an apparent spherical diameter. In a review of the available methods, Swift et al. (1972) found the Coulter counter method to be the most versatile method overall; however, it does not provide selling information. Another potential method for determining the particle size distribution of a very fine fraction is through the use of electron microscopy (Leppard et al., 1988). Collection techniques for very fine material can result in aggregation of larger colloidal structures (Leppard, 1986; Leppard et al., 1988). In general, particle settling methods are preferred to sediment sizing methods.

Percent Water or Moisture Content

Water content is a measurement of sediment moisture usually expressed as a percentage of the whole sediment weight. It is known to influence toxicity and is used to aid in the interpretation of sediment quality investigations. Sediment moisture content is measured as the difference between wet weight of the sediment and dry weight following oven drying at 50 to 105°C to a constant weight. Percent water is used to convert sediment concentrations of substances from wet weight to a dry weight. Methods for determining moisture content were described by Plumb (1981) and Vecchi (1999). Additional methods are provided in USEPA (1987).

Salinity of the Porewater (Marine Sediments)

Salinity is a measure of the mass of dissolved salt in a given mass of solution. The most reliable method to determine the true or absolute salinity is by complete chemical analysis; however, this is time consuming and costly. Therefore, indirect methods are more suitable. Indirect methods include conductivity, density, sound speed, or refractive index (APHA, 1995). Salinity is then calculated from the empirical relationship between salinity and the indirect measurement. Conductivity measurements have the greatest precision but respond only to ionic solutes (APHA, 1995). Density measurements respond to all solutes. APHA (1995) recommended the electrical conductivity method, because it is sensitive and easily performed, and the density method using a vibrating flow densitometer. USEPA (1986b) methods should also be consulted. A salinity refractometer can be used for quick readings of salt density in solutions such as seawater. These refractometers are easy to read, noncorrosive, and lightweight. They have dual scales and an adjustable focus. Temperature and non-temperature compensating refractometers are available. Most refractometers are accurate to 1 ppt and read specific gravity (1.000 to 1.070 in 0.001 divisions) and parts per thousand (0 to 100 in 1 ppt divisions).

Conductivity of the Porewater (Freshwater Sediments)

Conductivity is a measure of the ability of an aqueous solution to carry an electric current. This ability is dependent on the presence of ions in the solution, the concentration of the ions, their mobility and valance, and temperature. Solutions of inorganic compounds are usually good conductors, whereas those of organic compounds are usually poor conductors. Conductivity is enhanced by calcium, potassium, sodium, and magnesium chlorides and sulfides. Meters can be used to measure the degree to which electrical current can travel through water. The unit of measure is 1 mS/m (millisiemens/meter) or 1 μS/cm (microsiemens/centimeter). The reading indicates the amount of ions in the water. Although traditional chemical tests for hardness measure calcium and magnesium, they fail to provide an indication of other ions (e.g., sodium). The conductivity meter provides a much better measure of ionic strength.

Acid-Volatile Sulfides

Measurement of acid-volatile sulfides (AVSs) and simultaneously extracted metal (SEM) concentrations associated with AVS extraction can provide insight into the bioavailability of metals in anaerobic (anoxic) sediments (Ankley et al., 1996; DiToro et al., 1990). AVSs are the reactive solid-phase sulfide fraction that is extracted by cold hydrochloric acid. AVSs appear to affect the bioavailability of most divalent metal ions, as the sulfide ions have a high affinity for divalent metals. This affinity results in the formation of insoluble metal sulfides with greatly reduced bioavailability. AVS concentrations in freshwater and marine sediments can range between <0.1 and >50 μmol AVS per gram of sediment (DiToro et al., 1990). The bioavailability of metals in sediments has been predicted by comparing the molar concentration of

AVSs to the molar concentration of SEMs (methods described below). If the concentration of AVSs is greater than that of SEMs, the metals are bound in sulfide complexes with greatly limited bioavailability. However, if the concentration of AVSs is less than that of SEMs, metals may or may not be toxic due to other controlling factors (e.g., TOC). The easily extractable sulfide fraction can be measured using the acid purge and trap technique. The sample sulfide is solubilized in cold hydrochloric acid. The analytical method involves conversion of sulfides to aqueous H_2S. This may be measured with a sulfide probe or by following a wet chemistry method. In the latter method, silver sulfide is precipitated in a gas-tight assembly and flushed with nitrogen to eliminate oxidation. The precipitation is filtered, dried, and weighed. The weight is compared with the weight obtained from a non-acidified sample, and the difference is attributed to the AVS fraction (DiToro et al., 1990).

Simultaneously Extracted Metals

A model for predicting toxicity from divalent trace metals (DiToro et al., 1990) is based on the binding of these metals to AVSs. Where the sum of the moles of the SEMs, including Ag, Cd. Cu, Ni, Pb, and Zn, is exceeded by the molar concentration of AVS, the metals are insoluble and largely unavailable to biota. The extraction of AVS and metals should be achieved using a single methodology to ensure that recoveries associated with each measure are consistent. Simultaneous extraction improves the efficiency of the methodology. SEM can be measured in filtered aliquots by atomic absorption methods (DiToro et al., 1990). Recent SEM analysis methods use inductively coupled plasma atomic emission spectrometry (Berry et al., 1999).

Metals

Low levels of trace metals occur naturally in the environment, but highly elevated levels in sediment are generally associated with human-impacted contaminant loads. Metals are partitioned in sediments as soluble free ions, soluble organic and inorganic complexes, easily exchangeable ions, precipitates of metal hydroxides, soluble organic and inorganic complexes, easily exchangeable ions, precipitates organic complexes, insoluble sulfides, and residual forms (Gambrell et al., 1976).

Current instrument methods available for the analysis of trace metals include electrochemistry (e.g., differential pulse polarography), spectrophotometry (e.g., silver diethyldithiocarbamate), atomic absorption spectrophotometry, atomic emission spectrophotometry (AES), x-ray fluorescence (XRF), and neutron activation; the most commonly used instrumental method to analyze sediments for metals is atomic absorption spectrophotometry (Puget Sound Estuary Program, 1987c). Inductively coupled plasma mass spectrometry (ICP-MS) or ICP-AES allow for simultaneous determination of many metals at sub-part-per-billion levels with little pretreatment (Berry et al., 1999; Crecelius et al., 1987).

The concentration of salt in marine or estuarine samples may interfere with metals analyses (USEPA/USACE, 1998); therefore, acid digestion and atomic absorption spectroscopy should be coupled with an appropriate technique to control for this interference. Methods in USEPA (1986b) are recommended for the analysis of mercury in sediments, and EPRI (1986) methods are recommended for the analysis of selenium and arsenic. USEPA methods for cadmium, hexavalent chromium, copper, lead, mercury, nickel, selenium, silver, and zinc are described in USEPA (1986b). The Puget Sound Estuary Program (1987c) suggests that mercury can be extracted using vacuum distillation and analyzed by gas chromatography/mass spectrophotometry.

Synthetic Organic Compounds (Pesticides, PCBs, TCDD-Dioxin)

Analytical techniques for measuring organic compounds require five general steps: (1) drying the sample, (2) extraction, (3) drying the extract, (4) clean up of the extract, and (5) analysis of the extract. The Puget Sound Estuary Program (1987b) recommends centrifugation or sodium sulfate to dry the sample and a solvent extraction, with application of shaker/roller, or sonication (i.e., applying sound energy). Sample drying with sodium sulfate is recommended for samples weighing approximately 10 grams (after overlying water is decanted). The sediment and sulfate mixture is extracted and the extract is processed (MacLeod et al., 1985). Soxhlet® extraction (USEPA, 1986b) involves distillation with a solvent such as acetone, dichloromethane/methanol (2:1), dichloromethane/methanol (9:10), or benzene/methanol (3:2). Sonication with solvent mixtures and a 30-gram subsample of sediment has been recommended (USEPA, 1983).

Drying the extract can be accomplished through separatory funnel partitioning as needed to remove water and sodium sulfate or by using a Kuderna–Danish apparatus and rotary evaporation with purified nitrogen gas for concentration to smaller volumes (Puget Sound Estuary Program, 1987c). Using the separatory funnel partitioning method, the wet sample is mixed with methanol and centrifuged. The supernatant is decanted and extracted later. Extraction of the sample is continued using less polar solvents and the water/methanol and solvent extracts are combined and dried.

Elemental sulfur can be removed from the sediment sample with vigorous mechanical agitation using a Vortex or Genie® or using activated copper (Puget Sound Estuary Program, 1987c). Organic interferences can be removed with gel permeation chromatography (GPC) (USEPA, 1983), bonded octadecyl columns (Puget Sound Estuary Program, 1987c), high-performance liquid chromatography (HPLC) (Metro, 1981), silica gel (Puget Sound Estuary Program, 1987c), or alumina (USEPA, 1983). Instrumental analyses for volatiles and semivolatiles and for pesticides/PCBs are performed using gas chromatography/mass spectrophotometry (GC/MS) and gas chromatography/electron capture detection (GC/ECD), respectively (Burgess and McKinney, 1996; Puget Sound Estuary Program, 1987b).

Oil and Grease

Oil and grease tests for sediments measure material recovered that is soluble in a nonpolar solvent under acidic conditions. Oil and grease compounds are substances such as hydrocarbons, vegetable oils, animal fats, waxes, soaps, and greases. Many solvents can dissolve other substances (e.g., sulfur compounds, organic dyes, chlorophyll). Therefore, oil and grease are operationally defined by the solvent used and the analytical method used to perform the analysis. There are two basic methods used to analyze oil and grease: the gravimetric technique and the infrared (IR) spectrophotometer technique (Puget Sound Estuary Program, 1996).

Petroleum Hydrocarbons and PAHs

Petroleum hydrocarbons are oil and grease constituents that remain in solution after contact with silica gel. Petroleum distillates, also called hydrocarbons or petrochemicals, refer to a broad range of compounds that are extracted by distillation during the refining of crude oil. During the fractional distillation of petroleum, crude oil is heated to allow various compounds to turn from liquid into gas and is then captured as they rise, cool, and condense. Lighter, more volatile compounds rise higher before they condense and are collected on distillation trays. Heavier, less volatile compounds such as diesel fuel and oil are collected on lower distillation trays. Waxes and asphalts are collected from the bottom after the other products have volatized.

Petroleum distillates contain both aromatic hydrocarbons (carbon rings) and aliphatic hydrocarbons (straight carbon chains). The chemical structure of the hydrocarbon largely defines the nature and behavior of these compounds. Most aromatic hydrocarbons are chronic toxins and known carcinogens. Aliphatic compounds are found in all crude oils and most petroleum products. Many aromatic hydrocarbons have a pleasant odor and include such substances as naphthalene, xylene, toluene, and benzene. Aliphatic hydrocarbons are flammable and may be explosively flammable. Aliphatic hydrocarbons include methane, propane, and kerosene.

Aromatic and aliphatic hydrocarbons were analyzed in sediments by Page et al. (1995a,b). Sediment samples were spiked with the appropriate surrogates, mixed with equal amounts of sodium sulfate to dry the samples, and extracted with a methylene chloride–acetone mixture (Method 3550, USEPA, 1986b). The concentrated extracts were partitioned on an alumina column into saturated and unsaturated hydrocarbon fractions (Method 3611, USEPA, 1986b). The fractions were concentrated using the appropriate preinjection volume, spiked with the appropriate internal standards, and analyzed by gas chromatography with flame ionization detection (GC/FID) and gas chromatography with mass spectrometry (GC/MS) detection operating in the selected ion monitoring (SIM) mode. The method of internal standards (Method 8000, USEPA, 1986b) using the average relative response factors generated from the linear initial calibration was used to quantify the target compounds. All data were corrected for the recovery of the appropriate surrogate compound. Their relative abundances could then be used for identification and quantification purposes.

Total petroleum hydrocarbons (TPHs) and polycyclic aromatic hydrocarbons (PAHs) have also been analyzed by first acidifying the sample with concentrated hydrochloric acid and then extracting hydrocarbons with a mixture of methanol and hexane. The hexane extracts were then spiked with an internal standard and analyzed by GC/FID for TPH content and by GC/MS for PAH analysis. Kaplan et al. (1996) extracted hydrocarbons using anhydrous Na_2SO_4 with methylene chloride and sonication. The total solvent extract was then concentrated with Kuderna–Danish equipment. The concentrate was further concentrated using a gentle stream of dry nitrogen. An aliquot was then injected directly into the gas chromatography.

Total Sulfides

Total sulfides represent the combined amount of acid-soluble H_2S, HS^-, and S^{2-} in a sample. Sulfides are often measured because they are common in some sediments, particularly those that are anoxic, and they can be toxic to aquatic organisms. The Puget Sound Estuary Program (1996) described a method to measure total sulfides in sediments. Oxygen is removed from the sample using nitrogen gas, methyl orange and hydrochloric acid are added, and the mixture is heated. Amine solution and iron chloride are added to develop a colorimetric reaction product, and sample absorbance is measured spectrophotometrically. Methods for measuring sulfides in aqueous samples include potentiometric methods described by ASTM (2000e) and APHA (Method 4500, 1995). Sulfide ions are measured using a sulfide ion-selective electrode in conjunction with a double-junction, sleeve-type reference electrode (Phillips et al., 1997). Potentials are read using a pH meter or a specific ion meter having a direct concentration scale for the sulfide ion. Samples are treated with a sulfide antioxidant buffer, which fixes the solution pH at a high alkaline level and retards air oxidation of sulfide ion in solution. This ensures that the sulfide measured represents total sulfides as S^- ion and rather than the HS^- or H_2S found at lower pH values. APHA (Method 4500, 1995) provides qualitative as well as quantitative methods to determine aqueous sulfide concentrations. Qualitative methods include the antimony test, silver–silver sulfide electrode test, lead acetate paper test, and silver foil test. Quantitative methods include the photometric method, automated photometric methylene blue colorimetric method, and iodometric titration method for standardizing stock solutions.

Sediment Oxygen Demand

Sediment can exhibit significant rates of oxygen uptake attributable to (1) a benthic ecosystem supported by soluble organic substances in the water column, (2) naturally occurring sediments derived from aquatic plants and animals, or (3) detritus discharged into the water body by natural runoff. When numerical modeling is required to predict dissolved oxygen concentrations, the rate of dissolved oxygen consumed by the benthic ecosystem is defined as the sediment (benthic) oxygen demand (SOD) in grams O_2 per m^2 per day.

Two approaches for measuring SOD were reviewed by Truax et al. (1995), including *in situ* respiratory and laboratory respirometry methods. Numerous techniques have been developed for each approach. Generally, *in situ* methods are considered more credible than laboratory measurements, although both apply the same technique. A given amount of sediment is enclosed in a chamber with a known water volume, and oxygen uptake is measured over time. The SOD rate is then calculated based on the area of the enclosed sediment, the volume of water in the chamber, and the rate of uptake.

In situ sediment oxygen demand measurement methods were described by Uchrin and Ahlert (1985). A cylindrical respirometer, dissolved oxygen probe with a stirring mechanism, and dissolved oxygen meter were used. Ambient dissolved oxygen was measured using the probe/meter as well as by using the Winkler method (APHA,

1995) in the laboratory to determine the effect of respiration on total dissolved oxygen uptake. The respirometer was deployed in a level area at the bottom of the water body. Dissolved oxygen levels were recorded initially and at 15-minute intervals thereafter to determine the SOD rate.

Sediment Biochemical Oxygen Demand

Biochemical oxygen demand (BOD) is a measure of the dissolved oxygen consumed by microbial organisms while assimilating and oxidizing the organic matter in a sample (Puget Sound Estuary Program, 1996). The test is an empirical methodology in which standardized laboratory procedures are used to determine the relative oxygen uptake of environmental samples. The test measures the amount of molecular oxygen used during a specified incubation period to biochemically degrade organic material and to oxidize reduced forms of nitrogen (APHA, 1995). Plumb (1981) described a method to analyze BOD in sediments using freshwater bacteria as seed material and buffered distilled water as dilution water. The Puget Sound Estuary Program (1996) described an alternative procedure to analyze BOD in marine sediments using marine bacteria as the seed material and filtered, oxygenated seawater as the dilution water. USEPA (1987) methods should also be consulted.

Sediment Chemical Oxygen Demand

Chemical oxygen demand (COD) is a measure of the oxygen equivalent of organic matter content in a sample that is susceptible to oxidation by a strong chemical oxidant at elevated temperature and reduced pH. The test was devised to augment the biochemical oxygen demand test. Chemical oxygen demand can be related empirically to biochemical oxygen demand, carbon, or total volatile solids. The Puget Sound Estuary Program (1996) described a method for analyzing sediment COD using a closed reflux/colorimetric method. Dichromate (Cr_2O_7) ions are used to oxidize organic matter to carbon dioxide and water and to provide oxygen. The dichromate ions remaining after the reaction are measured by titration, and the amount of oxygen consumed is then calculated. Four standard procedures for measuring COD in water are (1) open reflux method, (2) closed reflux method, (3) titrimetric method, and (4) closed reflux/colorimetric method (APHA, 1995). USEPA procedures for the colorimetric and titrimetric methods are described in USEPA (1979), and semiautomated methods are described in USEPA (1993). Hach (Loveland, CO) modified the USEPA-approved dichromate reflux method and the reactor digestion method. The methods are photometric and are adapted for use with Hach photometers.

Cation Exchange Capacity of Sediments

Cation exchange capacity (CEC) is a parameter that provides information relevant to metal bioavailability studies. Cations or positively charged elements (such as calcium, magnesium, hydrogen, and potassium) are attached to negatively charged surfaces of clay and organic matter. There is a continuous exchange of cations between sediment and water. CEC is a measure of the sediment's ability to retain cationic elements. It is also a measure of clay activity and mineralogy, which are used to calculate mineralization and leaching rates and to predict interactions with contaminants. The degree of CEC is dependent on the kind and amount of suitable surfaces such as organic matter and clay. High cation exchange capacities are associated with high clay contents and high organic matter, and changes in CEC are typically associated with changes in organic carbon content and pH of the sediment. Organic matter generally supplies a greater number of exchange sites than clay particles.

Redox Potential of Sediments

Redox (Eh) is a measure of the oxidation–reduction potential of sediments. Measurements of Eh are particularly important for metal speciation and for determining the extent of sediment oxidation. Eh values below approximately –100 millivolts would indicate biologically important sulfide concentrations. Some trace metals form insoluble complexes with sulfides. These metal–sulfide complexes bind the metals in a form that is not bioavailable. Because free ionic metals are generally thought to possess the greatest toxicity potential, it is important to understand conditions that control binding dynamics, such as pH and Eh. Potentiometric measurements of Eh using a millivolt reader can be obtained with a platinum electrode relative to a standard hydrogen electrode (Plumb, 1981). APHA (1995) has not recommended the standard hydrogen electrode as it is fragile and impractical. Instead, their preferred method uses a silver–silver-chloride or calomel reference electrode. A graphite rather than platinum electrode for sediments is recommended (APHA, 1995). Once the Eh equilibrium is reached, the difference between the platinum or graphite electrode and the reference electrode is equal to the redox potential of the system.

Total Inorganic Carbon

Inorganic carbon has been measured as a complement to microbial activity to determine the fate of an organic contaminant in biodegradation studies (West and Gonsior, 1996) and to determine the percent carbon unaccounted for in fate transport predictions of hydrophobic contaminants (Tye et al., 1996). Often the total inorganic carbon (TIC) fraction in samples is many times greater than the TOC fraction and presents an interference in the measurement of TOC. There are several options to eliminate TIC interferences when trying to measure TOC. One option is to compensate for the IC interference by measuring total carbon (TC) and total inorganic carbon. The difference between the two is the TOC. TIC is determined by acidifying the sample to convert the inorganic carbon (i.e., carbonates, bicarbonates, and dissolved CO_2) to carbon dioxide. Carbon dioxide is purged from the sample and then detected by a nondispersive infrared detector (NDIR) calibrated to directly display the mass of carbon dioxide measured. This mass is proportional to the mass of the TIC (USEPA, 2001).

Total Volatile Solids

Total volatile solids represent the fraction of total solids that are lost on ignition at a higher temperature than that used to determine total solids. Total volatile solids are used as a crude estimate of the amount of organic matter in total solids (Puget Sound Estuary Program, 1996). In this regard, total volatile solids are often measured instead of, or in addition to, organic carbon content. Total volatile solids are operationally defined by ignition temperature. Total volatile solids content does not always represent the organic content of a sample because some organic material may be lost at the drying temperature and some inorganic material (e.g., carbonates, chlorides) may be lost at the ignition temperature. Because of the temperature dependency of total volatile solids, valid interstudy comparisons require the use of standardized drying and ignition temperatures (Puget Sound Estuary Program, 1996). Total volatile solids measurements are generally made by igniting the sediments at 550 ± 10°C until a constant weight is achieved and reporting the percent ash free dry weight (APHA, 1995; Keilty et al., 1988; McLeese et al., 1980).

Dissolved Organic Carbon in Porewater

Dissolved organic carbon (DOC) often consists of humic substances and is the fraction of the organic carbon pool that is dissolved in water and passed through a 0.45-μm glass-fiber filter. DOC is an indicator of the chemically reactive organic fraction and accurately measures the dissolved organic load. Sediment pore waters can be rich in humic acids. As much as 50 to 90% of the porewater DOC can be colloidal, which is a significant factor because organic chemicals will preferentially partition to porewater DOC (Burgess and McKinney, 1996; Resendes et al., 1992). Hermann (1996) and Gilek et al. (1996) measured DOC using a TOC apparatus and infrared detection of CO_2. Borga et al. (1996) measured DOC using atomic emission spectrometry. The APHA methods for total organic carbon that can be applied to the measurement of DOC are (1) the combustion–infrared method, (2) the persulfate–ultraviolet oxidation method, and (3) the wet oxidation method (Method 5310, APHA, 1995). Adjustments for inorganic carbon interference may be required.

Alkalinity and Hardness of the Porewater (Freshwater Sediments)

Alkalinity is defined as the acid-neutralizing (i.e., proton-accepting) capacity of water. It is the sum of all the table bases and a measure of the quality and quantity of constituents in the pore water that result in a shift in the pH toward the alkaline side of neutrality. The measured value may vary significantly with the pH endpoint used. Studies have shown that effects of certain contaminants such as metals are influenced by alkalinity as it alters speciation and bioavailability. APHA (1995) recommended a color-change titration method to measure alkalinity that was also described by ASTM (2000h). The sample is titrated with standard alkali or acid to a designated pH and the endpoint is determined electrometrically or by the color change of an internal standard. In addition, ASTM (2000b) described two additional methods:

1. A titration curve is developed to identify inflection points, a standard acid of alkali is added to the sample by small increments, and pH is recorded after each addition. The total volume of acid or alkali is plotted against the observed pH values.
2. The pH is determined and standard acid is added to lower the pH to 4.0 or less. The solution is boiled with hydrogen peroxide and titrated while hot to the phenolphthalein endpoint or, when cooled electrometrically, with standard alkali to pH 8.2, the desired endpoint.

The color-change titration method is most commonly used. Hach (Method 8202) has developed a portable water chemistry kit based on the APHA (1995) color-change titration method and an additional method using sulfuric acid with a digital titrator (Hach Method 8203). Hardness is the concentration of metallic cations, with the exception of alkali metals, present in water samples. Generally, hardness is a measure of the concentration of calcium and magnesium ions in water. Hardness is usually expressed as calcium carbonate equivalent in mg/L. Like alkalinity, hardness alters speciation and the bioavailability of certain contaminants, particularly many metals.

REFERENCES AND RECOMMENDED READING

Adams, D.D. (1991). Sampling sediment pore water. In: *CRC Handbook of Techniques for Sediment Sampling* (Mudroch, A. and MacKnight, S.D., Eds.), pp. 117–202. Boca Raton, FL: CRC Press.

Allen, H.E., Fu, G., and Deng, B. (1993). Analysis of acid-volatile sulfide (AVS) and simultaneously extracted metals (SEM) for the estimation of potential toxicity in aquatic sediments. *Environmental Toxicology and Chemistry*, 12: 1441–1453.

Allen, T., (1975). *Particle Size Measurement.* New York: John Wiley & Sons.

Ankley, G.T., Schubauer-Berigan, M.K., and Dierkes, J.R. (1981). Predicting the toxicity of bulk sediments to aquatic organisms with aqueous test fractions: porewater vs. elutriate. *Environmental Toxicology and Chemistry*, 10: 925–939.

Ankley, G.T., Katko A., and Arthur, J.W. (1990). Identification of ammonia as a major sediment-associated toxicant in the lower Fox River and Green Bay, Wisconsin. *Environmental Toxicology and Chemistry*, 9: 313–322.

Ankley, G.T., DiToro, D.M., Hansen, D.J., and Berry, W.J. (1996). Technical basis and proposal for deriving sediment quality criteria for metals. *Environmental Toxicology and Chemistry*, 15: 2056–2066.

APHA. (1971). *Standard Methods for the Examination of Water and Wastewater*, 17th ed. Washington, DC: American Public Health Association.

APHA. (1995). *Standard Methods for the Examination of Water and Wastewater*, 18th ed. Washington, DC: American Public Health Association.

APHA. (1998). *Standard Methods for the Examination of Water and Wastewater*, 20th ed. Washington, DC: American Public Health Association.

ASTM. (2000a). Standard guide for collection, storage, characterization, and manipulation of sediments for toxicological testing, E1391-94. In: *ASTM Standards on Environmental Sampling.* West Conshohocken, PA: American Society for Testing and Materials International.

ASTM. (2000b). Standard guide for general planning of waste sampling, D4687. In: *ASTM Standards on Environmental Sampling.* West Conshohocken, PA: American Society for Testing and Materials International.

ASTM. (2000c). Standard guide for selecting grab sampling devices for collecting benthic macroinvertebrates, D4387. In: *ASTM Standards on Environmental Sampling*. West Conshohocken, PA: American Society for Testing and Materials International.

ASTM. (2000d). Standard guide for core-sampling submerged, unconsolidated sediments, D4823-95. In: *ASTM Standards on Environmental Sampling*. West Conshohocken, PA: American Society for Testing and Materials International.

ASTM (2000e). Standard practice for preparation of sediment samples for chemical analysis, D3976-92. In: *ASTM Standards on Environmental Sampling*. West Conshohocken, PA: American Society for Testing and Materials International.

ASTM. (2000f). Standard guide for conducting acute toxicity tests with fishers, macroinvertebrates, and amphipods, E729-96. In: *ASTM Standards on Environmental Sampling*. West Conshohocken, PA: American Society for Testing and Materials International.

ASTM. (2000g). Standard test methods for ammonia nitrogen in water, D1426-93. In: *ASTM Standards on Environmental Sampling*. West Conshohocken, PA: American Society for Testing and Materials International.

ASTM. (2000h). Standard test methods for acidity and alkalinity of water, D1067-92. In: *ASTM Standards on Environmental Sampling*. West Conshohocken, PA: American Society for Testing and Materials International.

AWRI. (2000). *Plankton Sampling*. Allendale, MI: Robert B. Annis Water Resource Institute, Grand Valley State University.

AWWA. (1995). *Water Treatment*, 2nd ed. Denver, CO: American Water Works Association.

Axelman, J., Carina, N., Broman, D., and Kristoffer, N. (1999). Accumulation of polycyclic aromatic hydrocarbons in semipermeable membrane devices and caged mussels (*Mytilus edulis* I.) in relation to water column phase distribution. *Environmental Toxicology and Chemistry*, 18(11): 2454–2461.

Bahls, L.L. (1993). *Periphyton Bioassessment Methods for Montana Streams*. Helena: Montana Water Quality Bureau, Department of Health and Environmental Science.

Barbour, M.T., Garretson, J., Snyder, B.D., and Stabling, J.B. (1997). *Revision to Rapid Bioassessment Protocols for Use in Streams and Rivers: Periphytons, Benthic Macroinvertebrates, and Fish*. Washington, DC: U.S. Environmental Protection Agency.

Baudo, R. (1990). Sediment sampling, mapping and data analysis. In: *Sediments. Chemistry and Toxicity of In-Place Pollutants* (Giesy, J.P. and Muntau, H., Eds.), pp. 15–60. Chelsea, MI: Lewis Publishers.

Bender, M., Martin, W., Hess, J., Sayles, F., Ball, L., and Lambert, C. (1987). A whole-core squeezer for interfacial pore-water sampling. *Limnology and Oceanography*, 32: 1214–1225.

Bennett, E.R., Metcalfe, C.D., and Metcalfe, T.L. (1996). Semi-permeable membrane devices (SPMDs) for monitoring organic contaminants in the Otonabee River, Ontario. *Chemosphere*, 33: 363–375.

Berry, W.J., Cantwell, M.G., Edwards, P.A., Serbst J.R., and Hansen, D.J. (1999). Predicting toxicity of sediments spiked with sliver. *Environmental Toxicology and Chemistry*, 18: 40–48.

Bischoff, J.L., Greet, R.E., and Luistro, A.O. (1970). Composition of interstitial waters of marine sediments: temperature of squeezing effect. *Science*, 167: 1245–1246.

Bly, T.D. and Smith, G.F. (1994). *Biomonitoring Our Streams: What's It All About*? Nashville, TN: U.S. Geological Survey.

Bollinger, R., Bandl, H., Hohener, P., Hanselmann, K.W., and Bachofen, R. (1992). Squeeze-water analysis for the determination of microbial metabolites in lake sediments-comparison of methods. *Limnology and Oceanography*, 37: 448–455.

Borga, P., Elowson T., and Liukko, K. (1996). Environmental loads from water-sprinkled softwood timber. 2. Influence of tree species and water characterizations on wastewater discharges. *Environmental Toxicology and Chemistry*, 15: 1445–1454.

Botkin, D.B. (1990). *Discordant Harmonies*. New York: Oxford University Press.

Bottomly, E.Z. and Bayly, I.L. (1984). A sediment pore water sampler used in root zone studies of the submerged macrophyte, *Myriophyllum spicatum*. *Limnology and Oceanography*, 29: 671–673.

Bray, J.T., Bricker, O.P., and Troup, B.N. (1973). Phosphate in interstitial waters of anoxic sediments: oxidation effects during sampling procedure. *Science*, 180: 1362–1364.

Brinkman, A.G., van Raaphorst, W., and Lijklema, L. (1982). *In situ* sampling of interstitial water from lake sediments. *Hydrobiologia*, 92: 659–663.

Brumbaugh, W.G., Ingersoll, C.G., Kemble, N.E., May, T.W., and Zajicek, J.L. (1994). Chemical characterization of sediments and porewater from the Upper Clark Fork River and Milltown Reservoir, Montana. *Environmental Toxicology and Chemistry*, 13: 1971–1973.

Bufflap, W.E. and Allen, H.E. (1995). Sediment pore water collection methods: a review. *Water Research*, 29: 165–177.

Burgess, R.M. (1996). Enrichment of marine sediment colloids with polychlorinated biphenyls: trends resulting from PCB solubility and chlorination. *Environmental Science & Technology*, 30(8): 2556–2566.

Burgess, R.M. and McKinney, R.A. (1996). Effects of sediment homogenization on interstitial water PCB geochemistry. *Archives of Environmental Contamination and Toxicology*, 33: 125–129.

Burgess, R.M., Schweitzer, K.A., McKinney R.A., and Phelps, D.K. (1993). Contaminated marine sediments: water column and interstitial toxic effects. *Environmental Toxicology and Chemistry*, 12: 127–138.

Burton, Jr., G.A. (1991). Assessment of freshwater sediment toxicity. *Environmental Toxicology and Chemistry*, 10: 1585–1627.

Burton, Jr., G.A., Rowland C.D., Greenberg M.S., Lavoie, D.R., Nordstrom, J.F., and Eggert, L.M. (2003). A tiered, weight-of-evidence approach for evaluating aquatic ecosystems. In: *Sediment Quality Assessment and Management: Insight and Progress* (Munawar, M., Ed.). The Netherlands: Backhuys Publishers.

Call, D.J., Christine, N., Polkinghorne, T.P., Markee, L.T., Brooke, D.L., Geiger, J.W., Gorsuch, K., and Robillard, N. (1999). Silver toxicity to *Chironomus tentans* in two freshwater sediments. *Environmental Toxicology and Chemistry*, 18(1): 30–39.

Camann, M. (1996). *Freshwater Aquatic Invertebrates: Biomonitoring*. Arcata, CA: Humboldt State University.

Carignan, R. (1984). Interstitial water sampling by dialysis: methodological notes. *Limnology and Oceanography*, 29: 667–670.

Carignan, R. and Lean, D.R.S. (1991). Regeneration of dissolved substances in a seasonally anoxic lake: the relative importance of processes occurring in the water column and in the sediments. *Limnology and Oceanography*, 36: 683–703.

Carignan, R., Rapin, F., and Tessier, A. (1985). Sediment porewater sampling for metal analysis: a comparison of techniques. *Geochimica et Cosmochimica Acta*, 49: 2493–2497.

Carignan, R., St. Pierre, S., and Gachter, R. (1994). Use of diffusion samples in oligotrophic lake sediments: effects of free oxygen in sampler material. *Limnology and Oceanography*, 39: 468–474.

Carins, Jr., J. and Dickson, K.L. (1971). A simple method for the biological assessment of the effects of waste discharges on aquatic bottom-dwelling organisms. *Journal of Water Pollution Control Federation*, 43: 755–772.

Carlton, R.G., and R.G. Wetzel. (1985). A box corer for studying metabolism of epipelic microorganisms in sediment under *in situ* conditions. *Limnology and Oceanography*, 30: 422.

Carr, R.S. (1998). Marine and estuarine porewater toxicity testing. In: *Microscale Testing in Aquatic Toxicology: Advances, Techniques, and Practice* (Wells, P.G., Lee, K., and Blaise, C., Eds.), pp. 523–538. Boca Raton, FL: CRC Press.

Carr, R.S. and Chapman, D.C. (1992). Comparison of solid-phase and pore-water approaches for assessing the quality of marine and estuarine sediments. *Chemical Ecology*, 7: 19–30.

Carr, R.S. and Chapman, D.C. (1995). Comparison of methods for conducting marine and estuarine sediment porewater toxicity tests: extraction, storage, and handling techniques. *Archives of Environmental Contamination and Toxicology*, 28: 69–77.

Carr, R.S., Williams, J.W., and Fragata, C.T.B. (1989). Development and evaluation of a novel marine sediment porewater toxicity test with the polychaete *Dinophilus gyrociliatus*. *Environmental Toxicology and Chemistry*, 8: 533–543.

Chin, Y. and Gschwend, P.M. (1991). The abundance, distribution, and configuration of pore water organic colloids in recent sediment. *Geochimica et Cosmochimica Acta*, 55: 1309–1317.

Crecelius, E.A., Jenne E.A., and Anthony, J.S. (1987). *Sediment Quality Criteria for Metals: Optimization of Extraction Methods for Determining the Quantity of Sorbents and Adsorbed Metals in Sediments*. Washington, DC: U.S. Environmental Protection Agency, Criteria and Standards Division.

Day, P.R. (1965). Particle fractionation and particle-size analysis. In: *Methods of Soil Analysis*. Part 1. *Agronomy* (Black, C.A. et al., Eds.), pp. 545–567. Madison, WI: American Society of Agronomy.

DiToro, D.M., Mahony, J.H., Hansen, D.J., Scott, K.J., Hicks, M.B., Mayr, S.M., and Redmond, M. (1990). Toxicity of cadmium in sediments: the role of acid volatile sulfides. *Environmental Toxicology and Chemistry*, 9: 1487–1502.

DiToro, D.M., Zarba, C.S., Hansen, D.J., Berry, W.J., Swartz, R.C., Cowan, C.E., Pavlou, S.P., Allen, H.E., Thomas, N.A., and Paquin, P.R. (1991). Technical basis for establishing sediment quality criteria for nonionic organic chemicals using equilibrium partitioning. *Environmental Toxicology and Chemistry*, 10: 1541–1583.

Doig, L. and Liber, K. (2000). Dialysis minipeeper for measuring porewater metal concentrations in laboratory sediment toxicity and bioavailability tests. *Environmental Toxicology and Chemistry*, 19: 2882–2889.

Duncan, G.A. and Lattaie, G.G. (1979). Size analysis procedures used in the sedimentology laboratory. In: *National Water Research Institute Manual*. Burlington, Ontario: National Water Research Institute, Canada Centre for Inland Waters.

Environment Canada. (1994). *Guidance Document on Collection and Preparation of Sediments for Physicochemical Characterization and Biological Testing*, Report EPS 1/RM/29. Gatineau, Quebec: Environment Canada, Environmental Protection Services.

EPRI. (1986). *Speciation of Selenium and Arsenic in Natural Waters and Sediments*, Vol. 2, EPRI EA-4641. Palo Alto, CA: Electrical Power Research Institute.

EPRI. (1999). *Review of Sediment Removal and Remediation Technologies at MGP and Other Contaminated Sites*, TR-113106. Palo Alto, CA: Electrical Power Research Institute; Berlin, CT: Northeast Utilities.

Frazier, B.E., Naimo, T.J., and Sandheinrich, M.B. (1996). Temporal and vertical distribution of total ammonia nitrogen and un-ionized ammonia nitrogen in sediment porewater from the upper Mississippi River. *Environmental Toxicology and Chemistry*, 15: 92–99.

Froelich, P.M., Klinkhammer, G.P., Bender, M.L., Luedtke, N.A., Heath, G.R. et al. (1979). Early oxidation of organic matter in pelagic sediments of the eastern equatorial Atlantic: suboxic diagenesis. *Geochimica et Cosmochimica Acta*, 43: 1075–1090.

Gambrell, R.P., Khalid R.A., and Patrick, Jr., W.H. (1976). Physicochemical parameters that regulate mobilization and immobilization of toxic heavy metals. In: *Proceedings of the Specialty Conference on Dredging and Its Environmental Effects*, Mobile, AL, January 26–28, pp. 418–434.

Gee, G.W. and Bauder, J.W. (1986). Particle size analyze. In: *Methods of Soil Analysis*, 2nd ed. Part I. *Physical and Mineralogical Methods* (Klute, A., Ed.), pp. 383–411. Madison, WI: American Society of Agronomy.

Gilek, M., Bjork, M., Broman, D., Kautsky, N., and Naf, C. (1996). Enhanced accumulation of PCB congeners by Baltic Sea blue mussels, *Mytilus edulis*, with increased algae enrichment. *Environmental Toxicology and Chemistry*, 15: 1597–1605.

GLNPO. (1994). *Assessment and Remediation of Contaminated Sediments (ARCS) Program*, EPA 905-B94-002. Chicago, IL: U.S. Environmental Protection Agency, Great Lakes National Program Office.

Golterman, H.L., Sly P.G., and Thomas, R.L. (1983). *Study of the Relationship between Water Quality and Sediment Transport*. Mayenne, France: UNESCO.

Gonzalez, A.M. (1996). A laboratory-formulated sediment incorporating synthetic acid-volatile sulfide. *Environmental Toxicology and Chemistry*, 15(12): 2209–2230.

Gustafsson, O., Haghesta, F., Cahn, C., MacFarlane, J., and Gschwend, P.M. (1997). Quantification of the dilute sedimentary soot phase: implication for PAH speciation and bioavailability. *Environmental Science and Technology*, 31(1): 203–209.

Guy, H.P. (1969). Laboratory theory and methods for sediment analysis. In: *Techniques of Water-Resources Investigations*, Book 5, Chapter C1. Reston, VA: U.S. Geological Survey.

Hermann, R. (1996). The daily changing pattern of hydrogen peroxide in New Zealand surface waters. *Environmental Toxicology and Chemistry*, 15: 652–662.

Hesslein, R.H. (1976). An *in-situ* sampler for close internal pore water studies. *Limnology and Oceanography*, 21: 912–914.

Hill, B.H. (1997). The use of periphyton assemblage data in an index of biotic integrity. *Bulletin of the North American Ethological Society*, 14: 158.

Ho, K.T., McKinney, R., Kuhn, A., Pelletier, M., and Burgess, R. (1997). Identification of acute toxicants in new Bedford Harbor sediments. *Environmental Toxicology and Chemistry*, 16(3): 551–558.

Ho, K.T., Kuhn, A., Pelletier, M.C., Burgess, R.M., and Helmstetter, A. (1999). Use of *Ulva lactuca* to distinguish pH-dependent toxicants in marine waters and sediments. *Environmental Toxicology and Chemistry*, 18(2): 207–212.

Hoffman, E. (1998). *Tributyltin Analysis: Clarification of Interstitial Water Extraction and Analysis Methods—Interim*, EPA, 206-553-0038. Seattle, WA: U.S. Environmental Protection Agency, Dredged Material Management Program.

Hoss, S., Haitzer, M., Traunspurger, W., and Steinberg, C.E.W. (1999). Growth and fertility of *Caenorhabditis elegans* (Nematoda) in unpolluted freshwater sediments: response to particle size distribution and organic content. *Environmental Toxicology and Chemistry*, 18(12): 2921–2925.

Howes, B.L., Dacey, J.W.H., and Wakeham, S.G. (1985). Effects of sampling technique on measurements of porewater constituents in salt marsh sediments. *Limnology and Oceanography*, 30: 221–227.

Huckins, J.N., Ruvwefwn, M.Q., and Mnuqwwe, F.K. (1990). Semipermeable membrane devices containing model lipid: a new approach to monitoring the bioavailability of lipophilic contaminants and estimating their bioconcentration potential. *Chemosphere*, 20: 533–552.

Huff, W.R. (1993). Biological indices define water quality standard. *Water Environment & Technology,* 5: 21–22.

Hulbert, M.H. and Brindel, M.P. (1975). Effects of sample handling on the composition of marine sedimentary pore water. *Geological Society of America Bulletin*, 86(1): 109–110.

Jackson, M.L. (1958). *Soil Chemical Analysis*. Englewood Cliffs, NJ: Prentice-Hall.

Jahnke, R.A. (1988). A simple, reliable, and inexpensive pore-water sample. *Limnology and Oceanography*, 33: 483–487.

Jenne, E.A. and Zachara, J.M. (1984). *Factors Influencing the Sorption of Metals: Fate and Effects of Sediment-Bound Chemicals in Aquatic Systems*. New York: Pergamon Press.

Kalil, E.K. and Goldhaker, M. (1973). A sediment squeezer for removal of pore waters without air contact. *Journal of Sediment Petrology*, 43: 554–557.

Kaplan, I., Lu, S.T., Lee R.P., and Warrick, G. (1996). Polycyclic hydrocarbon biomarkers confirm selective incorporation of petroleum in soil and kangaroo rat liver samples near an oil well blowout site in the western San Joaquin Valley, California. *Environmental Toxicology and Chemistry*, 15: 696–707.

Karr, J.R. (1981). Assessment of biotic integrity using fish communities. *Fisheries*, 66: 21–27.

Karr, J.R., Bausch, K.D., Hagermeier, P.L., Yant, P.R., and Schlosser, I.J. (1986). *Assessing Biological Integrity in Running Waters: A Method and Its Rationale*, Special Publ. 5. Champaign: Illinois Natural History Survey.

KDOW. (1993). *Methods for Assessing Biological Integrity of Surface Waters*. Frankfort: Kentucky Department of Environmental Protection, Division of Water.

Keilty, T.J., White, D.S., and Landrum, P.G. (1988). Short-term lethality and sediment avoidance assays with endrin-contaminated sediment and two oligochaetes from Lake Michigan. *Archives of Environmental Contamination and Toxicology*, 17: 95–101.

Kemble, N.E., Dwyer, F.J., Ingersoll, C.G., Dawson, T.D., and Norberg-King, T.J. (1999). Tolerance of freshwater test organisms to formulated sediments for use as control materials in whole-sediment toxicity tests. *Environmental Toxicology and Chemistry*, 18: 222–230.

Kittrell, F.W. (1969). *A Practical Guide to Water Quality Studies of Streams*. Washington, DC: U.S. Department of Interior.

Klemm, D.J., Lewis, P.A., Fulk, F., and Lazorchak, J.M. (1990). *Macroinvertebrate Field and Laboratory Methods for Evaluating the Biological Integrity of Surface Waters*, EPA-600-4-90-030. Cincinnati, OH: U.S. Environmental Protection Agency, Environmental Monitoring and Support Laboratory.

Knezovich, J.P. and Harrison, F.L. (1987). A new method for determining the concentration of volatile organic compounds in sediment interstitial water. *Bulletin of Environmental Contamination and Toxicology*, 38: 937–940.

Knezovich, J.P. and Harrison, F.L. (1988). The bioavailability of sediment-sorbed chlorobenzenes to larvae of the midge *Chironomus decorus*. *Ecotoxicology and Environmental Safety*, 15: 226–241.

Kristensen, E. and Blackburn, T.H. (1987). The fate of organic carbon and nitrogen in experimental marine sediment systems: influence of bioturbation and anoxia. *Journal of Marine Research*, 45: 231–257.

Kriukov, P.A. and Manheim, F.T. (1982). Extraction and investigative techniques for study of interstitial waters of unconsolidated sediments: a review. In: *The Dynamic Environment of the Ocean Floor* (Fanning, K.A. and Manheim, F.T., Eds.), pp. 3–26. Washington, DC: Lexington Books.

Leonard, E. (1991). *Standard Operating Procedures for Total Organic Carbon Analysis of Sediment Samples*. Duluth, MN: U.S. Environmental Protection Agency, Office of Research and Development, Environmental Research Laboratory.

Leppard, G.G. (1986). The fibrillar matrix component of lacustrine biofilms. *Water Research*, 20: 697–702.

Leppard, G.G., Buffle, J., De Vitore, R.R., and Pereet, D. (1988). The ultrastructure and physical characteristics of a distinctive colloidal iron particulate isolated from a small eutrophic lake. *Archives of Hydrobiology*, 113: 405–424.

Long, E.R., Buchman, M.F., Bay, S.M., Breteler, R.H., Carr, R.S. et al. (1990). Comparative evaluation of five toxicity tests with sediments from San Francisco Bay and Tomales Bay, California. *Environmental Toxicology and Chemistry*, 9: 1193–1214.

Lyons, W.B., Gaudette, J., and Smith, G. (1979). Porewater sampling in anoxic carbonate sediments: oxidation artifacts. *Nature*, 277: 48–49.

MacLeod, Jr., W., Brown, D.W., Friedman, A.J., Burrows, D.G., Maynes, O., Pearce, R.W., Wigren, C.A., and Bogar, R.G. (1985). *Standard Analytical Procedures of the NOAA National Analytical Facility, 1984–1985: Extractable Toxic Organic Compounds*, NOAA Technical Memorandum NMFS F/NWC-92. Silver Spring, MD: National Oceanic and Atmospheric Administration.

Mangelsdorf, P.C. and Wilson, T.R.S. (1969). Potassium enrichments in interstitial waters of recent marine sediments. *Science*, 165: 171.

Mayer, L.M. (1976). Chemical water sampling in lakes and sediments with dialysis bags. *Limnology and Oceanography*, 21: 909.

McCabe, I.M. and Jarvis, J. (1973). Use of the Model T Coulter Counter in size analysis of fine to coarse sand. *Sedimentology*, 20: 305–315.

McLeese, D.W., Metcalfe, C.D., and Pezzack, D.S. (1980). Uptake of PCBs from sediment by *Nereis virens* and *Crangon septemspinosa*. *Archives of Environmental Contamination and Toxicology*, 9: 507–518.

Metro. (1981, rev. 1983). *Analytical Support and Data Validity: Organics*. Seattle, WA: Toxicant Pretreatment Planning Study, Municipality of Metropolitan Seattle.

Mudroch, A. and Azcue, J.M. (1995). *Manual of Aquatic Sediment Sampling*. Boca Raton, FL: Lewis Publishers.

Mudroch, A. and MacKnight, S.D. (1994). *CRC Handbook of Techniques for Aquatic Sediment Sampling*, 2nd ed. Boca Raton, FL: CRC Press.

Murray, J.W., Grunamanis, V., and Smethie, Jr., W.M. (1978). Interstitial water chemistry in the sediments of Saanich Inlet. *Geochimica et Cosmochimica Acta*, 42: 1011–1026.

O'Toole, C., Ed. (1986). *The Encyclopedia of Insects*. New York: Facts on File.

Ohio EPA. (2001). *Ohio EPA Sediment Sampling Guide and Methodologies*. Columbus: State of Ohio Environmental Protection Agency.

Page, A.L., Miller, R.H., and Keeney, D.R., Eds. (1982). *Methods of Soil Analysis*, Parts 1 and 2. Madison, WI: American Society of Agronomy.

Page, D.S., Boehm, P.D., Douglas, G.S., and Bence, A.E. (1995a). Identification of hydrocarbon sources in the benthic sediments of Prince William Sound and the Gulf of Alaska following the *Exxon Valdez* oil spill. In: *Exxon Valdez Oil Spill: Fate and Effects in Alaskan Waters*, ASTM STP 1219 (Wells, P.G., Butler, J.N., and Hughes, J.S., Eds.), pp. 41–83. Philadelphia, PA: American Society for Testing and Materials International.

Page, D.S., Gilfillan, E.S., Boehm, P.D., and Harner, E.J. (1995b). Shoreline ecology program for Prince William Sound, Alaska, following the *Exxon Valdez* oil spill. Part I. Study design and methods. In: *Exxon Valdez Oil Spill: Fate and Effects in Alaskan Waters*, ASTM STP 1219 (Wells, P.G., Butler, J.N., and Hughes, J.S., Eds.), pp. 263–295. Philadelphia, PA: American Society for Testing and Materials International.

Patrick, R. (1973). Use of algae, especially diatoms, in the assessment of water quality. In: *Biological Methods for the Assessment of Water Quality*, ASTM STP 528 (Carins, Jr., J. and Dickso, K.L., Eds.), pp. 76–95. Philadelphia, PA: American Society for Testing and Materials International.

Patrick, Jr., W.H. (1958). Modification of method particle size analyses. *Proceedings of the Soil Science Society of America*, 4: 366–367.

Phillips, B.M., Anderson, B.S., and Hunt, J.W. (1997). Measurement and distribution of interstitial and overlying water ammonia and hydrogen sulfide in sediment toxicity tests. *Marine Environmental Research*, 44(2): 117–126.

Plumb, R.H. (1981). *Procedures for Handling and Chemical Analysis of Sediment and Water Samples*. Buffalo, NY: State University College at Buffalo, Great Lakes Laboratory.

Puget Sound Estuary Program. (1987a). *Recommended Guidelines for Measuring Metals in Puget Sound Water, Sediment, and Tissue Samples*. Olympia, WA: U.S. Environmental Protection Agency; Olympia, WA: Puget Sound Water Quality Authority.

Puget Sound Estuary Program. (1987b). *Recommended Guidelines for Measuring Metals in Puget Sound Water, Sediment, and Tissue Samples.* Olympia, WA: U.S. Environmental Protection Agency; Olympia, WA: Puget Sound Water Quality Authority.

Puget Sound Estuary Program. (1987c). *Recommended Guidelines for Measuring Organic Compounds in Puget Sound Water, Sediment, and Tissue Samples.* Olympia, WA: U.S. Environmental Protection Agency; Olympia, WA: Puget Sound Water Quality Authority.

Puget Sound Estuary Program. (1987d). *Recommended Guidelines for Sampling Marine Sediment, Water Column, and Tissue in Puget Sound.* Olympia, WA: Puget Sound Estuary Program, Puget Sound Water Quality Action Team.

Puget Sound Estuary Program. (1996). *Recommended Guidelines for Measuring Selected Environmental Variables in Puget Sound.* Olympia, WA: U.S. Environmental Protection Agency; Olympia, WA: Puget Sound Water Quality Authority.

Reeburgh, W.S. (1967). An improved interstitial water sampler. *Limnology and Oceanography*, 12: 163–165.

Resendes, J., Shiu, W.Y., and Mackay, D. (1992). Sensing the fugacity of hydrophobic organic chemicals in aqueous systems. *Environmental Science and Technology*, 26: 2381–2387.

Rodgers, Jr., J.H., Dickson, K.L., and Cairns, Jr., J. (1979). A review and analysis of some methods used to measure functional aspects of periphyton. In: *Methods and Measurements of Periphyton Communities: A Review*, ASTM STP 690 (Weitzel, R.L., Ed.), pp. 142–167. West Conshohocken, PA: American Society for Testing and Materials International.

Rosenfeld, J.K. (1979). Ammonia absorption in nearshore anoxic sediments. *Limnology and Oceanography*, 24: 356–364.

Rukavina, N.A. and Duncan, G.A. (1970). F.A.S.T—fast analysis of sediment texture. In: *Proceedings of the 13th Conference of the International Association for Great Lakes Research.* Ann Arbor, MI: International Association for Great Lakes Research, pp. 274–281.

Saager, P.M., Sweerts, J.-P., and Ellermeijer, H.J. (1990). A simple pore-water sampler for coarse, sandy sediments of low porosity. *Limnology and Oceanography*, 35: 747–751.

Sanford, R.B. and Swift, D.J.P. (1971). Comparisons of sieving and settling techniques for size analysis, using a benthos rapid sediment analyzer. *Sedimentology*, 17. 257–264.

Santschi, P.H., Lenhart, J., and Honeyman, B.D. (1997). Heterogeneous processes affecting trace contaminant distribution in estuaries: the role of natural organic matter. *Marine Chemistry*, 58: 99–125.

Sarda, N. and Burton, Jr., G.A. (1995). Ammonia variation in sediments: spatial, temporal and method-related effects. *Environmental Toxicology and Chemistry*, 14: 1499–1506.

Sasson-Brickson, G. and Burton, Jr., G.A. (1991). *In situ* and laboratory sediment toxicity testing with *Ceriodaphnia dubia*. *Environmental Toxicology and Chemistry*, 10: 201–207.

Sayles, F.L., Wilson, T.R.S., Hume, D.N., and Mangelsdorf, Jr., P.C. (1973). *In situ* sampler for marine sedimentary porewaters: evidence for potassium depletion and calcium enrichment. *Science*, 180: 154–156.

Schubauer-Berigan, M.K., and Ankley, G.T. (1991). The contribution of ammonia, metals and nonpolar organic compounds to the toxicity of sediment interstitial water form an Illinois River tributary. *Environmental Toxicology and Chemistry*, 10: 925–939.

Schults, D.W., Ferraro, S.P., Smith, L.M., Roberts F.A., and Poindexter, C.K. (1992). A comparison of methods for collection interstitial water for trace organic compounds and metals analyses. *Water Research*, 26: 989–995.

SETAC. (2001). *Summary of a SETAC Technical Workshop: Porewater Toxicity Testing: Biological, Chemical, and Ecological Considerations with a Review of Methods and Applications, and Recommendations for Future Areas of Research.* Pensacola, FL: Society of Environmental Toxicology and Chemistry.

Simon, N.S., Kennedy, M.M., and Massoni, C.S. (1985). Evaluation and use of a diffusion controlled sampler for determining chemical and dissolved oxygen gradients at the sediment water interface. *Hydrobiologia*, 126: 135–141.

Singer, J.K., Anderson, J.B., Ledbetter, M.T., McCabe, I.N., Jones, K.P.N., and Wright, R. (1988). An assessment of analytical techniques for the size analysis of fine-grained sediments. *Journal of Sediment Petrology*, 58: 534–543.

Skalski, C. and Burton, G.A. (1991). Laboratory and *In Situ* Sediment Toxicity Evaluations Using Early Life Stages of *Pimephales promelas*, master's thesis, Wright State University, Dayton, OH.

Spellman, F.R. (2015). *The Science of Water: Concepts and Applications*, 3rd ed. Boca Raton, FL: CRC Press.

Sternberg, R.W. and Creager, J.S. (1961). Comparative efficiencies of size analysis by hydrometer and pipette methods. *Journal of Sediment Petrology*, 31: 96–100.

Stevenson, R.J. (1996). An introduction to algal ecology in freshwater benthic habitats. In: *Algal Ecology: Freshwater Benthic Ecosystems* (Stevenson, R.J., Bothwell, M., and Lowe, R.L., Eds.), pp. 3–30. San Diego, CA: Academic Press.

Stevenson, R.J. (1998). Diatom indicators of stream and wetland stressors in a risk management framework. *Environmental Monitoring and Assessment*, 51: 107–108.

Stevenson, R.J. and Pan, Y. (1999). Assessing ecological conditions in rivers and streams with diatoms. In: *The Diatoms: Application to the Environmental and Earth Sciences* (Stoermer, E.G. and Smol, J.P., Eds.), pp. 11–40. Cambridge, U.K.: Cambridge University Press.

Steward, A.R. and Malley, D.F. (1999). Effect of metal mixture (Cu, Zn, Pb, and Ni) on cadmium partitioning in littoral sediments and its accumulation by the freshwater macrophyte *Eriocailon septangilare*. *Environmental Toxicology and Chemistry*, 18(3): 436–447.

Swift, D.J.P., Schubel, J.R., and Sheldon, R.W. (1972). Size analysis of the fine-grained suspended sediments: a review. *Journal of Sedimentary Petrology*, 42: 122–134.

Tchobanoglous, G. and Schroeder, E.D. (1985). *Water Quality*. Reading, MA: Addison-Wesley.

Thurston, R.V., Russo, R.C., and Vinogradov, G.A. (1981). Ammonia toxicity to fishes: effect of pH on the toxicity of the unionized ammonia species. *Environmental Science and Technology*, 15: 837–840.

Troup, B.N., Bricker, O.P., and Bray, J.T. (1974). Oxidation effect on the analysis of iron in the interstitial water of recent anoxic sediments. *Nature*, 249: 237–239.

Truax, D.D., Shindala, A., and Sartain, H. (1995). Comparison of two sediment oxygen demand measurement techniques. *Journal of Environmental Engineering*, 121(9): 619–624.

Tye, R., Jepsen R., and Lick, W. (1996). Effects of colloids, flocculation, particle size, and organic matter on the adsorption of hexachlorobenzene to sediments. *Environmental Toxicology and Chemistry*, 15: 643–651.

Uchrin, C.G. and Ahlert, W.K. (1985). *In situ* sediment oxygen demand determinations in the Passaic River (NJ) during the late summer/early fall 1983. *Water Resources*, 19: 1141–1144.

USEPA. (1979). *Chemistry Laboratory Manual for Bottom Sediments and Elutriate Testing*, EPA-095-4-79-014 (NTIS PB 294596). Chicago, IL: U.S. Environmental Protection Agency, Region V.

USEPA. (1983). *Technical Support Manual: Waterbody Surveys and Assessments for Conducting Use Attainability Analyses*, EPA 440486037. Washington, DC: U.S. Environmental Protection Agency.

USEPA. (1986a). *Occupational Health and Safety Manual*. Washington, DC: U.S. Environmental Protection Agency.

USEPA. (1986b). *Test Methods for Evaluating Solid Waste (SW-846): Physical/Chemical Methods.* Washington, DC: U.S. Environmental Protection Agency, Office of Solid Waste.

USEPA (1987). *Quality Assurance/Quality Control (QA/QC) for 301(h) Monitoring Programs: Guidance on Field and Laboratory Methods*, EPA 430/9-86-004. Washington, DC: U.S. Environmental Protection Agency.

USEPA. (1991). *Methods for Measuring the Acute Toxicity of Effluents and Receiving Waters to Freshwater and Marine Organisms*, 4th ed., EPA-600/4-90/027F. Cincinnati, OH: U.S. Environmental Protection Agency.

USEPA. (1995). *QA/QC Guidance for Sampling and Analysis of Sediments, Water, and Tissues for Dredged Material Evaluations (Chemical Evaluations)*, EPA 832-B-95-002. Washington, DC: U.S. Environmental Protection Agency, Office of Water.

USEPA. (1998). *Contaminated Sediment Management Strategy*, EPA 823-R-98-001. Washington, DC: U.S. Environmental Protection Agency, Office of Water.

USEPA. (1999). *1999 Update of Ambient Water Quality Criteria for Ammonia*, EPA-823-F-99-013. Washington, DC: U.S. Environmental Protection Agency, Office of Water.

USEPA. (2000). *Monitoring Water Quality: Intensive Stream Bioassay.* Washington, DC: U.S. Environmental Protection Agency.

USEPA. (2001). *Methods for Collection, Storage and Manipulation of Sediments for Chemical and Toxicological Analyses: Technical Manual*, EPA-823-B-01-002. Washington, DC: U.S. Environmental Protection Agency, Office of Water.

USEPA/USACE. (1998). *Evaluation of Dredged Material Proposed for Discharge in Waters of the U.S.: Testing Manual*, EPA-823-B-98-004. Washington, DC: U.S. Environmental Protection Agency, Office of Water.

Vanderpleog, H.A. (1981). Effect of the algal length/aperture length ratio on Coulter analyses of Lake Seston. *Canadian Journal of Fisheries and Aquatic Science*, 38: 912–916.

Veechi, M., Reynoldson, T.B., Pasteris, A., and Bonomi, G. (1999). Toxicity of copper-spiked sediments to *Tubifex tubifex* (Oligochaeta, Tubificidae): comparison of the 28-day reproductive bioassay with an early-life-stage bioassay. *Environmental Toxicology and Chemistry*, 18(6): 1144–1148.

Velz, C.J. (1970). *Applied Stream Sanitation.* New York: Wiley Inter-Science.

Warren, Jr., M.L. and Burr, B.M. (1994). Status of freshwater fishes of the United States: overview of an imperiled fauna. *Fisheries* 19(1): 6–18.

Watson, P.G. and Frickers, T.E. (1990). A multilevel, *in situ* pore-water sample for use in intertidal sediments and laboratory microcosms. *Limnology and Oceanography*, 35: 1381–1389.

Watson, P.G., Frickers, P., and Goodchild, C. (1985). Spatial and seasonal variations in the chemistry of sediment interstitial waters in the Tamar estuary. *Estuarine, Coastal and Shelf Science*, 21: 105–119.

Weitzel, R.L. (1979). Periphyton measurements and applications. In: *Methods and Measurements of Periphyton Communities: A Review*, ASTM STP 690 (Weitzel, R.L., Ed.), pp. 3–33. West Conshohocken, PA: American Society for Testing and Materials.

West, R.J. and Gonsior, S.J. (1996). Biodegradation of triethanolamine. *Environmental Toxicology and Chemistry*, 15: 72–480.

Whitfield, M. (1978). The hydrolysis of ammonium ions in sea water: experimental confirmation of predicted constants at one atmosphere pressure. *Journal of the Marine Biological Association of the United Kingdom*, 58: 781–787.

Whitman, R.L. and Nevers, M.B. (2003). Foreshore sand as a source of *Escherichia coli* in nearshore water of a Lake Michigan beach. *Applied and Environmental Microbiology*, 69: 5555–5562.

Winger, P.V. and Lassier, P.J. (1991). A vacuum-operated porewater extractor for estuarine and freshwater sediments. *Archives of Environmental Contamination and Toxicology*, 21: 321–324.

Word, J.Q., Ward, J.A., Franklin, L.M., Cullinan, V.I., and Kiesser, S.L. (1987). *Evaluation of the Equilibrium Partitioning Theory for Estimating the Toxicity of the Nonpolar Organic Compound DDT to the Sediment Dwelling Amphipod Rheopoxynius abronius*. Sequim, WA: Battelle Marine Research Laboratory.

Glossary

Abiotic: Occurring without the direct involvement of organisms.

Absorb: To take in; many things absorb water.

Absorption: The uptake of water, other fluids, or dissolved chemicals by a porous material, a cell, or an organism.

Accuracy: The closeness of an individual measurement or of the average of a number of measurements to the true value. Deviation from the true value is a measure of bias in the individual measurement or averaged value.

Acid rain: Acidic rainfall that results when rain combines with sulfur oxide emissions from the combustion of fossil fuels (e.g., coal).

Acid-volatile sulfides (AVSs): Sulfides removed from sediment by cold arc extraction, consisting mainly of iron sulfide. AVSs are the principal binding phase in sediment for divalent metals.

Acre-feet (acre-foot; ac-ft): An expression of water quantity. One acre-foot covers 1 acre of ground 1 foot deep. An acre-foot contains 43,560 cubic feet, 1233 cubic meters, or 325,829 gallons (U.S.).

Activated carbon: Carbon derived from vegetable or animal materials by roasting in a vacuum furnace. Its porous nature gives it a very high surface area per unit mass—as much as 1000 square meters per gram, which is 10 million times the surface area of 1 gram of water in an open container. Used in adsorption (see definition), activated carbon adsorbs substances that are not or are only slightly adsorbed by other methods.

Activated sludge: The solids formed when microorganisms are used to treat wastewater using the activated sludge treatment process. It includes organisms, accumulated food materials, and waste products from the aerobic decomposition process.

Actual retention (*F*): Retention after runoff begins, expressed in inches.

ACWF: America's Clean Water Foundation.

Adsorption: Electrochemical attraction of positively or negatively charged molecules onto solids with an opposite charge.

Advanced wastewater treatment: Treatment technology used to produce an extremely high-quality discharge.

Advection: Transport of a substance by a fluid (e.g., groundwater) through the fluid's bulk motion in a particular direction (i.e., convection).

AED: Aerodynamic equivalent particle diameter.

AER: Allowable emission rate.

Aeration: The process of bubbling air through a solution, sometimes cleaning water of impurities by exposure to air.

Aerobic: Conditions in which free, elemental oxygen is present; also used to describe organisms, biological activity, or treatment processes that require free oxygen.

Agglomeration: Floc particles colliding and gathering into a larger settleable mass.

Aggradation: The geological process of building up a surface by the accumulation of deposits.

Air gap: The air space between the free-flowing discharge end of a supply pipe and an unpressurized receiving vessel.

Albedo: The percentage of incoming radiation that is reflected by a natural surface such as the ground, ice, snow, or water. The term is often used for the reflectivity of snow in particular.

Algae bloom: A phenomenon whereby excessive nutrients within a river, stream, or lake causes an explosion of plant life that results in depletion of the oxygen in the water needed by fish and other aquatic life. Algae blooms are usually the result of urban runoff (lawn fertilizers, etc.). A potential consequence is a fish kill, where the stream life dies *en masse*.

Aliquot: A sample or portion of a total amount of a liquid.

Alum: Aluminum sulfate, a standard coagulant used in water treatment.

Ambient: The expected natural conditions that occur in water unaffected or uninfluenced by human activities.

Anaerobic: Conditions in which no oxygen (free or combined) is available. Also used to describe organisms, biological activity, or treatment processes that function in the absence of oxygen.

Animal feeding operation (AFO): Any facility that feeds livestock or poultry in confinement such that the animals are not sustained on forages growing in the confinement area or that relies on imported feed. As defined by the USEPA (40 CFR 122.23): a "lot or facility" where animals "have been, are, or will be stabled or confined and fed or maintained for a total of 45 days or more in any 12 month period and crops, vegetation, forage growth, or post-harvest residues are not sustained in the normal growing season over any portion of the lot or facility."

Animal feedlot: A lot or building or combination of contiguous lots and buildings intended for the confined feeding, breeding, raising, or holding of animals and specifically designed as a confinement area where manure may accumulate or where the concentration of animals is such that vegetative cover cannot be maintained within the enclosure. Open lots used for the feeding and rearing of poultry (poultry ranges) are considered animal feedlots, but pastures are not considered animal feedlots.

Animal unit (AU): A unit of measure that is used to compare different animal species:

USEPA (66FR 2960-3138)—1 cattle excluding mature dairy and veal cattle; 0.7 mature dairy cattle; 2.5 swine weighing over 55 pounds; 10 swine weighing 55 pounds or less; 55 turkeys; 100 chickens; 1 veal calf.

USDA—1000 pounds of live animal weight.

Animal waste area: A holding area or lagoon used or intended to be used for the storage or treatment of animal manure and other waste products associated with an animal feedlot.

Annual flood: The maximum instantaneous peak discharge in a water year.

Annual flood series: A list of the maximum flood peak discharges occurring in each year for the period of record.

Annual runoff: The total natural discharge of a stream for a year, usually expressed in inches depth or acre-feet.

Annual yield: The total amount of water obtained in a year from a stream, spring, well, etc. Usually expressed in inches depth, acre-feet, millions of gallons, or cubic feet.

Anoxic: Conditions in which no free, elemental oxygen is present. The only source of oxygen is combined oxygen, such as that found in nitrate compounds. Also used to describe biological activity of treatment processes that function only in the presence of combined oxygen.

Antecedent runoff conditions (ARCs): The average condition of a watershed when flooding occurs; formerly called *antecedent moisture conditions.*

Antidunes: Bed forms that occur at a water velocity higher than the velocity that forms dunes and plane beds. Antidunes commonly move upstream and are accompanied by and are in phase with waves on the water surface.

Aquifer: A water-bearing stratum of permeable rock, sand, or gravel.

Aquifer system: A heterogeneous body of introduced permeable and less permeable material that acts as a water-yielding hydraulic unit of regional extent.

Area–depth curve: Graph showing the change in average rainfall depth as the size of the area receiving the rainfall changes.

Areal rainfall: Average rainfall over an area.

Armor: A layer of particles, usually gravel size, that covers the bed as a coarse residue after erosion of the finer bed materials.

Artesian water: A well tapping a confined or artesian aquifer in which the static water level stands above the top of the aquifer. The term is sometimes used to include all wells tapping confined water. Wells with water levels above the water table are said to have positive artesian head (pressure), and those with water levels below the water table have negative artesian head.

Artifact: An undesirable, detectable feature (e.g., chemical or physical change) in a sample that has resulted from sampling, sample handling or storage, or manipulations of the sample.

Available water capacity (available moisture capacity): Capacity of soils to hold water available for use by most plants. It is commonly defined as the difference between the amount of soil water at field capacity and the amount at the wilting point. It is commonly expressed as inches of water per inch of soil. The capacity, in inches, in a 60-inch profile is expressed as:

Very low	0–3 inches
Low	3–6 inches
Moderate	6–9 inches
High	9–12 inches
Very high	>12 inches

Average monthly discharge limitation: The highest allowable discharge over a calendar month.

Average weekly discharge limitation: The highest allowable discharge over a calendar week.

Backflow: Reversal of flow when pressure in a service connection exceeds the pressure in the distribution main.

Backwash: Fluidizing filter media with water, air, or a combination of the two so that individual grains can be cleaned of the material that has accumulated during the filter run.

Bacteria: Any of a number of one-celled organisms, some of which cause disease.

Bar screen: A series of bars formed into a grid used to screen out large debris from influent flow.

Base: A substance that has a pH value between 7 and 14.

Base flow: The sustained or fair-weather discharge that persists after storm runoff and associated quick return flow are depleted. It is usually derived from groundwater discharge or gradual snow or ice melt over extended periods of time, but it need not be continuous flow. It can be based on annual or seasonal periods depending on when major floods usually occur. It may also be defined as the stream discharge derived from groundwater sources. It is sometimes considered to include flow from regulated lakes or reservoirs.

Basin: A groundwater reservoir defined by the overlying land surface and underlying aquifers that contain water stored in the reservoir.

Bed form: Generic term used to describe a sand streambed; includes ripples, dunes, plane bed, and antidunes.

Bed-material load: The part of the total load of a stream that is composed of particle sizes present in appreciable quantities in the shifting parts of the streambed.

Bedload: Material moving on or near the streambed by rolling, sliding, and making brief excursions into the flow a few diameters above the bed.

Beneficial use of water: The use of water for any beneficial purpose. Such uses include domestic use, irrigation, recreation, fish and wildlife, fire protection, navigation, power, and industrial use. The benefit varies from one location to another and by custom. What constitutes beneficial use is often defined by statute or court decisions.

Benthic: Associated with the bottom of a waterbody.

Bioaccumulation: The net accumulation of a substance by an organism as a result of uptake from all environmental sources.

Bioavailability: The degree to which a chemical is taken up by aquatic organisms.

Biochemical oxygen demand (BOD): The oxygen used to meet the metabolic needs of aerobic microorganisms in water rich in organic matter.

Biosolids: Solid organic matter recovered from a sewage treatment process and used especially as fertilizer [or soil amendment]—usually used in plural (*Merriam-Webster's Collegiate Dictionary*, 10th ed., 1998). *Note:* In this text, the term *biosolids* is used in many places (activated sludge being the exception) to replace the standard term *sludge*. It is the opinion of the author that the term *sludge* is an ugly four-letter word inappropriate to describe biosolids. Biosolids are a product that can be reused; they have some value. Because biosolids have value, they certainly should not be classified as a waste product, and when the topic of biosolids for beneficial reuse is addressed it is made clear that they are not a waste product.

Biota: All the species of plants and animals indigenous to a certain area.

Boiling point: The temperature at which a liquid boils; the temperature at which the vapor pressure of a liquid equals the pressure on its surface. If the pressure of the liquid varies, the actual boiling point varies. The boiling point of water is 212°F (100°C).

Breakpoint: Point at which chlorine dosage satisfies chlorine demand.

Breakthrough: In filtering, when unwanted materials start to pass through the filter.

Buffer: A substance or solution that resists changes in pH.

Calcium carbonate: Compound principally responsible for hardness.

Calcium hardness: Portion of total hardness caused by calcium compounds.

Carbonaceous biochemical oxygen demand (CBOD): The amount of biochemical oxygen demand that can be attributed to carbonaceous material.

Carbonate hardness: Caused primarily by compounds containing carbonate.

Chain-of-custody: Documentation that establishes the control of a sample between the time it is collected and the time it is analyzed. It usually applies to legal samples to demonstrate that there was no tampering with, or contamination of, the sample during this time.

Channel flow: Water flow in a defined channel, either natural or human-made.

Chemical oxygen demand (COD): The amount of chemically oxidizable materials present in the wastewater.

Chlorination: Disinfection of water using chlorine as the oxidizing agent.

Clarifier: A device designed to permit solids to settle or rise and be separated from the flow; also known as a settling tank or sedimentation basin.

Clean: Refers to sediment or water test samples determined not to contain concentrations of contaminants that cause apparent and unacceptable harm (or effects) to the test organisms.

Coagulation: Neutralization of the charges of colloidal matter.

Coefficient of kinematic viscosity: Ratio of the coefficient of viscosity to the density of a fluid.

Coefficient of viscosity: Ratio of shear stress to the velocity gradient perpendicular to the direction of flow of a Newtonian fluid or the ratio of shear stress in a moving liquid to the rate of deformation.

Coliform: A type of bacteria whose presence indicates possible human or animal contamination of water.

Combined sewer: A collection system that carries both wastewater and stormwater flows.

Comminution: A process to shred solids into smaller, less harmful particles.

Composite sample: A combination of individual samples taken in proportion to flow.

Concentration: The ratio of weight or volume of test materials to the weight or volume of sediment or water.

Connate water: Pressurized water trapped in the pore spaces of sedimentary rock at the time it was deposited. It is usually highly mineralized.

Connected impervious area: Occurs where runoff from an impervious area flows directly to storm drains, street gutters, or stream channels.

Consumptive use: (1) The quantity of water absorbed by crops and transpired or used directly in the building of plant tissue, together with the water evaporated from the cropped area. (2) The quantity of water transpired and evaporated from a cropped area or the normal loss of water from the soil by evaporation and plant transpiration. (3) The quantity of water discharged to the atmosphere or incorporated in the products of the process in connection with vegetative growth, food processing, or an industrial process.

Contaminated sediment: Sediment containing chemical substances at concentrations that pose a known or suspected threat to environmental or human health.

Contamination (water): Damage to the quality of water sources caused by sewage, industrial waste, or other material.

Control sediment: A sediment that is essentially free of contaminants and is used routinely to assess the acceptability of a test. Any contaminants in control sediment may originate from the global spread of pollutants and do not reflect any substantial input from local or nonpoint sources. Comparing test sediments to control sediments is a measure of the toxicity of a test sediment beyond inevitable background contamination.

Core sample: A sediment sample collected to obtain a vertical profile using a variety of instruments.

Correlation: A statistical index that measures linear variation between variables.

Cover: Vegetation and vegetational debris, such as mulch and residue, that exist on the soil surface.

Crest: The elevation of the uppermost surface of a dam or an earth spillway.

Crest staff gauge: A gauge used to make a quick visual observation of water surface levels in reservoirs, rivers, streams, irrigation channels, weirs, and flumes.

Cross-connection: A connection between a storm drain system and a sanitary collection system; a connection between two sections of a collection system to handle anticipated overloads of one system; or a connection between drinking (potable) water and an unsafe water supply or sanitary collection system.

Cross-section: The shape of a channel, stream, or valley determined by a line approximately perpendicular to the mean path of water flow, along which measurements of distance and elevations are determined.

Curve number (CN): A dimensionless number of 98 or less that relates runoff to the soil–cover complex of a watershed. The curve number indicates the runoff potential of a soil–cover complex during periods when the soil is not frozen. Higher numbers mean greater runoff.

Daily discharge: The discharge of a pollutant measured during a calendar day or any 24-hour period that reasonably represents a calendar day for the purposes of sampling. Limitations expressed as weight are total mass (weight) discharged over the day. Limitations expressed in other units are average measurements of the day.

Daily maximum discharge: The highest allowable values for a daily discharge.

Dam: An artificial barrier, together with any associated spillways and appurtenant works, across a watercourse of natural drainage area that does or may impound or divert water.

Damage reach: A length of floodplain or valley selected for damage evaluation.

Darcy's law: An equation for computing the quantity of water flowing through porous media. Darcy's law assumes that the flow is laminar and that inertia can be neglected. The law states that the rate of viscous flow of homogeneous fluids through isotropic porous media is proportional to, and in the direction of, the hydraulic gradient.

Decontamination: A process of washing or rinsing that removes chemicals adhering to equipment and supplies.

Degree-day: A day with an average temperature 1 degree above a defined basis or threshold. The average is usually obtained by averaging the maximum and minimum temperatures for the day. Depending on usage, the threshold temperature may vary. For example, in snowmelt studies, a degree-day is defined as a day with an average daily temperature above 32°F, so a day with an average temperature of 40°F (40° – 32°) = 8 degree-days.

Degrees of freedom: The number of independent pieces of information, or parameters, required to form a statistical estimate.

Depression storage: The volume of water stored on the soil surface.

Depth–area curve: A graph showing the change in average rainfall depth as size of area changes.

Design storm: A specified rainfall depth and rainfall distribution used to estimate runoff in the design of hydraulic structures.

Detention time: The theoretical time water remains in a tank at a given flow rate.

Dewatering: The removal or separation of a portion of water present in a sludge or slurry.

Digital elevation model: A digital representation of a topographic surface.

Diffusion: The process by which both ionic and molecular species dissolved in water move from areas of higher concentration to areas of lower concentration.

Discharge monitoring report (DMR): The monthly report required by the treatment plant's National Pollutant Discharge Elimination System (NPDES) discharge permit.

Disinfection: Water treatment process that kills pathogenic organisms.

Disinfection byproducts (DBPs): Chemical compounds formed by the reaction of disinfectants with organic compounds in water.

Dissolved oxygen (DO): The amount of oxygen dissolved in water or sewage. Concentrations of less than 5 parts per million (ppm) can limit aquatic life or cause offensive odors. Excessive organic matter present in water because of inadequate waste treatment and runoff from agricultural or urban land generally causes low DO levels.

Dissolved solids: The total amount of dissolved inorganic material contained in water or wastes. Excessive dissolved solids make water unsuitable for drinking or industrial uses.

Domestic consumption (use): Water used for household purposes such as washing, food preparation, and showers. The quantity (or quantity per capita) of water consumed in a municipality or district for domestic uses or purposes during a given period, it sometimes encompasses all uses, including the quantity wasted, lost, or otherwise unaccounted for.

Drawdown: Lowering the water level by pumping. It is measured in feet for a given quantity of water pumped during a specified period or after the pumping level has become constant.

Drinking water standards: Standards established by state agencies, U.S. Public Health Service, and U.S. Environmental Protection Agency for drinking water in the United States.

Dunes: Bed forms with a triangular profile having a gently upstream slope. Dunes advance downstream as sediment moves up the upstream slope and is deposited on the steeper downstream slope. Dunes move downstream much more slowly than the stream flow.

Ecotox thresholds (ETs): Benchmark values in ecological risk assessments defined as media-specific contaminant concentrations above which there is sufficient concern regarding adverse ecological effects to warrant further site investigation.

Effluent: Something that flows out, usually a polluting gas or liquid discharge.

Effluent limitation: Any restriction imposed by a regulatory agency on quantities, discharge rates, or concentrations of pollutants discharged from point sources into state waters.

Elutriate: An aqueous solution obtained after adding water to a solid substance or loose material (e.g., sediment, tailings, drilling mud, dredge spoil), shaking the mixture, then centrifuging or filtering it or decanting the supernatant.

Energy: In scientific terms, the ability or capacity of doing work. Various forms of energy include kinetic, potential, thermal, nuclear, rotational, and electromagnetic. One form of energy may be changed to another, such as when coal is burned to produce steam to drive a turbine, which produces electric energy.

Equilibration: The condition in which a material or contaminant is at a steady state between the solid or particulate sediment and the interstitial water.

Erosion: The wearing away of the land surface by wind, water, ice, or other geologic agents. Erosion occurs naturally from weather or runoff but is often intensified by human land use practices.

Eutrophication: The process of enrichment of water bodies by nutrients. Eutrophication of a lake normally contributes to its slow evolution into a bog or marsh and ultimately to dry land. Eutrophication may be accelerated by human activities, thereby speeding up the aging process.

Evaporation: The process by which water becomes a vapor at a temperature below the boiling point.

Facultative organisms: Organisms that can survive and function in the presence or absence of free, elemental oxygen.

Fall diameter or standard fall diameter: The diameter of a sphere that has a specific gravity of 2.65 and the same terminal velocity as a particle of any specific gravity when each is allowed to settle alone in quiescent distilled water of infinite extent and at a temperature of 24°C. A particle reaches terminal velocity when the water resistance is equal to the force of gravity.

Fecal coliform: The portion of the coliform bacteria group that is present in the intestinal tracts and feces of warm-blooded animals.

Field capacity: The capacity of soil to hold water. It is measured as the ratio of the weight of water retained by the soil to the weight of the dry soil.

Filtration: The mechanical process that removes particulate matter by separating water from solid material, usually by passing it through sand.

Floc: Solids that join to form larger particles that will settle better.

Flocculation: Slow mixing process in which particles are brought into contact to promote their agglomeration.

Flume: A flow rate measurement device.

Fluoridation: Chemical addition to water to reduce the incidence of dental caries in children.

Food-to-microorganism ratio (F/M): An activated sludge process control calculation based on the amount of food (BOD or COD) available per pound of mixed liquor volatile suspended solids.

Force main: A pipe that carries wastewater under pressure from the discharge side of a pump to a point of gravity flow downstream.

Formulated sediment: Mixtures of material used to mimic a natural sediment.

Global positioning system (GPS): A navigation system that relies on satellite information. It can give continuous position reports (i.e., latitude and longitude) that vary in accuracy depending on the sophistication of the receiving unit.

Grab sample: An individual sample collected at a randomly selected time.

Graywater: Water that has been used for showering, clothes washing, and faucet uses. Kitchen sink and toilet water is excluded. This water has excellent potential for reuse as irrigation for yards.

Grit: Heavy inorganic solids, such as sand, gravel, eggshells, or metal filings.

Groundwater: The supply of freshwater found beneath the surface of the Earth (usually in aquifers) often used for supplying wells and springs. Because groundwater is a major source of drinking water, concern is growing over areas where leaching agricultural or industrial pollutants or substances from leaking underground storage tanks (USTs) are contaminating groundwater.

Groundwater hydrology: The branch of hydrology that deals with groundwater, its occurrence and movements, its replenishment and depletion, the properties of rocks that control groundwater movement and storage, and the methods of investigation and use of groundwater.

Groundwater recharge: The inflow to a groundwater reservoir.

Groundwater runoff: A portion of runoff that has passed into the ground, has become groundwater, and has been discharged into a stream channel as spring or seepage water.

Hardness: The concentration of calcium and magnesium salts in water.

Head loss: Amount of energy used by water to move from one point to another.

Head space: The space in the storage container between the top of the sample and the lid of the container.

Heavy metals: Metallic elements with high atomic weights (e.g., mercury, chromium, cadmium, arsenic, lead). They can damage living things at low concentrations and tend to accumulate in the food chain.

Holding pond: A small basin or pond designed to hold sediment-laden or contaminated water until it can be treated to meet water quality standards or used in some other way.

Holding time: The period of time for which a sediment or water sample can be stored after collection and before analysis or being used in a biological test. Changes that occur in sediments or water should be minimal during this period, and the integrity of the sample should not be compromised to any substantial degree with respect to its physical, chemical, or biological characteristics.

Homogenization: The complete mixing of sediment, either by hand or mechanical means, until physical, chemical, or biological homogeneity of the sample is achieved.

Hydraulic cleaning: Cleaning pipe with water under enough pressure to produce high water velocities.

Hydraulic gradient: A measure of the change in groundwater head over a given distance.

Hydraulic head: The height above a specific datum (generally sea level) that water will rise in a well.

Hydrologic cycle (water cycle): The cycle of water movement from the atmosphere to the Earth and back to the atmosphere through various processes. These processes include precipitation, infiltration, percolation, storage, evaporation, transpiration, and condensation.

Hydrology: Science dealing with the properties, distribution, and circulation of water.

Index period: Specific time period in which sampling or *in situ* analyses are conducted; generally pertains to an ecologically important season or desired environmental conditions under which sampling is performed.

Impoundment: A body of water such as a pond that is confined by a dam, dike, floodgate, or other barrier and is used to collect and store water for future use.

***In situ*:** Refers to the original location from which test samples are being collected or at which organisms are exposed to undisturbed water or sediments for extended periods.

Industrial wastewater: Wastes associated with industrial manufacturing processes.

Infiltration: The gradual downward flow of water from the surface into soil material.

Inflow: Extraneous flows in sewers; simply, inflow is water discharged into sewer pipes or service connections from such sources as foundation drains, roof leaders, cellar and yard area drains, cooling water from air conditioners, and other clean water discharges from commercial and industrial establishments. It is defined as follows:

Infiltration—Water entering the collection system through cracks, joints, or breaks.

Steady inflow—Water discharged from cellar and foundation drains, cooling water discharges, and drains from springs and swampy areas. This type of inflow is steady and is identified and measured along with infiltration.

Direct flow—Type of inflow that has a direct stormwater runoff connection to the sanitary sewer and causes an almost immediate increase in wastewater flows. Possible sources are roof leaders, yard and areaway drains, manhole covers, cross-connections from storm drains and catch basins, and combined sewers.

Total inflow—The sum of the direct inflow at any point in the system plus any flow discharged from the system upstream through overflows, pumping station bypasses, and the like.

Delayed inflow—Stormwater that may require several days or more to drain through the sewer system. This category can include the discharge of sump pumps from cellar drainage as well as the slowed entry of surface water through manholes in ponded areas.

Influent: Wastewater entering a tank, channel, or treatment process.

Inorganic chemicals or compounds: Chemical substances of mineral origin, not of a carbon structure. These include metals such as lead, iron (ferric chloride), and cadmium.

Interferences: Characteristics of sediments or sediment test systems that can potentially affect analytical results or test organism responses aside from responses related to sediment contamination. Types of interferences include non-contaminant characteristics (e.g., sediment texture or grain size, lighting); changes in chemical bioavailability due to sample handling or storage (e.g., ammonia generation); and the presence of indigenous organisms. Also referred to as *confounding factors*.

Interstitial water: Water occupying the spaces between sediment or soil particles.

Ion exchange process: Process used to remove hardness from water.

Jar test: Laboratory procedure used to estimate proper coagulant dosage.

Laminar flow: Low-velocity flow in which layers of fluid slip over contiguous layers without appreciable mixing.

Langelier saturation index (LI): A numerical index that indicates whether calcium carbonate will be deposited or dissolved in a distribution system.

Leaching: The process by which soluble materials in the soil such as nutrients, pesticide chemicals or contaminants are washed into a lower layer of soil or are dissolved and carried away by water.

License: A certificate issued by the State Board of Waterworks/Wastewater Works Operators authorizing the holder to perform the duties of a wastewater treatment plant operator.

Lift station: A wastewater pumping station designed to "lift" the wastewater to a higher elevation. A lift station normally employs pumps or other mechanical devices to pump the wastewater and discharges into a pressure pipe called a force main.

Maximum contaminant level (MCL): An enforceable standard for protection of human health.

Mean cell residence time (MCRT): The average length of time a mixed liquor suspended solids particle remains in the activated sludge process; may also be referred to as *sludge retention time*.

Mechanical cleaning: Cleaning pipe through the use of equipment (bucket machines, power rodders, or hand rods) that scrapes, cuts, pulls, or pushes material out of the pipe.

Membrane process: A process that draws a measured volume of water through a filter membrane with small enough openings to take out contaminants.

Metering pump: A chemical solution feed pump that adds a measured amount of solution with each stroke or rotation of the pump.

Milligrams/liter (mg/L): A measure of concentration equivalent to parts per million (ppm).

Mixed liquor: Suspended solids concentration of the mixed liquor.

Mixed liquor volatile suspended solids (MLVSS): The concentration of organic matter in the mixed liquor suspended solids.

Nephelometric turbidity unit (NTU): Indicates the amount of turbidity in a water sample.

Nitrogenous oxygen demand (NOD): A measure of the amount of oxygen required to biologically oxidize nitrogen compounds under specified conditions of time and temperature.

Nonpoint source (NPS) pollution: Pollution caused by sediment, nutrients, and organic and toxic substances originating from land use activities being carried to lakes and streams by surface runoff. Nonpoint source pollution occurs when the rate of materials entering these water bodies exceeds natural levels.

NPDES permit: National Pollutant Discharge Elimination System permit, which authorizes the discharge of treated wastes and specifies the conditions that must be met for discharge.

Nutrients: Substances required to support living organisms; usually refers to nitrogen, phosphorus, iron, and other trace metals.

Organic chemicals/compounds: Animal- or plant-produced substances containing mainly carbon, hydrogen, and oxygen; examples include benzene and toluene.

Overlying water: The water placed over sediment in a test chamber during a test.

Parts per million (ppm): The number of parts by weight of a substance per million parts of water. This unit is commonly used to represent pollutant concentrations. Large concentrations are expressed in percentages.

Pathogenic: Disease causing; a pathogenic organism is capable of causing illness.

Peepers: Devices that collect interstitial water by diffusion through membranes attached to collection chambers. The chambers are typically placed in the sediment for extended periods of time to allow for equilibration between the internal water environment of the peeper and the surrounding ambient sediment/interstitial water matrix.

Percolation: The movement of water through subsurface soil layers, usually continuing downward to the groundwater or water table reservoirs.

pH: An expression of acidity and alkalinity on a scale of 0 to 14, with 7 representing neutrality; numbers less than 7 indicate increasing acidity, and numbers greater than 7 indicate increasing alkalinity.

Photosynthesis: A process in green plants in which water, carbon dioxide, and sunlight combine to form sugar.

Piezometric surface: An imaginary surface that coincides with the hydrostatic pressure level of water in an aquifer.

Plane bed: A sedimentary bed with irregularities no larger than the maximum size of the bed material.

Point source pollution: A type of water pollution resulting from discharges into receiving waters from easily identifiable points. Common point sources of pollution are discharges from factories and municipal sewage treatment plants.

Pollution: Alteration of the physical, thermal, chemical, or biological quality of, or the contamination of, any water to a state that renders the water harmful, detrimental, or injurious to humans, animal life, vegetation, property, or public health, safety, or welfare, or impairs the usefulness or the public enjoyment of the water for any lawful or reasonable purpose.

Porosity: Measure of the part of a rock that contains pore spaces without regard to size, shape, interconnection, or arrangement of openings. It is expressed as a percentage of the total volume occupied by spaces.

Potable water: Water satisfactorily safe for drinking purposes from the standpoint of its chemical, physical, and biological characteristics.

Precipitate: A deposit of hail, rain, mist, sleet, or snow. The process by which atmospheric water becomes surface or subsurface water is *precipitation*, a term that is commonly used to designate the quantity of water precipitated.

Preventive maintenance (PM): Regularly scheduled servicing of machinery or other equipment using appropriate tools, tests, and lubricants. This type of maintenance can prolong the useful life of equipment and machinery and increase its efficiency by detecting and correcting problems before they cause a breakdown of the equipment.

Purveyor: An agency or person that supplies potable water.

Radon: A radioactive, colorless, odorless gas that occurs naturally in the earth; when radon is trapped in buildings, concentrations build up that can cause health hazards such as lung cancer.

Recharge: The addition of water into a groundwater system.

Reference sediment: A whole sediment, collected near an area of concern, that is used as a point of comparison to assess sediment conditions exclusive of the materials or activities of interest. The reference sediment may be used as an indicator of localized sediment conditions exclusive of the specific pollutant input of concern. The sediment would be collected near the site of concern and would represent the background conditions resulting from any localized pollutant inputs as well as global pollutant input. Program-specific guidance documents should be consulted, as some USEPA programs have specific definitions and requirements for reference sediment.

Reservoir: A pond, lake, tank, or basin (natural or human-made) where water is collected and used for storage. Large bodies of groundwater are called *groundwater reservoirs*; water behind a dam is also called a *reservoir of water*.

Return activated sludge solids (RASS): The concentration of suspended solids in the sludge flow being returned from the settling tank to the head of the aeration tank.

Reverse osmosis: Process in which almost pure water is passed through a semipermeable membrane.

Ripples: Bed forms that have a triangular profile and are similar to dunes but much smaller.

River basin: A term used to designate the area drained by a river and its tributaries.

Sampling platform: A working space, such as the deck of a boat, from which all sample collection activities are conducted.

Sanitary wastewater: Wastes discharged from residences and from commercial, institutional, and similar facilities that include both sewage and industrial wastes.

Schmutzdecke: Layer of solids and biological growth that forms on top of a slow sand filter, allowing the filter to remove turbidity effectively without chemical coagulation.

Scum: The mixture of floatable solids and water removed from the surface of the settling tank.

Sediment: Transported and deposited particles derived from rocks, soil, or biological material.

Sediment concentration: Ratio of the weight of the sediment in a water–sediment mixture to the total weight of the mixture.

Sediment discharge: Usually the mass but sometimes the volume of sediment passing a stream transect in a unit of time.

Sediment load: The quantity of sediment that passes a cross-section of a stream or river in a specified period of time.

Sedimentation: A process that reduces the velocity of water in basins so that suspended material can settle out by gravity.

Seepage: The appearance and disappearance of water at the ground surface. Seepage refers to the movement of water in saturated material; it differs from percolation, which is predominantly the movement of water in unsaturated material.

Septic tanks: Tanks that are used to hold domestic wastes when a sewer line is not available to carry them to a treatment plant. The wastes are piped to underground tanks directly from a house. Bacteria in the wastes decompose some of the organic matter, the sludge settles on the bottom of the tank, and the effluent flows out of the tank into the ground through drains.

Settleability: A process control test used to evaluate the settling characteristics of the activated sludge. Readings taken at 30 to 60 minutes are used to calculate the settled sludge volume (SSV) and the sludge volume index (SVI).

Settled sludge volume (SSV): The volume (in percent) occupied by an activated sludge sample after 30 to 60 minutes of settling. It is normally written as SSV with a subscript to indicate the time of the reading used for calculation (e.g., SSV_{60}, SSV_{30}).

Sieving: Selectively removing certain size fractions of a sediment sample by processing the sediment through selected mesh sizes. Sediment samples are sieved to remove unrepresentative material, such as shells, stones, trash, and twigs; to increase homogeneity and replicability of samples; to remove indigenous organisms prior to toxicity testing; and to facilitate organism counting, sediment handling, and subsampling.

Silt: Detritic sedimentary rock that has noncemented grains of particle size between fine sand and clay.

Site: A study area that can be comprised of multiple sampling stations.

Sludge: The mixture of settleable solids and water removed from the bottom of the settling tank.

Sludge retention time (SRT): See *mean cell residence time.*

Sludge volume index (SVI): A process control calculation used to evaluate the settling quality of activated sludge; the SSV_{30} and mixed liquor suspended solids test results are required to calculate the SVI.

Soil moisture (soil water): Water diffused in the soil. It is found in the upper part of the zone of aeration from which water is discharged by transpiration from plants or by soil evaporation.

Specific heat: The heat capacity of a material per unit mass. It is expressed as the amount of heat (in calories) required to raise the temperature of 1 gram of a substance 1°C; the specific heat of water is 1 calorie.

Spiking: Addition of a known amount of test material to a sediment, a technique often used as a quality control check for bias due to interference or matrix effects.

Standing waves: Water waves that are in phase with antidunes.

Station: A sampling location within a study area or site where physical, chemical, or biological sampling or testing occurs.

Storm sewer: A collection system designed to carry only stormwater runoff.

Stormwater: Runoff resulting from rainfall and snowmelt.

Stream: A general term for a body of flowing water. In hydrology, the term is generally applied to the water flowing in a natural channel as distinct from a canal. More generally, it is applied to the water flowing in any channel, natural or artificial. Some types of streams include (1) *ephemeral*, a stream that flows only in direct response to precipitation and whose channel is at all times above the water table; (2) *intermittent* or *seasonal*, a stream that flows only at certain times of the year when it receives water from springs, rainfall, or from surface sources such as melting snow; (3) *perennial*, a stream that flows continuously; (4) *gaining*, a stream or reach of a stream that receives water from the zone of saturation (an effluent stream); (5) *insulated*, a stream or reach of a stream that neither contributes water to the zone of saturation nor receives water from it and is separated from the zones of saturation by an impermeable bed; (6) *losing*, a stream or reach of a stream that contributes water to the zone of saturation, an influent stream; and (7) *perched*, a losing stream or an insulated stream that is separated from the underlying groundwater by a zone of aeration.

Supernatant: The liquid standing above a sediment or precipitate.

Surface tension: The free energy that is produced in a liquid surface by the unbalanced inward pull exerted by molecules underlying the layer of surface molecules.

Surface water: All water naturally open to the atmosphere (e.g., rivers, lakes, reservoirs, streams, impoundments, seas, estuaries) and all springs, wells, or other collectors directly influenced by surface water.

Suspended load: The part of the total sediment load that moves above the bed layer. The weight of suspended particles is continuously supported by the fluid.

Thermal pollution: The degradation of water quality by the introduction of a heated effluent. It is primarily the result of the discharge of cooling waters from industrial processes (particularly from electrical power generation); waste heat eventually results from virtually every energy conversion.

Titrant: A solution of known strength of concentration; used in titration.

Titration: A process whereby a solution of known strength (titrant) is added to a certain volume of treated sample containing an indicator. A color change shows when the reaction is complete.

Titrator: An instrument, usually a calibrated cylinder (tube-form), used in titration to measure the amount of titrant being added to the sample.

Total dissolved solids: The amount of material (inorganic salts and small amounts of organic material) dissolved in water and commonly expressed as a concentration in terms of milligrams per liter.

Total suspended solids (TSS): Total suspended solids in water, commonly expressed as a concentration in terms of milligrams per liter.

Toxicity: The occurrence of lethal or sublethal adverse affects on representative sensitive organisms due to exposure to toxic materials. Adverse effects caused by conditions of temperature, dissolved oxygen, or nontoxic dissolved substances are excluded from the definition of toxicity.

Transpiration: The process by which water vapor escapes from the living plant—principally the leaves—and enters the atmosphere.

Turbulent: A state of flowing where the fluid is agitated by cross-currents and eddies.

Uniform flow: A flow in which the velocity is the same in both magnitude and direction from point to point along a reach.

Vaporization: The change of a substance from a liquid or solid state to a gaseous state.

Volatile organic compound (VOC): Any organic compound that participates in atmospheric photochemical reactions except for those designated by the USEPA administrator as having negligible photochemical reactivity.

Washload: The part of the sediment load of a stream composed of fine particles (usually smaller than 0.062 mm) found only in relatively small quantities in the streambed. Almost all of the washload is carried in nearly permanent suspension, and its magnitude depends primarily on that amount of fine material available to the stream from sources other than the bed.

Waste activated sludge solids (WASS): The concentration of suspended solids in the sludge being removed from the activated sludge process.

Wastewater: The water supply of a community after it has been contaminated by use.

Water cycle: The process by which water travels in a sequence from the air (condensation) to the Earth (precipitation) and returns to the atmosphere (evaporation). It is also referred to as the *hydrologic cycle*.

Water quality: A term used to describe the chemical, physical, and biological characteristics of water with respect to its suitability for a particular use.

Water quality standard: A plan for water quality management containing four major elements: water use, criteria to protect uses, implementation plans, and enforcement plans. An antidegradation statement is sometimes prepared to protect existing high-quality waters.

Water supply: Any quantity of available water.

Waterborne disease: A disease caused by a microorganism that is carried from one person or animal to another by water.

Watershed: The area of land that contributes surface runoff to a given point in a drainage system.

Weir: A device used to measure wastewater flow.

Whole sediment: Sediment and associated interstitial water that have had minimal manipulation.

Zone of aeration: A region in the earth above the water table. Water in the zone of aeration is under atmospheric pressure and would not flow into a well.

Zoogleal slime: The biological slime that forms on fixed-film treatment devices. It contains a wide variety of organisms essential to the treatment process.

Index

C

D

E

K

L

Q

R

S

T

X

Y

Z